Alfred Lewis Galabin

The Student's Guide to the Diseases of Women

Alfred Lewis Galabin

The Student's Guide to the Diseases of Women

ISBN/EAN: 9783744718219

Printed in Europe, USA, Canada, Australia, Japan

Cover: Foto ©berggeist007 / pixelio.de

More available books at **www.hansebooks.com**

THE

STUDENT'S GUIDE

TO THE

DISEASES OF WOMEN

BY

ALFRED LEWIS GALABIN, M.A., M.D., F.R.C.P.

LATE FELLOW OF TRINITY COLLEGE, CAMBRIDGE
OBSTETRIC PHYSICIAN AND LECTURER ON MIDWIFERY AND THE DISEASES OF WOMEN
TO GUY'S HOSPITAL; LATE EXAMINER IN OBSTETRIC MEDICINE TO THE
UNIVERSITY OF CAMBRIDGE, AND TO THE ROYAL COLLEGE OF PHYSICIANS OF LONDON

THIRD EDITION

LONDON

J. & A. CHURCHILL

NEW BURLINGTON STREET

—

1884

PREFACE TO THE THIRD EDITION.

In the present edition I have added sections on the operations for repair of ruptured perineum, and also on the operations for the cure of prolapse of the uterus and vagina, believing that it is possible to improve upon the mode of performing these operations, as described in most text-books of surgery. I have still, however, been unwilling to increase the size of the book by introducing an account of the operations for vesico-vaginal fistula, since practitioners in general are rarely called upon to perform this operation; and, moreover, it is described both in works on Surgery and in the larger works on Gynæcology.

For a similar reason, I have still omitted those subjects, such as extra-uterine fœtation, and retroversion of the pregnant uterus, which are to be found in text-books of Midwifery. Thirteen new figures have been introduced, the whole book has been revised, and other additions have been made, especially in the sections on anteflexion of the uterus, on the treatment of inflammation of the cervix uteri, and on the operative treatment of uterine myomata. The figures showing different pessaries in position, namely Figures 29, 34, 36, and 45, are drawn to a scale two-sevenths the size of nature. In Figure 3, showing the mode of introducing the uterine sound, and in Figures 40, 41, 42, and 43, illustrating the varieties of prolapse, the scale is two-ninths of the natural size.

December, 1883.

CONTENTS.

CHAPTER I.

CHAPTER II.

CHAPTER III.

A *

CHAPTER IV.

DISPLACEMENTS OF THE UTERUS AND PELVIC VISCERA.

CHAPTER V.

HYPERPLASIA AND ATROPHY OF THE UTERUS.

CHAPTER VI.

HYPERÆMIA AND INFLAMMATION OF THE UTERUS.

CHAPTER VII.

NEW GROWTHS OF THE UTERUS.

CHAPTER VIII.

DISEASES OF THE OVARIES.

CHAPTER IX.

DISEASES OF THE FALLOPIAN TUBES.

CHAPTER X.

DISEASES OF THE UTERINE LIGAMENTS AND OF THE ADJACENT PERITONEUM AND CELLULAR TISSUE.

CHAPTER XI.

DISEASES OF THE VAGINA AND VULVA.

CHAPTER XII.

LIST OF ILLUSTRATIONS.

DISEASES OF WOMEN.

CHAPTER I.

MEANS OF PHYSICAL DIAGNOSIS.

VAGINAL TOUCH AND BIMANUAL EXAMINATION.—
When a local investigation is considered desirable, the internal examination *per vaginam*, made by the index finger, or the index and middle fingers of either hand, will in most cases be the first exploratory measure which should be undertaken. It should, however, be invariably combined with abdominal palpation by the other hand placed externally over the pubes; for, if this be omitted, it is quite possible for the examiner to overlook the existence of tumours of considerable size, or of pregnancy of advanced duration. The position of the patient is of great importance. On the Continent and in America, the dorsal position is universally adopted, while in Britain it is more common to choose the left lateral position. Each position has its own advantages. The left lateral position, combined with the introduction into the vagina of the right index finger, has the disadvantage that the sensitive palmar surface of the finger can only be turned towards the posterior and lateral vaginal walls, and not towards the anterior wall, which it is most essential to explore. It has the still greater drawback that it does not allow of any effectual use of the conjoined or bimanual manipu-

B

lation. On the other hand, the lateral position allows
the perineum to be more fully retracted, so that the
finger can explore more deeply the posterior vaginal
cul-de-sac and posterior portion of the pelvis, while
its flexor surface has the most convenient direction
for this purpose. The dorsal position should always,
therefore, be employed first, but it is generally desirable to turn the patient afterwards into the lateral
position—left, if the right hand is being used for
vaginal examination, and conversely—to complete the
exploration.

By some it is preferred to introduce one or two
fingers of the *left* hand into the vagina, while the
patient is in the left lateral position. This plan allows
the bimanual examination to be effectually carried out,
but has the inconvenience that it requires the patient
to be placed somewhat transversely on the couch or
bed, and that the flexor surfaces of the fingers cannot
conveniently be turned to examine the posterior half
of the pelvis without changing hands.

For examination in the dorsal position, the patient
should lie upon a firm, flat surface, as a hard mattress,
the head, but not the shoulders, supported upon a low
pillow. The knees should be flexed and abducted.
When the skirts are tight, it may be necessary to slip
them above the knees, whether for examination in the
dorsal or in the lateral position. It is well, therefore,
to have a shawl at hand, which may be thrown over
the knees when required. The examining hand is
then passed beneath the thigh, and the index finger,
previously well lubricated,* is introduced into the
vulva from its posterior aspect, the perineum being
first sought for as a landmark. This is the readiest
mode of at once finding the vaginal outlet, and by
this means also the sensitive structures further for-

---

* Carbolic oil (1 in 20) may be used, but the following is a
more convenient antiseptic lubricant—Oil of eucalyptus ℥iss.,
paraffin ℥j., vaseline ℥j. : to be heated together, and mixed. It
has the advantage over oil that it is not liable to drop about.

ward, the clitoris and nymphæ, are avoided. The remaining fingers are flexed into the palm, and upon the extent to which they can be doubled back, even more than upon the length of the finger, depends the length of reach of the examiner. In most cases it is preferable to use the index finger alone at first. If two fingers are used, the vaginal spasm thereby excited more than counterbalances any advantage gained by the extra length of the middle finger. With a woman who has borne children, however, and whose vagina is capacious, two fingers may be used with advantage, especially in estimating the size and mobility of the uterus by the conjoined manipulation.

As it is carried up the vagina, the finger ascertains the perviousness and capacity of that canal, and also whether its mucous membrane is in a normal condition or otherwise as to smoothness, moisture, and temperature. Any undue sensitiveness or spasm at the vulval outlet is also noted as the finger enters, as well as any other abnormal condition of the perineum or the vulva itself, such as laceration or the presence of condylomata. The cervix is then examined with reference to size, hardness, position, and direction; and the os with reference to its size, the regularity, smoothness, and consistence of its lips, and the character of secretion, as to quantity and tenacity.

In ascertaining the position and size of the body of the uterus, the conjoined manipulation is brought to aid. In the normal condition nothing can be felt of the body of the unimpregnated uterus through the posterior vaginal cul-de-sac, but a portion of it can be reached by the finger in the vagina in front of the cervix. To carry out the bimanual method, the fingers of the left hand (or of the right hand, as shown in the figure, if the physician is standing at his patient's left side), passed beneath the clothes, and laid upon the abdomen, should be pressed firmly down into the pelvis, so as to push the fundus uteri downwards and forwards (*see* Fig. 1, p. 4). They should not be applied too close to the pubes, for, in that case, the fundus is apt to be pushed backward instead of forward. The manipulation is also facilitated if the cervix be at the same time pushed backwards by the finger in the vagina. The uterus is thus brought into a position somewhat of anteversion, and can be held firmly between the fingers of the two hands, and its size, shape, and any irregularities or prominences on its surface can be ascertained with much exactness. In

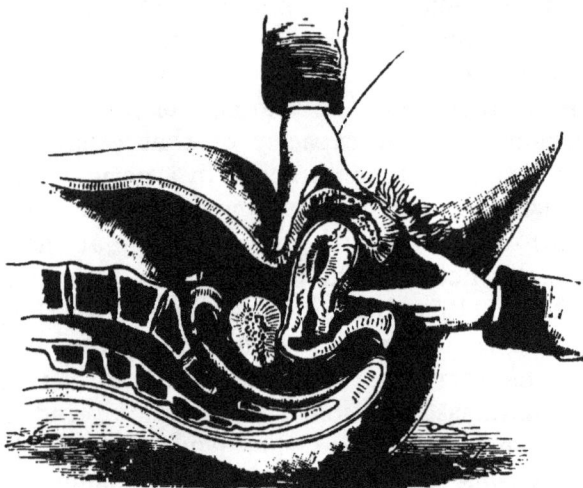

Fig. 1.—Method of Bimanual Examination. (After SIMS.)

carrying out this manipulation, it is essential to obtain the utmost possible relaxation of the abdominal muscles by causing the patient to look up to the ceiling, with her head firmly rested upon the pillow, and by distracting her attention with conversation. If she is directed to breathe deeply, the examiner may take advantage of each expiration to sink the hand gradually deeper into the pelvis without causing painful pressure. It is essential also that the bladder should be empty, or nearly so. A full bladder is generally readily detected as an elastic, fluctuating swelling. But in any case of important or difficult diagnosis, it is well, whenever the fundus uteri is not at once seized between the internal and external fingers, to pass a catheter, and so ensure the bladder being perfectly empty. If the fundus uteri be absent from its normal position, the external fingers may be brought down close upon that in the vagina. In nervous patients, when the abdominal muscles are held very rigid, the full advantage of this method sometimes cannot be obtained without the administration of an anæsthetic. A thick layer of fat in the abdominal walls also interferes with it. But even in such cases, although the uterus cannot be actually felt by the external hand, it is almost always possible to ascertain approximately its size and position by observing up to what level in the abdomen an impulse can be communicated to the finger resting upon the cervix.

While the uterus is thus balanced between the two hands, it is easy to estimate the mobility of the cervix, and of the whole organ, both to upward or downward, and to lateral displacements. At the same time any undue sensitiveness, either to pressure upon the fundus or cervix, or to either form of displacement, is noted. The examiner then quits the uterus, and explores in the same manner the rest of the pelvis. While the internal finger explores deeply the anterior vaginal wall, and all the vaginal culs-de-sac, and searches for any tumour or any abnormal resistance or tenderness, the external hand, at the same moment, defines the upper limits, and ascertains the size, shape, consistence, and mobility of any mass which may thus be detected If this can be fully carried out, it is scarcely possible for any swelling, however small, to escape detection. The *tactus eruditus* of the observer is called most fully into play in the estimation of slight deviations from the normal standard in the mobility of the uterus, and in the resistance of surrounding parts, which may be the only trace remaining of bygone inflammation. Thus there may be much significance in a slight difference of resistance in the two lateral culs-de-sac, or in a deviation of the cervix, the fundus, or the whole uterus to one side, the result of the contraction of old inflammatory material. In thin persons, when the abdominal walls are not too tense, the ovaries, if in their normal position, may be caught between the fingers at a point between the fundus uteri and crest of the ilium, and distant about one and a half inches from the former. The right ovary and right half of the pelvis are best explored by using the right index finger internally, the left ovary and left half of the pelvis by the left index finger.

As a final stage, the patient may be placed on the left side, with the head and shoulders low, the knees well drawn up, and the hips near the edge of the couch. In this position the posterior portion of the pelvis can be explored most deeply by the index finger of the right hand, and this method is especially serviceable in searching for a slightly prolapsed ovary, or a small tumour behind the uterus. The physician should accustom himself to use either hand with equal facility in both positions so that, in case of serious illness, a patient may not be needlessly disturbed.

ABDOMINAL PALPATION COMBINED WITH PERCUS-SION AND AUSCULTATION.—Abdominal palpation is in many cases not required. Frequently, the bimanual touch will assure the physician of the absence of any tumour or other condition upon which its employment could throw light, and thus, if the patient is dressed, the necessity of uncovering the abdomen will be avoided. If, however, the history of a case makes it seem possible that an abdominal tumour or pregnancy may exist, it is convenient to make abdominal palpation the first step of the examination. And if the bimanual touch have revealed the existence of any tumour or swelling, or any notable enlargement of uterus, a further examination will be necessary to ascertain the shape, size, consistence, and attachments of the mass. The examination may be made through a thin garment, but ocular inspection is often desirable to observe the appearance of the skin, the state of the veins, and the presence of any dark abdominal line. With palpation should be combined percussion — which is especially necessary for the distinguishing of phantom from real tumours, and the diagnosis of flaccid cysts or free fluid in the peritoneal cavity—and auscultation, which may reveal the sounds of a fœtal heart, the uterine souffle in pregnancy or in fibroid tumours, or friction sounds on respiration in the case of ovarian or other tumours.

EXAMINATION WITH THE UTERINE SOUND. — The uterine sound is a metallic staff, marked with notches at intervals of an inch (Fig. 2), so that if, in withdrawing it, the finger be kept upon the point corresponding to the os uteri, the distance to which it has penetrated into the uterus may be at once read off by the aid of figures marked upon the stem. For the terminal three or four inches the diameter of the instrument should be less, so that this portion of it may be readily bent to any desired curve, but is yet firm enough to retain its shape while being introduced, and to be used, if required, for the replacement of the uterus. A suitable combination of firmness and pliability is attained if the instrument is made of pure copper, plated. The sound should terminate in a smooth, slightly bulbous extremity, which,

Fig. 2.—Uterine Sound.

for ordinary use, should be about one-eighth of an inch in diameter—that is to say, should just pass through gauge No. 9 of the French scale. But for use in cases of stenosis of the cervical canal, it is necessary to have a sound with a diameter not greater than one-tenth, or even one-twelfth of an inch.

For the use of the sound the patient is placed upon the left side, with the hips near the edge of the couch, and knees well drawn up. There are two methods of

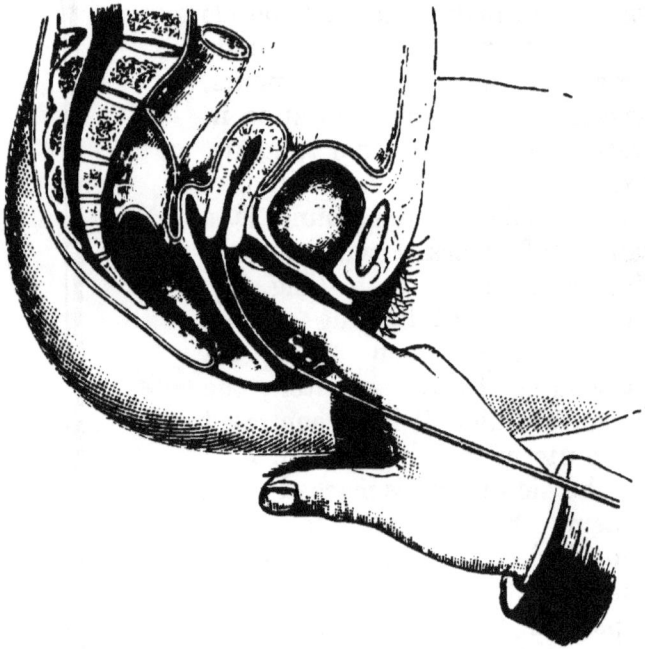

Fig. 3.—Mode of introducing Sound.

holding the instrument during its introduction. The one which I recommend is to introduce the index finger of the right hand into the vagina, and place it upon the os uteri, while the handle of the sound is held very lightly between the thumb and one or two fingers of the left hand, so that its stem rests between the thumb and index finger of the right hand, as shown in Fig. 3. If the vagina is moderately capacious, and the os has its normal direction, the concavity of the sound should, from the first, be directed anteriorly. The handle must at first be held well forward, close to the patient's thighs, and it is then easy, with the instrument in this position, to guide its point along the finger up to the os, and insinuate it gently into the cervical canal, and so onward to the fundus, the handle meanwhile being gradually carried backward. If, however, the vaginal orifice is narrow, and the perineum tight, as in the case of virgins, or if vaginal touch has shown that the os looks forward, instead of looking nearly in the axis of the pelvic brim, as it normally does, it is more convenient, holding the sound in the same way, to direct its concavity at first backward. As soon as it has been passed well into the vagina, in the former case, or as far as it will go into the cervical canal in the latter, its direction must be reversed by sweeping round the handle in a rather wide semi-circle, so that the stem of the instrument describes a semi-cone, while its point does not move, but its terminal portion of two and a half inches rotates nearly on its own axis. This manœuvre resembles the "tour de maitre" of a surgeon in introducing a catheter in the male, and is precisely the converse of that employed in introducing the sound into a retroflexed uterus (see Fig. 31, p. 88).

The sound is generally made with a projecting shoulder at its convex side, at a distance of two and a half inches from the end, to indicate the point which is normally just outside the os uteri, and the notches are also made upon its convex side. This shoulder interferes with flexibility, and is on the wrong side to be readily felt by the index finger of the right hand. For those, therefore, who introduce the sound in the way just described, it is far preferable to have the instrument made with a slightly prominent ring, readily felt from either side, in place of the shoulder, and to have the notches marked upon the concave side, as shown in Fig. 2, p. 7.

In the second method of introducing the sound, one

or two fingers of the left hand are introduced into the vagina, and placed upon the os, while the handle is held in the right hand, the concavity of the instrument being directed forward, and the point is thus guided into the cervix. This plan has the drawback that it cannot conveniently be carried out unless the patient is so placed that her trunk is nearly transverse to the couch, a position which it is often difficult to induce women to assume. Whichever method be adopted, the physician should be able, with equal dexterity, to make use of the other hand, placing the patient upon the opposite side. The sound should not be used, as a matter of routine, in every case, but only when it is likely to afford some additional information, or to clear up some point which previously remained doubtful. Its use is, as a rule, to be entirely avoided in cases of cancer, of acute uterine or periuterine inflammation, especially peritonitis, or when pregnancy is suspected, unless the diagnosis is of such extreme importance that it is desirable.to run the risk of inducing abortion. Even in chronic periuterine inflammation, it should be used only exceptionally, and with great caution. In all cases the direction of the uterine cavity should be previously ascertained, as far as possible, by bimanual touch, and the instrument should be warmed, that it may not, by its coldness, excite spasm of the cervix. If any great flexion of the uterus has been detected, the sound should first be bent to a corresponding curve, and its concavity turned in a suitable direction.

The first object in the use of the sound is *to measure the length of the uterine cavity.* If any obstacle be met with, it should be overcome by changing the direction of the point, or by very gentle and prolonged pressure, to which any temporary muscular spasm will gradually yield. It is to be remembered that a slight hindrance usually occurs at the internal os, and that the point of the sound is often arrested there in consequence of flexion, or, much more rarely, in consequence of stenosis. Some pain is often felt as the sound passes the internal os, and frequently a sudden pain indicates the moment when the point has reached the fundus, which is more sensitive than other parts, and may be excessively so in metritis or endometritis of the body of the uterus.

A second object is *to learn the direction and course of the uterine cavity.* In this respect the information to be gained is as positive as that which an autopsy could afford, and it is by verifying by the sound the inferences deduced from the vaginal and bimanual touch that the physician is best able to acquire the necessary *tactus eruditus.* The conditions in which this indication is most important are when there are tumours near the uterus, which might be mistaken for its fundus, when the uterus is embedded in inflammatory exudation, so that its position cannot be made out by palpation, or when it is distorted by fibroid or other tumours in its substance.

A third use is *to ascertain the permeability and diameter of the uterine canal.* The mode of doing this will be described under the head of stenosis of the cervix. Useful information is also obtained as to the *sensitiveness of the internal surface of the uterus at its different parts*, but for this purpose the sound must be used with much caution. A further application is *to decide upon the presence or absence of any foreign body*, such as a retained ovum, polypus, or other tumour in the cavity of the uterus, and to determine its attachments. For *testing the mobility of the uterus* bimanual touch is generally sufficient, but the sound may be used to great advantage to determine how intimately the uterus is connected with an ovarian or other tumour. In the case of fixation by inflammatory adhesions alone, the use of the sound as a test of mobility is not without danger, and other means are then generally sufficient.

The use of the sound *in conjunction with external palpation* is sometimes of great value, especially when the body of the uterus cannot be defined by the bimanual touch alone, or when it is required to distinguish it from other masses felt in the abdomen, and ascertain its connection with them. For this purpose the right hand may be used most conveniently for external palpation, and the left for holding the handle of the sound, while the patient remains on the left side. In some cases of difficulty, however, it is preferable to place her in the dorsal position. The handle of the sound being slightly rotated, the external hand detects the corresponding movement imparted to the fundus, and observes whether any other masses in the abdomen move with the fundus or not.

That the utmost gentleness is necessary in introducing the sound, is shown by the fact that it has not very unfrequently penetrated a soft uterus, so that its point could be felt beneath the abdominal wall. In some cases it may have passed along a dilated Fallopian tube, but there is no doubt that more frequently it has actually pierced the uterine wall, and sometimes an aperture has remained, through which it could be repeatedly passed. In most such cases no serious symptoms have followed, but the occurrence is not to be regarded as altogether without danger. It is most likely to occur when the uterus is softened by degeneration after parturition or abortion, or by the presence of cancer, or when its wall is extremely thin from super-involution.

The use of the sound for *replacement of the uterus* will be described under the heading of displacements of that organ (p. 87).

Dr. Marion Sims, followed by Dr. Thomas, and others in America, recommends as safer than the sound the uterine probe, which is only a little larger than the ordinary surgical probe, and is perfectly pliable, being made of pure silver or copper. This is used through a Sims' speculum, and the physician gives it the curve which he supposes the uterine canal to have, and keeps altering the curve, if necessary, until he can pass it without using the slightest force. This method has the drawback that the position of the uterus may be modified to an unknown extent by the introduction of the speculum, and the evidence derived from the probe thus rendered fallacious. Moreover, there are some cases of flexions in which there is great difficulty in passing the sound, and in which the operator may derive much assistance from lifting up the fundus with his finger, and so partially straightening the uterus, making due allowance in his mind for the change in its position so produced. This assistance is sacrificed by the use of the speculum, although there is some compensation in the fact that the cervix may be drawn downward or forward by a tenaculum. The dimensions of the vulva limit too much the movements of the handle of the probe to allow it to be passed through a speculum in a case of extreme flexion, the flexion remaining unreduced, while a properly made sound can be equally well bent to any desired curve. Again, when the vulva is at all narrow, and especially in the case of a virgin, the passing of a uterine sound by a skilful hand generally gives the patient far less discomfort than the introduction of a Sims' speculum.

RECTAL TOUCH.—In the case of tumours or inflammatory thickenings behind the uterus, the rectal touch is often the most valuable of all modes of exploration. The finger can reach *per rectum* to a higher level than *per vaginam;* the magnitude of any swelling, and its relation to the recto-vaginal septum and the posterior pelvic wall, can be accurately determined, and the ovaries can often be very exactly made out. The patient may be placed in the dorsal position, and the method combined with abdominal palpation, but for exploration of the posterior and lateral walls of the rectum, the lateral position is preferable. If the patient be directed to bear down as the finger is passing the sphincter, less discomfort is caused by its introduction. In the case of virgins with a very small hymeneal aperture, rectal may replace vaginal touch as a means of ascertaining the condition of the uterus, but as a general rule rectal proves much more disagreeable than vaginal exploration. An inexperienced person may be some-what puzzled in recognizing the cervix uteri as felt *per rectum*, but if the thumb be passed into the vagina, while the index finger is introduced into the rectum, the patient being in the dorsal position, the results of vaginal are at once brought into association with those of rectal touch. The uterus may also be grasped between the thumb in the vagina, and one or two fingers in the rectum, if the fundus is at the same time pushed down by the external hand. Rectal examination may be used in conjunction with a sound in the uterus to determine the connection of retro-uterine swellings with that organ, or in conjunction with a vesical sound in the bladder, in the case of absence or atresia of uterus or vagina, or to distinguish between a polypus and inversion of the uterus.

The scope of rectal exploration has been greatly extended by the method introduced by the late Professor Simon—namely, to place the patient under an anæsthetic, and introduce four fingers, or the whole hand, and, if necessary, a portion of the forearm into the rectum. Two or three fingers may even be passed into the commencement of the sigmoid flexure, and it is possible thus to reach as high as the lower portion of the kidneys. This method, when carried to its fullest extent, is not without danger, and has occasionally led to a fatal result. It should only be employed to establish a very important diagnosis as to the nature and connections of a tumour.

Certain special expedients, to aid the combination of vaginal and rectal touch with bimanual examination, are of use in difficult cases, especially for making out the attachments of a tumour. Thus, if the vagina is not sufficiently capacious, it may be stretched by preliminary plugging, or the use of an air-ball pessary. Another expedient is to place the patient on the left side, seize the cervix with tenaculum forceps and draw it down as far as is possible without using undue force. The handles of the forceps being then given to an assistant

to hold, one or two fingers of the left hand are introduced into the rectum, while the right hand, used externally, helps to push down the fundus if no tumour
intervenes. In this way the pedicle
of a tumour, or band of adhesion, may
often be put on the stretch and so detected. The fingers in the rectum may
also by this method reach as high as
the fundus, and any fault of development may be exactly made out. Dr.
Hegar, who specially recommends this
method, uses simple bullet forceps,
having a catch at the handle, to
draw down the cervix. The uterine
tenaculum forceps shown in Fig. 4,
give a more secure hold, the smaller
arm of the forceps being introduced
within the cervix. To carry out
this plan effectually, anæsthesia is
generally necessary.

EXPLORATION OF THE BLADDER.—
In a gynæcological examination it
may be desirable to empty the bladder
by catheter, or to pass a bladder
sound, in order to test whether the
uterus can be felt in its normal
position between the sound and a
finger in the rectum. The student
should acquire dexterity in performing either operation by the aid of
touch alone. In general, a male
gum-elastic catheter may be used
with quite as much advantage as the
silver female catheter, care being
taken not to push the instrument
too far into the bladder, so as to run the risk of
injuring the posterior bladder-wall. In some cases,
however, when the urethra is distorted, as by the
presence of tumours, the rigid metal catheter has an

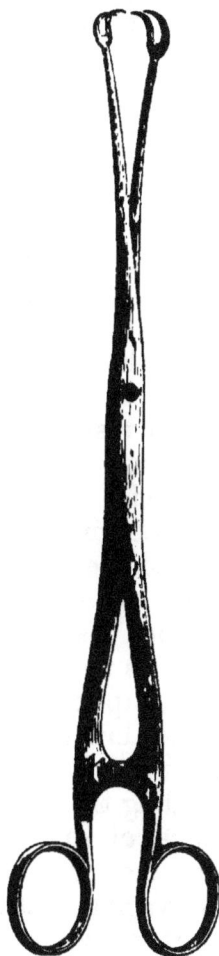

Fig. 4.—Uterine
Tenaculum Forceps.

advantage, from the fact that its course can be more
precisely directed. The use of the catheter, especially
if frequently repeated, is always liable to set up
cystitis, and one element in the production of this
result appears often to be the introduction of germs or
septic material, by means of the catheter, into the
bladder. Care should be taken, therefore, that the
catheter is perfectly clean, and previously disinfected
by carbolic solution, or other antiseptic. Either euca-
lyptic vaseline (*see* p. 2) or carbolic oil, not in too
great profusion, and not stronger than 1 in 20, may
be used to lubricate the instrument.

Mode of passing Catheter.—To pass the catheter,
the patient should be placed in the dorsal position,
with the knees flexed. A long elastic tube may be
fitted on to the catheter, in order to conduct the urine
into a vessel under the bed. It is generally preferable,
however, to have a small vessel close at hand, for the
physician can then instantly perceive as soon as the
urine begins to flow, and thus be warned that he has
passed the catheter far enough. A full-sized catheter,
from No. 10 to No. 12, should be chosen, for the point
is then less likely to catch in any depression of the
mucous membrane. The guide for finding the meatus
is the apex of the pubic arch. Supposing the physician
to be standing at the right side of his patient, he passes
his right hand beneath the thigh, and his left hand,
holding the catheter, above the thigh. With the index
finger of the right hand, he first finds the perineum,
and then introduces the tip of the finger just within
the vagina, that is to say within the circle of the hymen,
if there is one existing. The urethra can then be felt
as a cord against the apex of the pubic arch. The tip .
of the finger is slightly withdrawn to the extremity of
this cord, and feels, just in front of it, the orifice of
the urethra as an obvious depression. The catheter
being still held in the left hand, its point is then
guided into the orifice. If the upper part of the
urethra or neck of the bladder is pushed forward above

the pubes, as by a tumour, or by the presence of the fœtal head, it is often useful, as the catheter passes onward, to direct its point upward, through the medium of the urethral wall, by the finger passed into the vagina.

DIGITAL EXPLORATION OF THE BLADDER.—The anterior surface of the uterus and ovaries, and of any tumour in connection with them, may be very immediately reached by passing the finger into the bladder, after rapid dilatation of the urethra. For this purpose no instrument is more convenient than Bryant's urethral speculum dilator (Fig. 76). An anæsthetic is administered, and the urethra is then stretched by means of the dilator, until first the little finger, and afterwards the index finger, can be introduced. If necessary, the margins of the meatus may be slightly incised as a preliminary step. Some cystitis may be set up, and long-standing, if not permanent, incontinence of urine has occasionally followed: the plan, therefore, should only be adopted in order to make a diagnosis of great importance as regards the condition of the uterus. It is more frequently called for to ascertain the presence of growths or other diseased conditions in the bladder itself. As a rule, there is no permanent incontinence if the urethra be not dilated beyond the size of a moderately slender index finger.

THE SPECULUM.—The use of the speculum is less important for diagnosis than to facilitate the application of remedies and the introduction of instruments, as in the operation for the cure of fistulæ. In diagnosis, it serves chiefly to reveal the appearance of the cervix, especially as to the presence or absence of any erosion, the character and abundance of the secretion issuing from the os, and also the condition of the vaginal walls. Out of all the numerous varieties of specula there are four of special value, and of these each has such distinctive merits that three, at least, of them are essential to the gynæcologist for use under different circumstances.

Ferguson's Tubular Speculum. — The speculum

C

which concentrates far more light than any other
upon the os uteri, and one which commonly brings the
cervix readily into view, is Ferguson's speculum of
silvered glass, with bevelled extremity, and trumpet-
shaped entrance, whereby the rays of light are concentrated (Fig. 5). It has the further advantage that

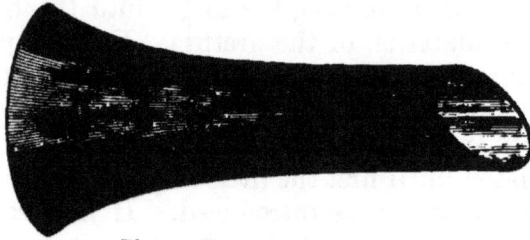

Fig. 5.—FERGUSON's Speculum.

it is readily cleaned, and is unaffected by acids or
other fluids, while its sides protect the vagina from
any application used, and a considerable quantity of
fluid can be conveniently poured into it, if such a
mode of application is desired. These specula can be
obtained of toughened glass, whereby the objection of
fragility is, in great measure, obviated.

For the introduction of the cylindrical or bivalve
speculum, it is more usual in Britain to place the
patient in the lateral, or, what is better, the semi-
prone position. This has the advantage in point of
delicacy, but is open to the drawback that it requires
a nearly horizontal light, such as is not easily obtained
in a ground floor room, and that the patient's legs,
feet, and dress are apt to interfere with the illumination. The dorsal position has the great advantage
that the effect of gravity then tends to bring the axis
of the uterus more nearly into coincidence with that
of the vagina, and so facilitates the exposure of the os.
It should always be adopted, therefore, if any great
difficulty is found in bringing the os into view, especially when this is due to anteversion of the uterus.
If the uterus be retroverted, the lateral position often

answers better, since the tip of the speculum can then be more easily directed forward to find the cervix. In either case the speculum is introduced without exposure of the patient. The position and direction of the cervix are first ascertained by the index finger : then by two fingers the labia are separated and perineum retracted so that the bevelled tip of the speculum can be passed beneath it. The speculum is then gradually pushed on in a backward direction, stretching the perineum still further back, while any painful pressure on the sensitive structures on the anterior wall of the vulva is avoided. The direction finally given is regulated by the position of the cervix as previously ascertained. If the os does not at once come into view, the speculum must be drawn back somewhat, and again pushed on in a different direction. Not unfrequently, when the uterus is anteverted, only the anterior surface of the cervix and anterior lip of the os are fully brought into view in this way, the whole circuit of the os not being fully seen. This difficulty may often be overcome by rotating the speculum till its projecting tip is anterior, in which position it tends to push up the fundus. Another plan is to draw the os into the centre of the field by means of a tenaculum hook, or by the sound passed just within the cervix. If this fails, the best plan is to use a bivalve or Sims' speculum.

A Ferguson's speculum is generally made about six inches long, and, when the vagina is long, or the woman very fat, some such length as this is necessary. For many purposes, however, a short Ferguson's speculum of full diameter has great advantages over a longer one, and it is well to have such an instrument in addition. It should be barely 4 inches long on the longer side, and $3\frac{1}{8}$ on the shorter, the bevelled end being less oblique than usual, the external diameter about $1\frac{1}{2}$ inches. The outer end should have only a moderate rim, not a wide, trumpet-shaped expansion. There are two special advantages in such a short speculum. First, it allows any point of the cervix to be felt through the speculum. This is very useful for guiding the bistoury or needle to distended cervical glands, which are often much more readily felt than seen. Secondly, by the lateral stretching of the vagina, it draws the cervix nearer the outlet, instead of pushing it further away, and, in consequence, a probe for intra-uterine medication can be much more easily passed into the canal, than with a longer speculum. This may, indeed, be done still more advantageously with Sims' speculum, but over that the short Ferguson's speculum has the great advantage that no assistant is required.

For illumination direct daylight is far superior to anything else, and, if the patient be in the dorsal position, a descending light, if the angle with the horizon be not greater than about 45°, answers excellently. If direct daylight cannot be obtained, it is often convenient to use a concave mirror, similar to a laryngoscopic mirror, having a rather large central aperture cut quite through the glass, and mounted upon a handle. This may be used to reflect either daylight or the rays of a lamp.

The Bivalve Speculum.—Of all valvular specula, the best is Cusco's bivalve speculum (Fig. 6). It is very easily introduced, and, in some respects, is the most convenient of all specula, especially in the fact that it is perfectly self-retaining. Its successful action depends upon a correct mode of introducing it. It is essential to ascertain first with the finger the exact direction and distance of the os. The speculum is tilted sideways to pass the vulva, then turned so that the blades are antero-posterior, and pushed on till their extremities are a little short of the os, but exactly in its direction, special care being taken that they do not pass beyond it into either cul-de-sac. The blades are then opened by the handles, the effect of which is that the fundus is pushed up by the anterior blade, and the antero-posterior stretching of the vagina at the same moment draws the cervix downward and forward, so that the axis of the uterus is brought nearly into coincidence with that of the vagina. The lips of the os are also drawn somewhat apart, so that the interior of the cervical canal can be seen. As soon as the os is fully in view, the speculum is at once fixed by the screw at the side. The essential points in a good speculum are that the blades should be capable of wide separation, that they should themselves be wide enough to prevent the lateral vaginal walls encroaching on the field of view (for which purpose a width of about $1\frac{1}{2}$ inches near the extremity is desirable), and that they should

Fig. 6.—CUSCO'S Bivalve Speculum.

not be too short. If one speculum only is used, the length of each blade should be about $4\frac{1}{2}$ inches, but a shorter instrument may be employed with advantage in a short vagina. The blades should never be placed laterally, for then the natural tendency of the vagina to become flattened antero-posteriorly causes its walls to encroach upon the field of view. The handles may be turned either toward the perineum or toward the pubes, as is most convenient. All the modifications, or so-called modern improvements, of valvular specula, in which three or four blades are employed, or the anterior blade is made much shorter than the posterior, interfere with this mechanism of bringing the uterus into a position of slight retroversion, and so do away with the special advantage of this form of speculum. In withdrawing the speculum, care must be taken not to allow the blades to close completely, and thereby pinch the vaginal walls.

Sims' Speculum.—Sims' univalve speculum (Fig. 7) is far superior to all others for many purposes, as

Fig. 7.—SIMS' Speculum.

when it is desired to introduce a tent or probe through the speculum, or to operate upon the cervix or vaginal walls. Its drawback is that it cannot be employed without an assistant, while a skilled assistant is necessary to give it its full value. The most important element in the use of this instrument is the position of the patient. To get the full benefit of the speculum, all dresses fastened round the waist should be loosened, as a preliminary step. The patient is placed on a high and firm couch, or table, and the light must be nearly horizontal. She lies on her left side, in a semi-prone position, with the head and shoulders low, and the left arm drawn behind her, so that the sternum is rotated forwards, coming very nearly into contact with the table. The legs are flexed at right angles to the trunk, and the right rather more than the left, so that the right knee lies just above the left, in contact with the table (Fig. 8). The nurse or assistant stands behind her, and pulls up the right side of the nates with the left hand. The physician then introduces the speculum, guiding it with the finger into its position behind the cervix, draws back the perineum so as to convert the vagina into a straight canal, and gives the instrument into the hand of the assistant, who holds it firmly in the desired position, maintaining the retraction of the perineum. In any long operation the hand of the assistant is apt to become fatigued, and therefore unsteady. Steadiness will be promoted if he can keep the speculum in position by fixing his hand as

Fig. 8.—Position for introduction of SIMS' Speculum.
(After SIMS.)

a wedge between its handle and the patient's sacrum, instead of depending solely upon muscular effort.

The object of this position is to make the vulva the highest point of the vaginal canal, and allow the effect of gravity on the abdominal viscera and walls to draw the anterior vaginal wall forward, and expand the canal into an air-containing cavity, almost as effectually as if the patient were in the knee-elbow position. When, however, the vagina or vulva is narrow, the anterior vaginal wall does not fall away sufficiently to allow the os to be seen, and it is then necessary to push it back, either by the finger, or by

a sound or similar instrument, or by a depressor made
for the purpose. This tends to draw the cervix forward. If, however, the cervix is still directed too
much backward to expose the os fully to view, or to
bring it into a convenient position for the introduction
of a probe or other manipulation, a small tenaculum
hook is to be fixed in the anterior lip of the os. By
this means the cervix is drawn forward until it is
nearly in the axis of the vagina (*see* Fig. 14, p. 30).
This measure causes very little pain or inconvenience,
and the shank of the hook serves at the same time for
a depressor of the anterior vaginal wall. In many
cases it answers still better to use, in place of the

Fig. 9.—Modified Sims' Speculum.

hook, either the uterine tenaculum forceps, shown in
Fig. 4, p. 15, or similar forceps having a single tooth,
instead of two to each blade. Various modifications
of Sims' speculum have been invented with the object
of attaching to the instrument a sacral plate and
depressor, and thereby rendering it self-retaining,
dispensing with the need of any assistant, and leaving
the operator's hands free. Another convenient modification is that shown in Fig. 9, in which the blade of
the speculum is split, and can be expanded by a screw,

according to the size of the vagina, and the amount of space required.

Neugebauer's Speculum.—A fourth speculum often of great service is Neugebauer's speculum (Fig. 10). This consists of two blades, each resembling a Sims' speculum, and introduced in a similar way, but so adjusted that one blade slides within the other in such manner that the two blades in combination form virtually a bivalve speculum. It is inferior to Cusco's speculum in self-retaining power and in efficacy for bringing the cervix into the line of the vagina. Its special advantage is that it can be guided exactly into position by the finger ; and thus it is generally superior

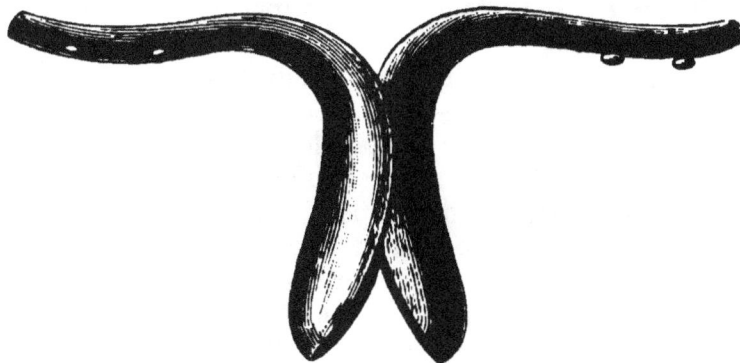

Fig. 10.—NEUGEBAUER'S Speculum, the blades of which may be united to form a SIMS' Speculum.

to all others if a speculum has to be used in a case of cancer of the cervix, other specula being liable to set up considerable hæmorrhage. Each blade should be about four inches long, and the handles may be so made as to clasp together in a reversed position to form a Sims' speculum (Fig. 7, p. 22). Dr. Barnes has introduced a modification of this instrument under the name of the "crescent speculum." For each handle another blade of different size is substituted. Thus the two pieces make a series—three different sizes of speculum—Nos. 1 and 3 being in one piece, and Nos. 2 and 4 in the other. For the use of Neugebauer's speculum the patient may be either in the semi-prone position or in the lithotomy position, with the nates overhanging the end of table. The larger blade should be introduced first, and guided by the finger into its position behind the cervix; the smaller blade will then slide into position within it.

DILATATION OF THE CERVIX BY MEANS OF TENTS. —The diagnosis of morbid conditions of the mucous membrane of the uterus, and of the presence of tumours or the products of conception within its cavity, is in many cases rendered impossible by the closure of the os. Dilatation of the cervix is then the only method of detecting the disease, and is of still greater importance in allowing access for thera-peutical means.

There are two substances commonly used for the manufacture of tents—compressed sponge and the laminaria digitata, or sea-tangle, each of which has special advantages under different circumstances, Sponge tents should be steeped in carbolic acid during their preparation, to render them antiseptic. They should be made of a uniformly conical shape, not bulging at the centre, and the string for their with-drawal should be attached to the upper extremity, and pass through the length of the tent, since it is other-wise liable to break away, and leave the greater part of the tent within the uterus. Laminaria tents (Fig. 12, p. 28), should be perforated from one end to the other, to allow them to be fixed upon a stylet for introduc-tion, and to render their expansion more rapid and com-plete. Tents made from the root of the tupelo tree have lately been introduced from America, and recom-mended as being more rapid in their action than those of laminaria, but they have not been found so efficacious in their dilating power.

Relative Advantages of Sponge and Laminaria Tents.—A sponge tent insinuates itself very closely into the interstices of the mucous membrane, and on this account it is less liable to slip out, and forms a more efficient plug in cases of hæmorrhage than the laminaria tent. The same property gives it an important therapeutic use in modifying the surface of the uterine mucous membrane, for which purpose it should be long enough to reach nearly to the fundus uteri (Fig. 11). It also causes less pain during its expansion than the harder laminaria tent. It has the

Fig. 11.—Sponge Tents.

disadvantage of more rapidly becoming offensive, but this is obviated, in great measure, by the preparation with carbolic acid, provided that the tent be not left in place longer than about twelve hours.

A laminaria tent is smoother, and can more conveniently be made of small size. It is, therefore, more easy to introduce, and is more suitable to commence with, when the cervix is small. In cases of flexion, it can be softened in warm water before introduction, and curved to suit the vaginal canal. It is capable of overcoming greater resistance in expansion than a sponge tent, and a wide dilatation may be effected by packing a number of laminaria tents side by side. As a rule, therefore, when the object is to explore the uterine cavity, laminaria are to be preferred to sponge tents, unless it is desired, at the same time, to arrest hæmorrhage.

Mode of Introducing Tents.—In most cases a laminaria tent is introduced most easily by the tent-introducer, contrived by Dr. Barnes (Fig. 13, p. 28). It consists of a wooden handle carrying a curved stem, at the extremity of which stylets of various sizes can be screwed in, and over which slides a gum-elastic

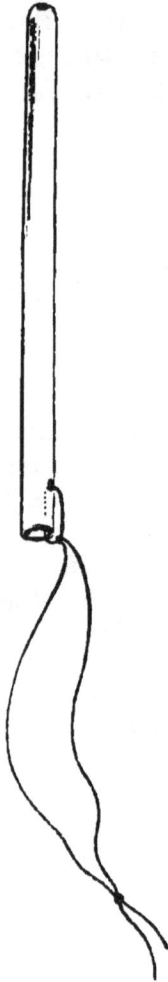

Fig. 12.
A Hollow Laminaria Tent.
(Actual size.)

Fig. 13.
BARNES' Tent Introducer.

tube. The tent being fixed firmly upon the stylet, the whole instrument is introduced exactly like the uterine sound, and the tube is then held steadily against the os, the disc at its lower extremity giving a point of resistance to the finger, while the stylet is withdrawn. The instrument can be extemporized by cutting off the end of a gum-elastic catheter, so that the stylet projects about an inch, and mounting the perforated tent upon this projecting end. In the case of a sponge tent, this method is less satisfactory, for the point is apt to become softened, by absorption of moisture, before it can be introduced. The difficulty may sometimes be overcome by smearing the point of the tent with a solid lubricant, lard or vaseline, and introducing it rather quickly, before it has time to soften. To keep the tent in place till it has time to swell, a tampon of cotton-wool, soaked in iodized glycerine or carbolic oil, should be placed beneath its extremity.

Another method of introducing a tent is to employ Sims' speculum and the semi-prone position. For a sponge tent, this plan is generally preferable, and it should be adopted in all cases of difficulty, even for the introduction of a laminaria tent. By the tenaculum hook or tenaculum forceps, the cervix is drawn nearly into the line of the vagina (Fig. 14, p. 30), and the direction of the uterine cavity is ascertained by the sound or probe. The tent is then guided into place either with a pair of forceps, or, more conveniently, by the tent-introducer. If laminaria tents are made about five inches long, instead of the usual length of about two inches, and are passed up nearly to the fundus, they are free from the risk of slipping out, but have the disadvantage that they hold the uterus forcibly in a position of retroversion, and hence cause more irritation. In some cases of fibroid tumour, in which the cervical canal is elongated, it is absolutely necessary to use a tent of extra length, in order to ensure its reaching within the internal os. If pain is produced during

the expansion of a laminaria tent, a morphia suppository should be administered.   If the internal os is
very rigid, pain may be great enough to call for more
decided opiate treatment.   In such case the tent may
be so tightly constricted at one point as to prevent any
great expansion, while it swells above and below.
The extraction may then be somewhat difficult, and
firm counter-pressure by the finger against the cervix
may be required to effect it.

Before the introduction of a tent, it is useful to
fasten a tape to the loop of thread which is passed
through it.   This tape serves to withdraw the tent.
It is also convenient, when a second tent is being

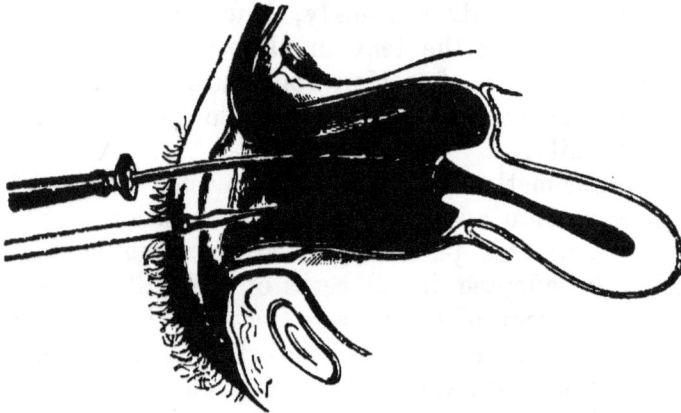

Fig. 14.—Mode of Introducing a Tent through SIMS' Spéculum.

introduced by the side of the first, to hold the first
steady by means of the tape, and so prevent its being
pushed up too far.   Care must be taken to leave the
ends of the tents projecting through the external
os, otherwise the tents are apt to expand in the canal,
leaving the os undilated, so that it may be impossible
to remove them except by incising or dilating the os.
For withdrawal the loop of thread is generally sufficient.   If this should break away, the tent being
firmly gripped at the internal os, the end of the tent

must be firmly seized with suitable forceps and drawn down.

Dangers from the Use of Tents, and Precautions required.—No small number of cases is on record in which the use of a tent has been followed by metritis, pelvic cellulitis or peritonitis, or even general and fatal septic peritonitis. The chief source of danger is the absorption of septic material by the lymphatics, a consequent rapid spread of inflammation along their course, and in some cases an almost immediate conveyance of septic contagion to the peritoneum. This danger may, to a very great extent, be avoided by suitable precautions. There are some cases, however, in which a risk is inevitable, as when a patient is exhausted by severe hæmorrhage, or is already the subject of septicæmia, but in which it may be justifiable to incur it on account of the still greater risk in non-interference. Serious effects have more frequently occurred when a series of tents have been used to effect progressive dilatation. The most important precaution, therefore, is not to employ tents more than twice in immediate succession, but if sufficient dilatation has not then been effected, to wait awhile before resuming the process. In dilatation for the purpose of diagnosis, or gaining access to a tumour, laminaria tents are the best to use. As many of these as can be introduced without force should be placed side by side. This may be done either at the first sitting, if the cervix is not small, or after preliminary dilatation by a single tent. An antiseptic vaginal injection should be used before the insertion of a tent and after its removal, and a sponge tent should not be left in place much more than twelve, or a laminaria tent more than twenty-four hours. If any rigor or rise of temperature occur, dilatation should at once be suspended. It is of the utmost importance also that the patient should be in bed when a tent is introduced, and should remain so until at least twenty-four hours after its removal. Tents should not be used, unless for extremely urgent cause, when any recent acute inflammation is present; and in cases of pelvic peritonitis, even of an old or chronic character, they should be avoided as a rule, since such an inflammation is apt to be rekindled on slight provocation.

INSTRUMENTAL DILATORS.—The most effective instrument for immediate dilatation of the cervical canal is that of Dr. Marion Sims (Fig. 15). This may sometimes be employed with advantage, the patient being placed under an anæsthetic, when, by previous use of tents, the cervix has been almost, but not quite, sufficiently dilated to allow the index finger to pass for exploration. It may also be used when spontaneous dilatation has occurred to a similar degree, as, for instance, in the case of abortion, and may then sometimes avoid the necessity for the use of tents.

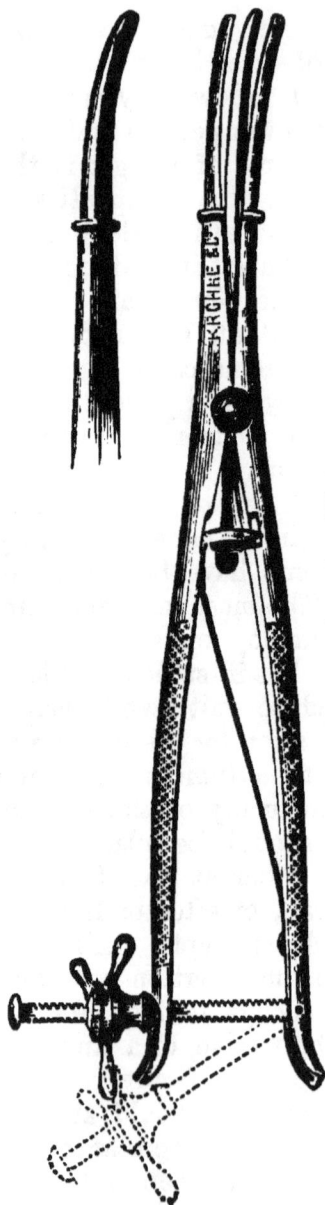

USE OF HYDROSTATIC DILATING BAGS. — In some cases, when the uterus is greatly enlarged by a

Fig. 15.—SIMS' Uterine Dilator.

tumour projecting into its cavity, and sufficient dilatation cannot be obtained by tents, the process may be completed by the aid of Dr. Barnes' hydrostatic dilators, which were designed especially for the gravid uterus. Except, however, in the case of such great enlargement, the cavity of the unimpregnated uterus is not large enough for even the smallest dilator.

CHAPTER II.

PHYSIOLOGY OF NORMAL MENSTRUATION.

By the term menstruation, or catamenia, is understood a hæmorrhage from the mucous membrane of the body of the uterus, which normally recurs at regular intervals of, approximately, one month, and continues throughout the whole period of sexual activity in women, except during pregnancy and lactation. In a menstrual period there are three phenomena intimately connected together. First, active hyperæmia of the uterus and ovaries, with engorgement of the erectile tissue surrounding those organs; second, rupture of one or more Graafian follicles, with escape of the contained ovules; third, disintegration of the surface of the mucous membrane lining the body of the uterus in degree sufficient to cause rupture of the vessels, and effusion of the menstrual blood. With the latter is associated an increased secretion from the cervix and vagina. There are several points in the physiology of menstruation, and in the relation between its several elements, as to which exact data are as yet wanting. Since the connection of menstruation with ovulation, first suggested by Power in 1821, was established by the researches of Négrier, Bischoff, Coste, Pouchet, Raciborski, and others, it has generally been believed that the mucous membrane becomes tumefied during the period, that the height of hyperæmia is coincident with the flow of menstrual blood, and that the follicle is ruptured at the same time or shortly after. It has

also been thought that conception is most frequent
shortly after the end of a period. Recent researches,
however, especially those of Kundrat, Engelmann,
John Williams, and Leopold, while differing in important points, have agreed in showing that the mucous
membrane attains its greatest thickness and development, and that hyperæmia is usually at its height,
immediately before the commencement of a period.
Anatomical evidence has also been adduced to show
that the follicle is commonly ruptured before the onset
of a period, though there are also cases recorded in
which it was found not yet ruptured, but apparently
on the point of rupture, during or immediately after
menstruation.   The view thus suggested by modern
observers—which is the same as that first supported
by Pouchet and Tyler Smith—is that in the inter-
menstrual epoch there is a growth of the uterine
mucous membrane, to render it a fit receptacle for the
ovum, and that the exfoliation of mucous membrane
and discharge of blood is already a retrogressive change,
analogous to the separation of the decidua in parturition, and denoting that the impregnation of that
particular ovum has not taken place.   From this view
would follow a conclusion contrary to that hitherto
general, namely, that the fertilized ovum commonly belongs, not to the last menstrual period which occurred,
but to the succeeding period which failed to appear.
The same conclusion has been supported by Löwen-
hardt, by evidence derived from the duration of
pregnancy.

Recent evidence has also compelled us to regard the
association of ovulation with menstruation as by no
means an invariable, although a general rule.   In women
whose ovaries are not developed, and in those who
have been spayed before puberty, menstruation never
appears.   Thus it is proved that a stimulus to the
nervous system, which originates in the ovaries, is
necessary for the establishment of that function.   A
considerable number of cases has been recorded, how-

ever, in which, after removal of both ovaries, menstruation is said to have continued more or less regularly for years. Anatomical evidence has also been found, on the one hand, of ovulation where no menstruation had ever taken place, and, on the other hand, of the absence of any sign of recent ovulation in women who had died during or immediately after menstruation. In rare cases, also, pregnancy has occurred in women who had never menstruated, though long past the age of puberty. Nevertheless, it remains true that the association of ovulation and menstruation is the general rule. After removal of both ovaries, menstruation does usually cease from the time of the operation, as was the case in eight out of nine instances observed by Thomas. Again, out of fifteen cases of oophorectomy, or extirpation of the functionally active ovaries, by Dr. Battey, there were nine in which both ovaries were completely removed. In all of these menstruation permanently ceased, while in the remainder it continued as before. Some of the apparent exceptions may be explained on the ground that the ovaries were removed piecemeal, and that some small portion of their tissue may have been left, or that a small supplementary ovary may, as is sometimes the case, have existed. Thus, in one case at least, not only menstruation, but pregnancy has occurred after the supposed complete removal of both ovaries. Again, amenorrhœa is a common result of cystic degeneration of both ovaries. Hence the probable conclusion is that the immediate source of the menstrual nisus, and of its periodical recurrence, lies rather in the nervous centres than in the ovaries, though the stimulus of the ovaries is necessary for its first establishment, and in most cases for its continuance. The final development and rupture of the Graafian follicle would then be rather the effect than the cause of the hyperæmia, and the exact period of its rupture might probably vary according to the stage which it had reached when the menstrual nisus commenced. It is also probable that Graafian follicles may occasionally be ruptured in the inter-menstrual intervals, especially under the influence of the hyperæmia induced by coitus; and it is certain that the menstrual period may pass without the rupture of any follicle, if there happen to be none sufficiently near to maturity.

The view that the immediate cause of the menstrual nisus lies in the nervous system appears to be supported by the fact that there are not merely local changes in menstruation, but a monthly cycle affecting the whole organism. Sphygmographic observations have shown that arterial tension is above the average for some time before the period, and becomes lowered during the period. Stevenson has shown that the curve of temperature rises about half a degree above the mean for about a fortnight before the period, falling to a similar extent below it during and after the period, and also that a curve representing the excretion of urea follows a similar course.

Source of the Menstrual Blood.—That the effusion of blood does not depend upon hyperæmia solely is shown by the fact that, when the uterine mucous membrane receives the stimulus in nutrition due to the implantation of a fecundated ovum, no hæmorrhage occurs, although the hyperæmia increases to a higher point than that of menstruation. The first step leading to rupture of the vessels is therefore a disintegration of the mucous membrane, and a fatty degeneration of this tissue, preceding the commencement of hæmorrhage, has been described by Williams. On careful microscopic examination of the menstrual blood, groups of cells belonging to the uterine mucous membrane may frequently be found, especially during the first two days of the period; and not unfrequently minute shreds of membrane, showing the apertures of the uterine glands, generally denuded of their epithelial lining, are also seen. The completeness of the disintegration appears to vary in different persons, but exfoliation in larger pieces is a morbid condition, which will be noticed under the head of membranous dysmenorrhœa. As to the depth of the normal exfoliation, final proof is yet wanting. Of recent observers, Williams maintains that the whole thickness of soft tissue, commonly regarded as mucous membrane, is thrown off every month, leaving only the extremities of the glands embedded in the muscular coat, the inner layer of which he regards as belonging, in development, to the mucous membrane, and as being, in fact, the muscularis mucosæ. The regeneration he describes as beginning at the internal os, and extending towards the fundus. The proof is incomplete from the fact that the instances in which complete exfoliation was found were cases of death by acute febrile diseases, so that the disintegration might have been morbid. Kundrat and Leopold adduce cases to show that, even near, or shortly after, the end of a period, no more than the most superficial layer of mucous membrane was found wanting, and attribute the decrease of the thickness to diminution of œdematous swelling rather than to loss of substance. Engelmann denies any exfoliation of even the surface.

The view .that permanent communications exist between the blood-vessels and the uterine glands, and that these are the source of the exudation of menstrual blood, may be regarded as now exploded. The mucous membrane of the cervix normally takes no part in the outpouring of blood, and its surface remains intact. The coagulation of menstrual blood is usually prevented in its admixture with the acid vaginal secretion. If the quantity of blood is excessive, or if it is retained long within the uterus, in consequence of stenosis or flexion, clots are formed. The quantity of blood normally lost is estimated at from three to seven ounces. The amount of loss depends in great measure upon the degree of active hyperæmia, as is shown by its increase from the effects of exercise, or in consequence of coitus. The natural duration of the flow is from three to five days, but in some women it lasts habitually for seven or eight. The period of recurrence, in women who are perfectly regular, usually varies from twenty-seven to thirty days.

Period of Possible Conception.—There are two considerations which render it very difficult to draw any positive conclusions as to the stage of the menstrual cycle at which it is possible, or usual, for conception to occur: first, that the life of spermatozoa within the uterus may be prolonged for certainly as much as eight days, and possibly for longer; and, secondly, that we have no evidence, in the human subject, as to the time occupied by the ovum in descending the Fallopian tube, or during which it may retain its vitality. There is no doubt that fruitful intercourse may occur at any part of the menstrual cycle, and that any method for preventing pregnancy by abstinence during any special period is unreliable. That abstinence shortly after the period has no such effect is shown by the case of the Jews, who are, if anything, more fertile than other nations. Strict observers of the Jewish law practise abstinence during five days for the period, and seven days for purification afterwards, reckoning from the end of the five days, or from the last appearance of blood, if the period lasted longer than five days, an interval which amounts to at least twelve days in all.* The converse fact that a single fruitful coitus may occur between four and ten days after the commencement of the flow is proved by cases recorded by Dr. Marion Sims, who considers the latter part of this period as the preferable time in order to ensure pregnancy. It can scarcely be doubted that menstruation is really analogous, in some measure, to the æstus or rut of animals; although there is the important contrast that, in animals, coitus takes place only at the time of æstus; but, in the human subject, usually at any other time except that of menstruation. The latter circumstance, however, is rather the result of civilization, and of a feeling of delicacy, for there is no doubt that an

* See Leviticus xv. 19 to end.

increase of sexual feeling does normally take place at the menstrual period as at that of æstus. Hence the common opinion that intercourse near the time of menstruation is more likely to prove fruitful is probably correct; but within what limits of time the uterine mucous membrane is capable of receiving an ovum remains as yet uncertain.

Commencement and Cessation of Menstruation.—The first appearance of menstruation usually coincides with the age of puberty, and the development of the breasts, the pelvis, and the hair on the pubes, as well as the mental changes which occur at the same time. The most frequent age is, in temperate climates, the fourteenth or fifteenth, or, somewhat less commonly, the sixteenth year; but variations between the tenth and twenty-first year are not very rare. The influence of climate is considerable, and in hot countries menstruation commences, on the average, about two years earlier, while it is, at the same time, more profuse. In arctic climes, on the other hand, its appearance is delayed to about an equal extent, and the quantity of blood lost is very small. Cases of precocious menstruation occasionally occur in childhood, and even infancy, and are then associated with premature development of breasts and pelvis, and probably with premature ovulation. In such a case, pregnancy has occurred at the age of eight years.

The time of cessation of menstruation (climacteric period, menopause, or change of life) is, on the average, about the age of forty-five. Women who menstruate early do not generally reach the menopause early, but the contrary, unless some diseased condition of ovaries or uterus has supervened; and, when menstruation is established late, the same ovarian inactivity often leads to an early cessation. In very rare cases true menstruation may continue, and pregnancy be possible, as late as the age of sixty.

Symptoms and Concomitants of Menstruation.—In women of robust health no premonitory signs are noticed, but, in those of more impressible nervous system, for some days before the period is due the breasts often become firm, or even painfully hard, and may be the seat of neuralgic pain, a condition which generally disappears within a day or two after the commencement of the flow. At the same time there is an increased irritability of nerve-centres, which, in women subject to hysteria, epilepsy, or migraine, is shown by the greater frequency of attacks at this period. If congestion of uterus or ovaries is present, pelvic pain precedes menstruation by some days. In cases of hernia of the ovaries, these organs have been found to become swollen and tender a little before menstruation, and continue so during the period. Vaginal touch during a menstrual period shows the uterus as well as the vagina to be turgid and soft. The soft condition of the uterus, however, is alternated with contraction, especially if any obstacle to the flow exists, and, if death occurs during a period, the muscular wall of the uterus is often found pale, from expulsion of the blood, while the mucous membrane, ovaries, and surrounding parts are highly congested. The cervix uteri, vagina, and vulva participate in the engorgement, and increased secretion from them precedes, accompanies, and follows the flow of blood. There is often a tendency to constipation shortly before the period is due, just as there is in early pregnancy, even before the uterus has enlarged sufficiently to produce any effect by pressure, and this constipation may be succeeded by relaxation of the bowels after the flow has commenced. Even in health, some degree of fulness, and of general lassitude, is usually felt just before, and for the first two or three days of the period—a condition expressed by the saying of women that they are "unwell."

CHAPTER III.

MALFORMATIONS OF THE UTERUS AND VAGINA.

THE Fallopian tubes, with the uterus and vagina, are developed from two distinct tubes, called Müller's ducts, which coalesce about the eighth week of fœtal life throughout that portion which forms the uterus and vagina, the point where junction should begin being marked by the insertion of the round ligaments. The graver congenital deformities of these organs depend upon a complete or partial failure either in the development of one or both of these ducts, or in their junction, and it will therefore be convenient to consider such deformities, both of uterus and vagina, in conjunction.

ABSENCE OR RUDIMENTARY DEVELOPMENT OF UTERUS. —The uterus may be completely absent, or may be a rudimentary membranous body with or without an enclosed cavity. Frequently in such cases there is a single solid cervix, and separate horns containing small cavities, a condition which constitutes the *uterus bipartitus.* The ovaries may be absent or present, the vagina absent or short, while the external genital organs are normal. When the ovaries are present, distress may arise from an unrelieved menstrual molimen. In one such case Dr. Battey has performed the operation of oophorectomy with a good result. The diagnosis is generally to be made by rectal touch in conjunction with bimanual examination, and may be aided by the introduction of a catheter, or of the finger, into the bladder. When the vagina is entirely absent, women may marry in ignorance of their deformity, and may afterwards be anxious for operative assistance. The attempt to make an artificial vagina, however, involves in these cases a risk of opening the peritoneal cavity.

The duct of Müller, on one side, may be formed normally, while that on the other is absent or imperfectly developed, and fails to coalesce fully with its fellow. This condition constitutes the *uterus unicornis*. The uterus is curved to one side and terminates in a point, from which the round ligament and other appendages take their origin. The rudimentary cornu, if present, is commonly attached about the position of the internal os, and may be pervious or not. Menstruation is usually normal. Pregnancy may occur in the developed horn, and proceed naturally. It may also take place in the rudimentary horn, and is then likely to lead to rupture, commonly before the end of the fourth month, and usually with a fatal result.

If both ducts are developed, but fail to coalesce completely, the *uterus bicornis*, or *uterus septus*, may be formed. In the former the body of the uterus is more or less bifid, as is the case in many animals; in the latter the externally normal uterus is divided by a septum into two halves. The septum may be incomplete, or may extend to the external os, and the vagina may be either single or double. Some recorded cases of superfœtation are explained by pregnancy having occurred on the two sides of a *uterus bicornis*, or *uterus septus*, at an interval of some months. If there are two vaginæ, generally one only serves for coition, but sometimes the septum leads to difficulty in this respect and requires removal.

The uterus is often imperfectly developed, and then assumes one of two forms—(1) the *infantile uterus*, in which the cervix is naturally formed, but the body remains of the same relative size as during infancy, constituting only one-fourth or one-third of the whole length, and having relatively thin walls ; (2) the *generally ill-developed uterus*, in which the normal relative proportion is maintained, but the whole organ is atrophic. The latter condition is often associated with stenosis of the external os and anteflexion, and will be further discussed under those headings. When the uterus is infantile, menstruation is generally absent ; when it is generally ill-developed, it is either absent or ·scanty. The infantile uterus may be diagnosed by bimanual examination, which reveals the small size of the body, while the sound passes only to a length of from $1\frac{1}{4}$ to $1\frac{3}{4}$ inches, and can be felt through the thin fundus with unusual distinctness. The generally ill-developed uterus is distinguished from the infantile by the small size of the vaginal portion.

Treatment.—When the uterus is imperfectly developed, nutrition should be stimulated as much as possible by nourishing diet and the administration of iron, especially if there is any tendency to chlorosis. Of still greater importance is hygienic treatment by abundance of fresh air and a suitable amount of exercise, with the avoidance of too prolonged study or sedentary occupations, especially about the age of puberty. The question of employing any local stimulus will be discussed under the head of amenorrhœa.

ATRESIA OF THE UTERUS, VAGINA, OR VULVA.

All occlusions of the genital canal, at whatever point situated, and whether congenital or acquired, have a common effect in preventing the exit of the menstrual blood, when the body of the uterus itself is developed, and so lead to a similar group of symptoms. It is, therefore, convenient to consider the several varieties of atresia together.

CONGENITAL UTERINE ATRESIA, apart from atresia of the vagina, is very rare, and may affect the external os, or, still more rarely, the whole cervix.

Atresia may exist in one half of a double uterus. Diagnosis is then more difficult than usual, for there is generally menstruation from the patent side of the uterus, while, on the other side, the menstrual fluid is retained, and the fact that the uterus is double may easily escape detection.

CONGENITAL VAGINAL ATRESIA is much commoner, and may consist either in complete or partial absence of the vagina, in an imperforate condition of the hymen, or in closure of the vagina by a transverse septum, which is generally situated immediately behind the hymen, and may easily be mistaken for *atresia hymenalis*. In many cases in which there is apparently a total absence of the vagina, the lower part of the cavity distended by menstrual blood is irregular in shape, and appears to correspond partly to the cervix and partly to a portion of the summit of the vagina. No distinct external os is formed, and the cavity has thick muscular walls, like those of the uterus rather than those of the vagina.

ACQUIRED UTERINE ATRESIA usually affects some portion of the cervical canal. It may result from the application of the actual cautery, potassa fusa, strong acids, or even the solid nitrate of silver, from amputation of the vaginal cervix, especially when performed by the galvanic écraseur, from the presence of growths in the cervix, whether fibroid or cancer, or from any injury to the cervix. It may also be the effect of cervical catarrh, through adhesion of the granulations formed on opposite sides of the canal, especially when the passage is no longer kept patent by the flow of menstrual blood. It is not uncommon, therefore, in old women, especially when prolapse of the uterus exists. Even before the menopause, atresia may result from abrasion close to the os, produced by friction upon a prolapsed cervix.

ACQUIRED VAGINAL ATRESIA is usually the result of sloughing of the vaginal walls after protracted labour, or, in rare cases, after abortion. It may also be the effect of injuries, of sloughing of the vagina after fevers, or of venereal ulceration. In some cases it is combined with vesical or rectal fistulæ. The labia majora are not uncommonly adherent in little girls, but the vagina is not completely closed thereby, and the adhesion is easily separated without any need for incision. This condition is not a fault of development, but may arise either during fœtal life, or after birth.

Results and Symptoms.—Congenital atresia usually attracts no attention during childhood, but occasionally, even in early life, an accumulation of secretion has taken place behind an occluding septum. As soon as menstruation commences, the menstrual blood collects behind the occlusion, and begins to distend the genital canal from below upwards ; first the vagina, if that is present, then the cervix, then the body of the uterus, and, lastly, the Fallopian tubes. Thus, in atresia of the hymen or at the lower portion of the vagina, the uterus does not participate in distension until quite a late stage. If, however, the atresia is about the situation of the external os, the whole uterus becomes dilated into a single cavity from the first, and the internal os is obliterated by distension, while the Fallopian tubes are much earlier affected than in the former case. During the inter-menstrual intervals a considerable portion of the fluid part of the blood is re-absorbed, and thus the swelling formed diminishes during such intervals, while the retained fluid acquires a thick, treacly consistence and dark appearance, but undergoes no putrefaction. The blood in the Fallopian tubes is not, in all cases, due to reflux from the uterus, but may be poured out into them under the stimulus of the morbid condition, as is proved by the fact that the uterine extremity of the distended tubes may be found quite narrow, or even occluded. Slight reflux of blood into the peritoneal cavity may occur, and the pavilions of the tubes often become adherent from this cause, but copious regurgitation does not often take place until the fluid has been partially evacuated. When the atresia is due to a thin membrane, a spontaneous termination, favourable or otherwise, may be brought about by rupture of the membrane under some sudden strain. Eventually the Fallopian tubes, or, less commonly, even the uterus itself, may rupture, and hæmatocele or fatal peritonitis be the result.

After the menopause, the uterus may be filled by mucous fluid (*hydrometra*), a condition usually resulting from acquired atresia of the cervical canal. I have met with one instance in which the uterus became largely distended by pus in consequence of an atresia produced by cancer about the internal os.

Attention is commonly attracted to congenital atresia either by amenorrhœa continuing beyond the age of puberty, by inability to perform the act of coition, or by the effects of menstrual retention. In the last case there will be spasmodic pain, recurring more or less regularly at monthly intervals, and eventually a tumour in the hypogastrium, enlarging in association with the pain, and subsiding somewhat in the intervals. Retention of urine, and other effects of pressure, may be produced. In some cases the atresia is not quite complete, and some slight escape of menstrual blood may occur.

Treatment.—When the occlusion consists only of a thin septum, the operation for evacuation of the retained fluid is extremely easy, but in these, as well as in more difficult cases, there is a grave peril of serious symptoms, and death has not unfrequently followed. The danger is in proportion to the degree of distension, and is especially great if the Fallopian tubes are involved ; while, if the collection of fluid is limited to the vagina, it is comparatively slight. The accidents most likely to occur are :—(1) Reflux of blood through the Fallopian tubes, due to spasmodic contraction of the uterus, or of the tubes, set up by sudden evacuation. (2) Rupture of some adherent portion of the

Fallopian tubes during the collapse of the tumour. (3) Decomposition of some of the retained fluid, which may lead to septic peritonitis, inflammation of the walls of the cavity, or, in some cases, rupture of these walls. (4) The walls of the cavity are also liable to become inflamed, even when no obvious decomposition has occurred. This is probably to be explained on the ground that, the cavity having been congenitally shut off from the outer surface, its walls have a susceptibility, like that of serous membranes, to the influence of germs commonly or occasionally present in the air.

When the accumulation of fluid is very considerable, so that the Fallopian tubes are distended, or the uterus very greatly enlarged, the danger appears to be best avoided by removing small quantities at a time by means of the aspirator, or a very fine trocar, which may be used under carbolic spray. The process should be commenced at the period of quiescence, shortly after a menstrual epoch, and the patient meanwhile kept perfectly at rest. If any considerable length of the vagina is affected by the atresia, the puncture may be made through the rectum. The method of gradual evacuation does not, however, invariably prevent the kindling of inflammation or septic change, antiseptic dressings being difficult to maintain at the vulva. If, therefore, there is any evidence of decomposition, or serious constitutional symptoms arise, a free opening should at once be made, and either immediately, or after a short interval, the cavity should be well washed out with an antiseptic fluid, especially a solution of sulphurous acid, or a weak solution of iodine,* and such injections should be continued at frequent intervals. Full doses of quinine, or other internal antiseptic, with opium, should also be given. If much distension still exist at the time of operation, it appears better to allow some hours for

---

* Acid. Sulphurosi, ℥ss. ad aq. Oj.
  Tinct. Iodi. ℨj. ad aq. Oj.

spontaneous gradual evacuation of the fluid, before the cavity is washed out, to lessen the risk of exciting violent contraction. If however, the fluid be decomposed, injection should not be deferred.

When the collection of fluid is comparatively small, or the vagina only is distended, a free opening may be made at once. It would seem to be worth while to make the incision under carbolic spray, and to attempt to maintain antiseptic dressings for the first day or two, the urine being drawn off at intervals. The benzoline cautery knife may be used to make the incision, so that the edges may be less inclined to close up, or to absorb any septic material. Experience has not yet fully decided whether it is better or not to wash out the cavity immediately. It appears preferable, however, as in the former case, to wait twelve or twenty-four hours for gradual evacuation before injecting, unless any sign of decomposition or febrile symptoms have previously appeared. Any pressure on the uterus should be carefully avoided, lest when the pressure is taken off air be sucked in. Dr. Emmet, however, prefers the plan of immediately washing out the cavity, whether large or small, and has obtained by this means a favourable result in a considerable number of cases. He is also opposed to puncturing *per rectum* under any circumstances.

When the whole or a considerable part of the vagina is absent, it is preferable, if possible, to make the permanent passage in the natural situation, rather than through the rectum ; and the making of an artificial vagina may be undertaken when any uterus can be detected, even though there is no collection of menstrual fluid. When, however, as is often the case, the septum is very thin between rectum and bladder, great care is required to avoid opening one of these cavities. The patient should be placed in the lithotomy position, and the knife or scissors used only to make a transverse incision through the mucous membrane, just in front of the fourchette. The rest of the passage should be torn

E

by the index-finger of the right hand, while the left index-finger is kept in the rectum, and a sound is held by an assistant in the bladder. The operator is thus guided by the sense of touch in making a passage equidistant from either cavity. If necessary, for enlargement of the canal, the finger may be removed from the rectum, and used to assist the other in the artificial passage, or a blunt instrument, such as the raspatory employed for scraping bones, may be used in conjunction with the finger. The uterus, when reached, may be pierced either by a trocar or knife, if there is no patent os externum. The artificial vagina should be made at first larger than required. A full-sized Sims' dilator of glass (Fig. 78) should be introduced at once, and must generally be worn continuously for a good many months to avoid the strong tendency to contraction which exists. This serves to check hæmorrhage in the first instance, and, under its unirritating pressure, an epithelium, like that of mucous membrane, may gradually spread over the artificial vagina. Eventually, it may be possible to substitute for the dilator a narrow Hodge's pessary. Marriage, if not already contracted, should not be advised until the patency of the new vagina has been tested for a considerable period. Should the attempt to make an artificial vagina fail, the only alternative, if menstrual fluid is poured out, is to endeavour to keep open a passage *per rectum*, and many operators have adopted this measure from choice in the first instance. The treatment of acquired atresia is similar to that of congenital, but the risk of evacuating retained fluid appears, in this instance, to be considerably less.

STENOSIS OF THE OS EXTERNUM.

Causation and Pathological Anatomy.—Congenital stenosis of the cervical canal is situated either at the external or internal os, and extreme stenosis is not uncommon at the former orifice, while it is rare at the latter. The intervening cervical canal is comparatively free, being somewhat spindle-shaped. A small external os is usually associated with a tapering, conical cervix, projecting more than usual into the vagina. Frequently also the cervix is flexed forward, so that the os looks in the direction of the vagina, or even still more anteriorly (Fig. 37), the posterior lip of the cervix being long, and the anterior lip short. More rarely the cervix is flexed backward. In many cases this form of cervix is associated with imperfect development of the whole uterus, or of the uterus and ovaries. The uterine cavity is then rather less than the normal length, and menstruation scanty. From some associated imperfection, sterility often persists after the stenosis has been cured. The vagina may partake in the same imperfect development, and be smaller than usual, and sexual feeling is often deficient. Dr. G. Roper has adduced a case to show that an infantile form of pelvis may also be an associated condition. Acquired stenosis may arise from gradual contraction of the os externum in old age, or after the use of caustics. It is also common in old cases of prolapse of the third degree.

Results and Symptoms.—The most marked results of stenosis of any part of the cervical canal are dysmenorrhœa and sterility. Dysmenorrhœa is, however, not invariable. If the menstrual flow is moderate and uniform, and the mucous membrane thrown off is completely disintegrated, no obstruction or pain may result ; but if the flow is more profuse, or if there are any clots or shreds of menstrual decidua (*see* p. 37) to pass, spasmodic pain is produced by the efforts of the uterus to overcome the difficulty. The extent both of the spasmodic contraction and of its painfulness depends in very great degree upon the irritability of the woman's nervous system and her sensibility to pain. Sterility is a more constant symptom than dysmenorrhœa ; nevertheless it does not imply an absolute hindrance, but only an increased difficulty in the access of spermatozoa to the uterus (*see* section on

sterility). I have met with several instances of
women whose os externum would not admit the
smallest surgical probe, but who had never suffered
the slightest dysmenorrhœa, although they were sterile.
In other cases, in addition to dysmenorrhœa, endometritis is produced by irritation, due to the retention
of menstrual and other secretions : and. this may lead
to hyperæmia and menorrhagia, although the primary
condition is usually that of scanty menstruation. The
uterus then becomes hypertrophied, partly from the
effect of hyperæmia, partly from the muscular efforts
to overcome obstruction. Women who have suffered
from symptoms of obstructive dysmenorrhœa, especially if menstruation is profuse, are more liable than
others to attacks of pelvic peritonitis, and even hæmatocele, in connection with menstruation. There is reason
to believe that an impediment to the outflow of menstrual blood may sometimes lead to dilatation of the
Fallopian tubes, and reflux into the peritoneal cavity,
possibly dependent upon spàsmodic contractions of the
uterus. Such a reflux of blood may be the starting
point of pelvic peritonitis, or form one of the milder
varieties of hæmatocele.*

Treatment.—If the contraction is only moderate,
and if the os points in the normal direction, dilatation
may be effected by mechanical dilators, such as graduated metallic bougies, Priestley's dilating sound (Fig.
21, p. 60), or any of the two-bladed or three-bladed
uterine dilators. Tents may also be used for dilatation, but their use appears to involve as much risk to
the patient as an incision limited to the os and lower
portion of the cervix, and the effect produced is not so
lasting. If, as is often the case, the os contracts up
again after dilatation, it is well to have recourse to the
method of incision; and, if the stenosis is very considerable, it is desirable to adopt this plan in the
first instance. The frequently associated condition of

* For anatomical evidence on this subject see Bernutz,
" Clinique Médicale sur les Maladies des Femmes."

cervical anteflexion, in which the cervix is curved forward, instead of being nearly in a line with the uterine cavity (*see* Figs. 20, 37), constitutes a reason in favour of performing the operation of posterior section of the cervix, in place of merely dilating the orifice, so that the new os may look in a more natural direction (*see* pp. 55, 57).

In the less common case in which the os looks in its normal direction, the incision, if required at all, should be bilateral. The incision may be made either by Kuchenmeister's scissors (Fig. 16), the blade of which has a point projecting at right angles to prevent retraction of the portion of cervix seized, or by any of the single or double-bladed metrotomes. If bilateral, the incision should not be made more than half-way up to the vaginal reflection, otherwise ectropion of the cervix, and its resulting evils, may be produced (*see* p. 176). Of the metrotomes, the most widely useful is the original metrotome of Simpson (Fig. 17). This is a bistoury caché, the amount of the projection of the blade of

Fig. 16.
KUCHENMEISTER'S Scissors.

which is regulated by a screw in the handle.

Of the two methods, the use of the scissors is most free from risk, the extent of tissue divided being less, but incision by the metrotome, more thoroughly lays open the lower part of the cervical canal (*see* Fig. 18), and the risk is but slightly greater, if care be taken not to cut as high as the internal os. The operation does not absolutely require an anæsthetic unless the patient is nervous, especially if scissors be used, as the pain is very brief. An anæsthetic, however, allows the incision to be made more deliberately, and its extent more exactly regulated. An antiseptic vaginal injection should be made before the operation, and instruments and fingers should be cleansed and disinfected with the utmost care. The incisions may be made by the sense of touch alone, without using any speculum, or by the aid of Sims' speculum. If the speculum is used, the most accurate method is first to cut the external os with the scissors to exactly the desired extent along the line *a b* (Fig. 18), and then to cut through the projecting angle of tissue *a b c*, by means of Sims' knife (*see* Fig. 20, p. 57). If the speculum is not employed, Simpson's metrotome may be used in the first instance. The instrument is set to cut pretty widely, its extremity passed up a little short of the internal os—that is to say, for something less than an inch into the cervical canal—and it is then gradually opened as it is withdrawn, so as to cut in the line *c a* (Fig. 18). It frequently happens that the resulting incision

Fig. 17.
SIMPSON'S Metrotome.

is not quite so wide externally as is desired, and the division of the external os may then be completed by the scissors to the exact extent wished, as shown in Fig. 18. If the case be one of posterior section for cervical anteflexion, the incision may be made up ·to, or nearly up to, the vaginal reflection, so as to throw the new aperture more nearly in a line with the upper part of the cervical canal (*see* Fig. 20, p. 57). If the incision is to be bilateral, Peaslee's metrotome (Fig. 23, p. 63), set to cut short of the internal os, affords the means of making symmetrical and precisely limited incisions.

For the double purpose of preventing primary union and avoiding the risk of septic absorption, it is a good

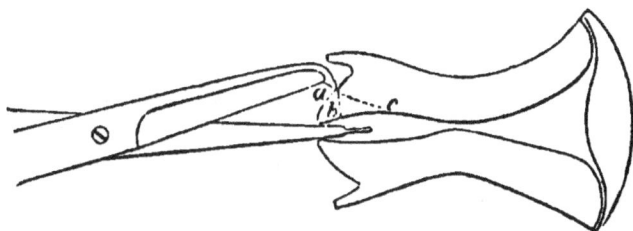

Fig. 18.
KUCHENMEISTER'S Scissors cutting Cervix.
a b, line of incision by scissors ; *a c*, line of incision by metrotome.

plan to swab the incision with a solution of chloride of zinc (gr. xxx. ad ʒj.). This has a mildly caustic effect upon the raw surface, and renders it less liable to absorb. If there is much bleeding, the wound may be swabbed also with a solution of perchloride or sub-sulphate of iron. If bleeding still continues, a small piece of absorbent cotton, dipped in the iron solution, should be placed in the incision as a plug. It may have a thread attached to draw it away after about twelve hours. A larger tampon soaked in carbolized or iodized glycerine may be placed in the vagina, and the vagina should be syringed several times a day with some antiseptic. For a few days occasional digital examinations may be made, to prevent adhesion taking place. After that time there is little tendency to close at the external os, if the incision at first be sufficiently free. Rest in bed should be· maintained at least ten days, and great caution, both with respect to movement and exposure to cold, should be enforced until the succeeding menstrual period has passed. This simple operation has occasionally produced severe cellulitis or peritonitis, and even death, but such a result appears to be due either to septic contamination at or after the operation, or to imprudence on the part of the patient, and is therefore avoidable. Out of twenty-five cases under my care in Guy's Hospital, disturbance followed the operation in one only. In this instance severe cellulitis appeared to result from contamination of the wound by a case of septicæmia in the same ward.

If the stenosis is so extreme that the metrotome will not pass, or the probe-pointed blade of the scissors cannot at first be introduced far enough to make an adequate incision, it is convenient first to expand the os partially by Priestley's dilating sound (Fig. 21, p. 60), or to dilate it by a small laminaria tent. In extreme cases it may be necessary to commence the dilatation by a sharp-pointed probe, or pointed bistoury.

The method of Marion Sims is to use his own speculum and a special knife, consisting of a small razor-shaped blade, which can be fixed at any angle at the end of a long handle (Fig. 19). The scissors are first used, and the uterus is then firmly held by a tena-

Fig. 19.—SIMS' Uterine Knife.

culum-hook, while the incision is made by the knife from below upwards, in a slanting line as far as the internal os, as shown in Fig. 20. In the case of ante-flexion of the cervix the incision is to be directly backward, otherwise it is to be bilateral.

In a considerable proportion of suitable cases of stenosis—that is to say, those associated with a dys-menorrhœa, shown by its characters to be obstructive rather than congestive or inflammatory*—incision of the cervix relieves the dysmenorrhœa more or less per-manently. It is comparatively rare for sterility to be

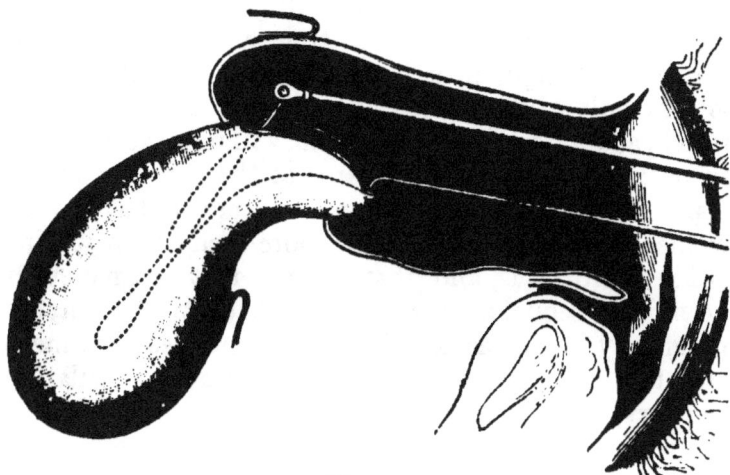

Fig. 20.
Posterior Section of the Cervix by SIMS' Knife. (After SIMS.)

cured by the operation. Dr. Pallen reports that out of 337 patients on whom he operated, thirteen or four-teen became pregnant afterwards, a proportion not at all greater than may be accounted for by mere coincidence.

STENOSIS OF THE OS INTERNUM.

Causation and Pathological Anatomy.—Opinions have differed widely as to the relative frequency of stenosis of the os internum, and some high authorities, as Barnes and Schroeder, have considered it so rare

* *See* section on dysmenorrhœa.

as seldom or never to require any operative interference. The majority, however, hold that a relative stenosis at least is not uncommon, and this is the result of my own experience. From autopsies made in a considerable number of nulliparous women, Dr. Peaslee has concluded that the average size of the internal os in them is equivalent to a circle ⅓-in. in diameter, a size which will allow the ordinary sound, whose extremity should be about ¼-in. in diameter, to pass pretty easily. In parous* women, who were neither sterile nor suffered from dysmenorrhœa, he found the average area to be nearly double that in nulliparous women, in the majority of cases admitting a sound ⅓-in. in diameter, though, in a large minority, one from ⅓-in. to ⅛-in. only could be passed. Hence, an internal os which, apart from flexion or spasm, will not readily admit an ordinary sound, not too large at the point, is abnormally small. Moreover, it is well known that parous women habitually menstruate more easily than virgins or the nulliparous, and that after a first pregnancy, if no morbid sequelæ remain, the probability of a further pregnancy is increased. It may be inferred that menstruation and conception may be facilitated by dilating the cervix to the average size of that in parous women. As regards dysmenorrhœa, this treatment is especially indicated, if the case is complicated by any flexion, by menorrhagia, leading to the formation of clots, by the discharge of shreds of membrane, by excessive hyperæmia leading to tumefaction of the cervical mucous membrane at menstrual periods, or by an irritable condition of the nervous system, owing to which a slight cause of obstruction sets up spasmodic and excessively painful uterine contractions.

It is to be remembered that the most definite sphincter in the uterine canal exists at the internal os, and that the size of the os therefore varies according to muscular tonicity. It appears, however,

* The word parous is used on the analogy of multiparous and nulliparous to denote one who has borne one or more children.

that, after being once completely dilated, as by parturition, the internal os does not usually contract up again quite to its original smallness.

Acquired Stenosis may affect the internal os, or other parts of the cervical canal. It may result from cicatricial contraction after the use of caustics, or other operative interference, from endometritis, with hyperplasia of the cervix, or from injuries received in parturition.

The **results and symptoms** of stenosis of the internal os resemble those of stenosis of the external os, as already enumerated, and are often combined with the effects of anteflexion.

Diagnosis.—The arrest of the sound near the internal os is much more frequently due to flexion than to stenosis. Stenosis can only be inferred when a full-sized sound is arrested, but a smaller sound having the same curve will pass. For this purpose a sound not more than $\frac{1}{10}$-in. or $\frac{1}{12}$-in. in diameter (equivalent to No. 2 or No. 1 bougie) may be required. It is very rare for the internal os to be too small to admit a sound of $\frac{1}{10}$-in. diameter, though flexion may render it very difficult to pass it. Temporary stenosis, due to spasmodic contraction of the internal os, is distinguished by its yielding after a while to very gentle pressure. A tendency to such spasm is often associated with some primary narrowness. For diagnosis of a degree of smallness which cannot be called in itself morbid in a nullipara, but yet may amount to a relative stenosis under the circumstances already mentioned, larger sounds are required. A convenient instrument for diagnosis, as well as for the purpose of effecting or maintaining dilatation, is a conical sound $\frac{1}{8}$-in. in diameter at the point, and enlarging to $\frac{1}{4}$-in. at the position corresponding to the internal os. If this can be passed with ease, the absence of any, even relative, stenosis is assured ; and if it is arrested, the point of arrest will afford an estimate of the size of the internal os, provided that it is ascertained by the finger that the arrest is not due to the external os.

Treatment.—The choice between tents, incisions, or instrumental dilators is more difficult in the case of the internal than in that of the external os. Incision is much more likely to be followed by adhesion and contraction than in the other case, but contraction is also likely to occur after dilatation. Nevertheless, the greater average size of the internal os in parous women shows that after full dilatation it does not usually so completely close again, and I therefore think it preferable first to make trial of dilatation. If symptoms of stenosis repeatedly recur after temporary improvement, or if the cervical canal is cicatricial in acquired stenosis, incisions may be used, but according to my experience, the necessity for incision of the internal os is extremely rare. Perhaps the safest mode of dilatation is to pass from time to time graduated metallic bougies, slightly conical, until the cervical canal is considerably larger than the required size, and will admit a No. 12 or No. 14 bougie, so that a considerable margin is allowed for subsequent contraction. In the case of virgins this method has the drawback that, to be effectual, it requires frequent manipulation. Dilatation by a laminaria tent avoids this difficulty, and, when flexion is superadded, it is advantageous from its effect of softening the walls of the uterus, and straightening it for the time being. It must be used with due precaution (*see* p. 31). A convenient mode of rather rapid dilatation is the use of Priestley's dilating sound (Fig. 21), formed of two blades joined at the extremity, and expanded by a screw at the handle, so that the external os is stretched to a considerable,

Fig. 21.—PRIESTLEY'S Dilating Sound.

the internal os to a moderate, size ; that is to say, the stretching is in proportion to the natural relative dimensions of the two orifices. The point should not be more than $\frac{1}{10}$-in. in diameter, and the blades should be capable of separation to a width of $\frac{1}{4}$-in. at the position of the internal os. The instrument should be used cautiously, and only partially expanded at first, with the view of gradually stretching the muscular fibres rather than causing any rupture. If the screw works easily, the degree of resistance in the cervix is readily estimated by the finger, and thus diagnosis as well as treatment is assisted. Other forms of mechanical dilators have been invented by Ellinger and others, in which the blades, two or three in number, are free at the extremity, and are separated by closing the handles. The most powerful is that of Dr. Marion Sims, previously described (Fig. 15, p. 32). This cannot generally be introduced into a small cervix, except after partial previous dilatation by bougies, laminaria tent, or other means. It has been recommended to effect immediate full dilatation by means of such an instrument with the aid of an anæsthetic. There is some risk, however, of exercising a dangerous degree of stretching upon the internal os or cavity of the uterus, and, with the weaker instruments, it is difficult to estimate exactly the expansion actually produced, on account of the elasticity of the blades. The more gradual mode of dilatation is therefore generally to be preferred.

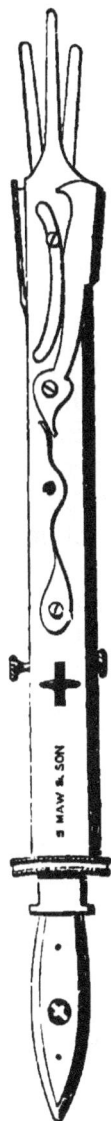

Incision may be performed by Simpson's single-bladed (Fig. 17, p. 54), or by any of the numerous two-bladed, metrotomes, introduced without any speculum. Much caution, however, is required in incising the internal os, since the large vessels which enter the uterus at this level are not far off, and alarming and even fatal hæmorrhage has sometimes occurred. Dr. Greenhalgh's metrotome (Fig. 22, p. 61) contains an ingenious mechanism by which two blades cut outwards and downwards in a definite curve, and an adjustment for regulating the width of the incision. The incision, however, so produced, even at its smallest, is dangerously wide at the internal os, when the instrument is fully introduced, and it is preferable only to expand the blades to a slight degree, and then cut by withdrawing the whole instrument. A simpler metrotome for use in this manner is that of Dr. Savage, in which each blade forms the shield for the other. Two-bladed metrotomes are liable to cut the two sides unequally from asymmetry of the uterus or from a difference of sharpness in the blades. With Simpson's metrotome the depth of the second incision is also uncertain, owing to the want of firm resistance to the back of the instrument. A graver objection to both forms of instrument is that, as usually made, they are so large that when incision of the internal os is required, they cannot be introduced without preliminary dilatation by a tent, after which it is difficult to judge exactly how much the canal will contract again, and therefore how deep an incision is required.

A safe and convenient instrument, though little known in Britain, is Dr. Peaslee's metrotome (Fig. 23). This consists of a flattened tube, narrowed for its terminal two inches, in which slides a single blade, lancet-shaped towards the point, but blunted at its extremity. There are two blades for each instrument, the cutting portion of one being $\frac{1}{4}$-in., of the other $\frac{3}{16}$-in. wide. A nut and screw on the handle of the blade regulate the extent of its passage into the uterus. The narrower blade is generally sufficient for incision of the internal os. I have used a modified form of this instrument, in which the tube is made round instead of flat, and its terminal portion of smaller size, being only $\frac{1}{12}$-in. in diameter near the extremity. It can then be passed through a very narrow cervical canal, being introduced like the ordinary sound. If the uterus be much flexed, it should first be straightened by means of a small sound, in the manner described under the head of flexions of the uterus (p. 105).

After incision, it is desirable to swab the cervical canal with solution of perchloride or subsulphate of iron, or with that of chloride of zinc (gr. xxx. ad ℨj.), to prevent primary union, and it is also necessary to maintain in some way its patency. The immediate introduction of an intra-uterine stem, as recommended by Barnes and Marion Sims, is not without considerable risk, and it is preferable to pass occasionally a large conical sound. If a stem be used at all, it is preferable to wait for two or three weeks before its introduction, and to use it for a limited number of weeks only. But the degree of risk which always attends the use of an intra-uterine stem, is greater after incision of the cervix than at other times. If adopted at all, an expanding stem (*see* p. 111), is convenient from its self-retaining quality.

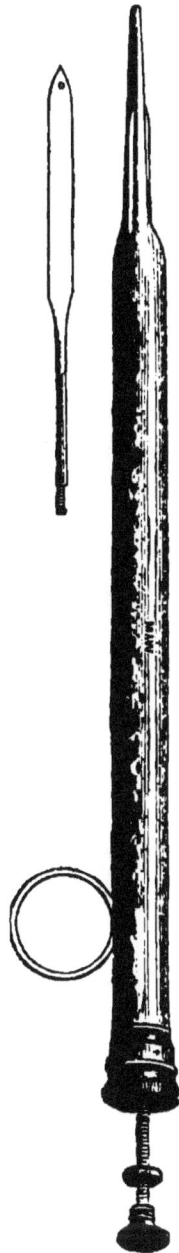

Fig. 23.—PEASLEE'S Metrotome

When stenosis of the external os exists, the internal os is not unfrequently also smaller than normal. It is preferable, however, not to incise both at a single operation, since the more limited incision may prove sufficient, and the subsequent dilatation necessary to keep the inner os patent, is unnecessary and undesirable after incision of the external os and cervical canal.

CHAPTER IV.

DISPLACEMENTS OF THE UTERUS AND PELVIC VISCERA.

RELATIVE IMPORTANCE OF DISPLACEMENTS OF THE UTERUS.

THE special attention which has been devoted of late years by some gynæcologists to the study of changes in the form and position of the uterus has led to much difference of opinion, and to some strength of feeling, with regard to the importance of these conditions. Thus, some authorities have gone so far as to maintain that in more than half of the patients suffering from symptoms referable to the uterus, the shape of that organ will be found to be materially altered or its position markedly changed; and, further, that in the great majority of those cases which were formerly regarded as chronic inflammation of the uterus an alteration in the shape of the organ is the principal and the really important feature. Other authorities are still found to retain the opinion which in former years was more common than it is of late, namely, that in all displacements of the uterus, except manifest prolapse, the truly scientific principle is, to the best of our power, to take care of the general symptoms, and to leave the displacement to take care of itself. The truth would seem to lie in the mean between these extreme opinions. Few will doubt that to find in a mechanical system of uterine pathology the key to the great majority of the maladies peculiar to women is as

F

one-sided a view as it would be to attribute a similar
importance to erosions or to lacerations of the cervix
uteri.   But, on the other hand, it is as erroneous to
regard as of little or no consequence all displacements
of the uterus, with the exception of external and obvious prolapse, as to overrate the importance of anteversion or anteflexion.   Of late years, according to the
varying fashion of the day, there has been a tendency
with many authorities to attach too much weight either
to changes in the position and shape of the uterus, or
to the minor lesions of the cervix, while depreciating
too much the relative importance of chronic inflammation in the pelvic disorders of women.   In the present
work, displacements are placed before inflammation on
account of their greater mechanical simplicity, and not
because I consider them to be the essential element
in the majority of cases.   A considerable proportion
of the engravings are devoted to their illustration,
because the subject can be explained in this way
much more readily than by words, while drawings
of the pathological appearances seen in inflammation and other diseases of the tissue of uterus and
ovaries find a more appropriate place in larger works
on gynæcology.

Displacements of the uterus may be either the cause
or the consequence of chronic hyperæmia, inflammation, or hyperplasia, and a further controversy has
taken place on the question as to which is the usual
sequence of events.   Its most important bearing is
the inference to be deduced from it as to treatment
when the two conditions are found combined.   There
is no doubt that even when a displacement is, in
the first instance, secondary to hyperæmia or inflammation, it has often a strong tendency to keep up and
intensify the condition which gave rise to it ; hence
the general principle of action is that if the displacement is important in its degree and effects, and can
be cured or alleviated by a pessary which is readily
tolerated, it is best to have recourse to early mechanical

treatment, in addition to other measures. This is usually the case in displacements of the uterus downward or backward. If, however, the displacement be but a slight departure from the normal condition, and if it can only be remedied by a pessary which is itself liable to cause irritation, then hyperæmia or inflammation, if present, should first be relieved, as far as possible, by general measures, and a pessary only tried, if tried at all, when other treatment has proved insufficient to relieve symptoms. This is more frequently the case in anterior displacements of the uterus.

There are some cases in which even a very acute flexion of the uterus is accidentally discovered, and in which no symptoms exist, just as there are some instances of extreme stenosis without any dysmenorrhœa. These are chiefly cases in which the flexion is primary, and has never become complicated by hyperæmia or inflammation, or in which, after the menopause, all such complications have long subsided. Minor degrees of displacement exist not unfrequently without producing any symptoms. When symptoms are absent, treatment is of course unnecessary; and even when symptoms are present, it must not be too hastily assumed, especially in the case of an anteversion or anteflexion, that the displacement is the cause of them, for it may be only an accidental concomitant. There is another class of cases in which mechanical treatment is, as a rule, forbidden, namely, that in which the displacement is secondary to inflammatory adhesions or deposits, by which the uterus is firmly fixed. Palliative treatment only is here admissible, for any attempt at immediate replacement is dangerous, and a pessary generally fails to remedy the displacement, while it excites irritation by pressure.

NORMAL POSITION OF THE UTERUS.—The uterus in a healthy and unimpregnated state is a very mobile organ, the whole of its body being free from any attachment except a very lax one by means of the broad and round ligaments, a provision necessary to allow of its expansion during pregnancy. The axis of the normal uterus varies from a straight line to a curve whose concavity looks forward, and whose angle does not

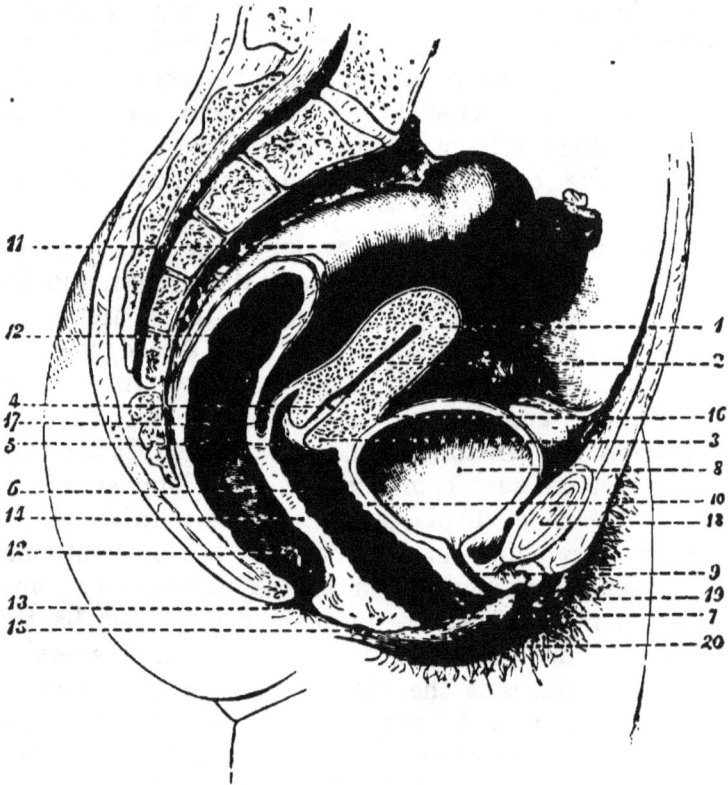

Fig. 24.
Longitudinal Section of the Pelvic Organs. (After SAPPEY.)

1, body of uterus; 2, its cavity; 3, the vaginal portion; 4, canal of cervix; 5, os uteri externum; 6, the vagina; 7, orifice of vulva; 8, interior of bladder; 9, urethra; 10, vesico-vaginal septum; 11, rectum; 12, its cavity; 13, anus; 14, recto-vaginal septum; 15, perineum, forming the lower border of the triangular perineal body; 16, vesico-uterine fossa of peritoneum; 17, recto-vaginal or Douglas's fossa of peritoneum; 18, os pubis; 19, labium minus; 20, labium majus.

exceed 45°. The maintenance of this axis during the movements of the uterus depends solely upon the firmness of the uterine tissue itself. The chief supports of the uterus are, anteriorly, its attachment to the bladder, and, by its means, intermediately to the pubes, and, posteriorly, the utero-sacral ligaments. By these attachments its centre is rendered comparatively a fixed point, while the broad ligaments place scarcely any restraint upon backward or forward displacements of the uterine body, and only partially limit lateral displacements. The round ligaments have a certain

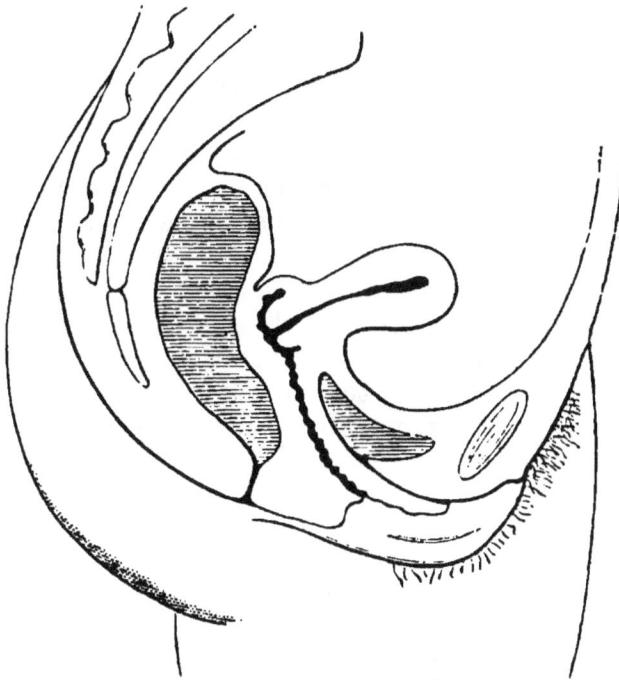

Fig. 25.—Diagrammatic Vertical Section, to illustrate the relations of the Uterus and Vagina in the virgin, the Bladder being nearly empty.

function in drawing the fundus forward after displacement by distension of the bladder or otherwise, but are usually not on a stretch, and have little efficacy in preventing displacements. The mean direction of the axis of the uterus is generally regarded as being coincident with that of the pelvic brim. It varies, however, considerably under different circumstances, and in

different positions of the body, being inclined more anteriorly in the erect position, when the bladder is empty
(Fig. 25, p. 69), and more posteriorly when the bladder
is full, especially when the rectum is at the same time
empty. Besides these movements upon a transverse
axis passing nearly through its centre, the uterus as a
whole is capable of a certain amount of upward and
downward movement, in which movement the base of
the bladder necessarily partakes. In Fig. 24, p. 68, and
in the other figures illustrating displacements of the
uterus and the position of pessaries, the vagina is represented, for the sake of clearness, as forming an actual
cavity. This is not, however, its condition in the living
woman, except when its walls have been separated by
the introduction of the finger or an instrument, or
unless air has entered in the prone or semi-prone
position. It is normally flattened, so that the anterior
and posterior walls are in contact, as indicated in
Fig. 25. Its direction is nearly at right angles to the
axis of the pelvic brim, and thus the anterior vaginal
wall, receiving through the bladder the intra-abdominal
pressure, which acts somewhat in the direction of the axis
of the brim, is supported by the posterior vaginal wall.
In the case of rupture of the perineal body, the lower
half of the anterior vaginal wall becomes unsupported,
and a tendency to bulge commences at this part.

The pressure of the intestines has an important
influence in maintaining or modifying the position of
the uterus. Being made up of that of the individual
coils, it is not equable on all sides, like a fluid pressure,
but is apt to be greatest where the coils are largest or
most numerous. This is usually the case in the retro-
uterine fossa of the peritoneum, which is more capacious
than the space in front of the fundus (*see* Fig. 25), and
hence the intestinal pressure is an important element
in maintaining the normal slight anteversion of the
uterus, the bladder being empty, in reference to the
axis of the brim. When in a healthy state the vagina,
as a muscular column in the form of a flattened cylin-

der, has also an influence in supporting the uterus, but when it has become excessively relaxed this function is lost, especially when by more or less complete laceration of the perineal body (Fig. 24, *14*, *15*, p. 68) in parturition, the cylinder has lost its base of support. The vagina then frequently becomes an active agent in producing prolapse.

Causation of Displacements in General.—Displacement of the uterus may be produced by any influence which tends to increase the weight of the organ, to weaken its supports, to push or to drag it out of place, or to diminish the firmness of the uterine tissue itself, the last element coming into play specially in the causation of flexion. Increased weight is most commonly due to the presence of fibroid or other tumours, to sub-involution or hyperplasia of the whole or a part of the uterus, to hyperæmia, or to pregnancy. Causes tending to weaken the uterine supports may be a general want of nutrition and laxity of tissue, associated with feeble health, especially if combined with a deficiency of fat. The most important, however, are the effects of pregnancy and parturition, including not only a stretching and loss of tone of all the uterine ligaments and of the vagina, but frequently also more or less damage to the perineal body. These are conjoined with excessive weight of the uterus so long as involution of that organ remains incomplete ; and thus a too early getting up, or undue exertion too soon after delivery, is the commonest of all causes of serious displacement. Among external causes are the effects of muscular efforts or falls, which may produce sudden displacement of an organ previously healthy. This, however, is comparatively rare, while a gradual effect, produced by prolonged muscular exertion in the standing position (as in the case of laundresses), or by repeated efforts of any kind, as chronic cough or straining in habitual constipation, is much more common. Excessive intra-abdominal pressure is also produced by tight-lacing, or the suspending of heavy skirts from the waist. Among the most irresistible forces tending to displacement are those exerted by tumours, effusions of fluid, or inflammatory deposits which push the uterus, or contracting adhesions which pull it, from its place.

Undue softness of uterine tissue is an important cause of flexions. It may result simply from imperfect nutrition, a condition most common in girls about the age of puberty as the effect of insufficient or unsuitable diet, or imperfect digestion, especially when associated with a too sedentary life, and lack of sufficient air and exercise. In the earlier stage of uterine hyperæmia or chronic metritis, the uterus is soft as well as increased in bulk, and therefore prone to flexion, while in the later stage it becomes indurated. Softness of the uterus also exists after parturition, and any cause which interferes with involution also prolongs the softened state of the organ.

RETROVERSION AND RETROFLEXION OF THE UTERUS.

Pathological Anatomy.—In retroversion the shape of the uterine axis is unaltered, but the whole organ is tilted backward, so that the fundus is inclined toward the sacrum, and the os toward the pubes. Retroversion is possible through a very large angle, and is not unfrequent up to one of about 135° (Fig. 26). In the case of the gravid uterus, at the third or fourth month, even this may be exceeded, and the angle of retroversion almost reach 180°, so that the fundus presses down upon the perineum, bringing down with it the retro-uterine pouch of peritoneum, and distending the recto-vaginal septum.

In retroflexion the axis of the uterus is bent upon itself, so as to create a curve with its concavity looking backward. The curve is generally not uniform, but has a point of maximum curvature usually near the internal os. In pure retroflexion the direction of the os uteri may be unaltered, but more frequently retroflexion is combined with more or less of retroversion, so that the axis of the uterus is carried backwards, while at the same time the os is tilted forwards (Fig. 27, p. 74). In primary retroflexion, on the other hand, the os may look too much backwards, as in anteversion. In a recent flexion of the uterus, whether backwards or forwards, the uterine wall on the convex side of the curve becomes the thinner as the result of stretching, exactly as would be the case with an india-rubber tube.

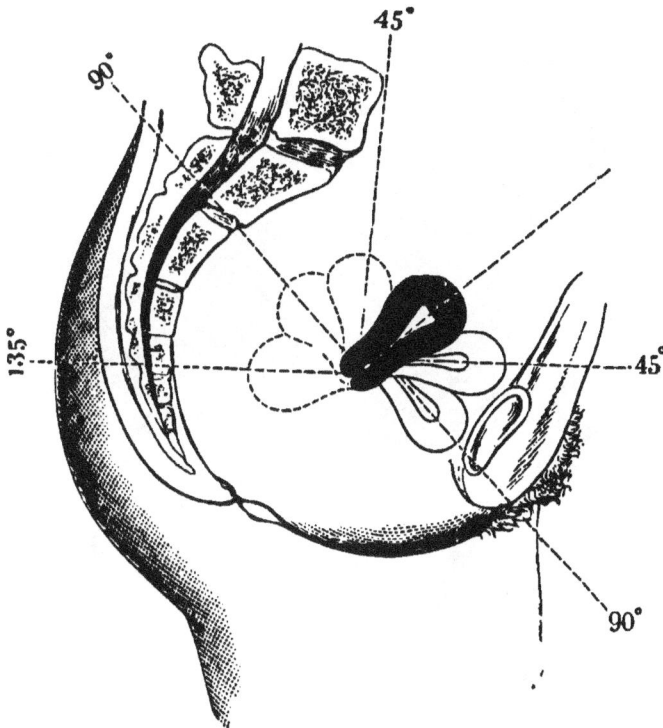

Fig. 26.
The Degrees of Retroversion and Anteversion.

In pathological specimens from old cases of flexion, however, it is often found that it is the wall on the concave side which is attenuated. This may be the result of atrophy from the prolonged effect of pressure and interference with the circulation, or may depend upon the flexion having been in the first instance due to a failure of development in the anterior or posterior wall of the uterus. The occasional existence of such a condition explains the great difficulty sometimes found in preventing the return of the uterus to its former shape after it has been straightened.

Retroversion and Retroflexion of the Gravid Uterus

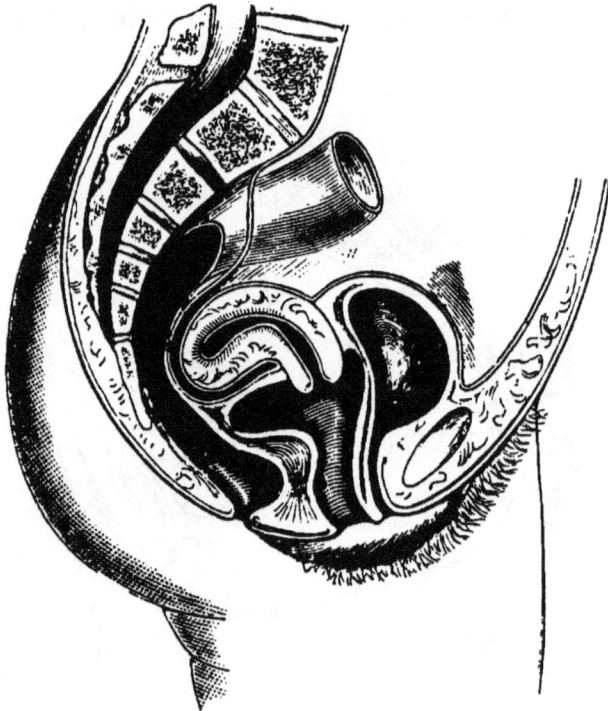

Fig. 27.—Retroflexion of the Uterus with associated Retroversion.

will not be discussed here, since they are considered in works on midwifery.

Causation.—The predisposing causes are the same as those enumerated for displacements in general. Retroversion is especially associated with prolapse, since the uterus as it descends tends to follow the curved axis of the pelvis, the cervix moving downward in the line of the vagina, the direction of least resistance (Fig. 40,

p. 113). The causes of prolapse, therefore, almost invariably produce at the same time more or less retroversion. Retroversion may also be brought about by the effect of gravity if the dorsal, or dorsal reclining position (as in an easy chair), be too persistently maintained, especially when the uterus is heavy, as after parturition. Such an effect may be increased by over-tight bandaging. A similar effect may result from the prolonged and excessive distension of the bladder to which women are especially liable, a temporary and partial retroversion being a necessary result of this distension. If a muscular effort is made under these conditions the retroversion may be suddenly increased.

Retroflexion is, in rare cases, a primary affection, being due to defective development of the posterior uterine wall, either in foetal life or at the time of puberty, when the organ is the subject of rapid growth. In the great majority of cases, however, it is secondary, and it is generally developed out of a partial retroversion. This may be partly the effect of gravity and partly that of the intra-abdominal pressure acting either gradually or during muscular efforts. When retroversion exceeds an angle of about 55° the weight of the body of the uterus itself, in the standing position, begins to tend to increase the retroversion or convert it into a retroflexion, instead of tending to bring the fundus forward as in the normal condition (*compare* Figs. 26, p. 73, 40, p. 113, *with* Figs. 24, p. 68, 25, p. 69). In the sitting or reclining position this effect comes into play at a less angle than 55°, the pelvic inclination being then much diminished. Again, when partial retroversion exists, there is more room for coils of intestine in front of the fundus than behind it, and thus the intestinal pressure, which normally should keep the fundus forward (*see* p. 70), comes to act upon its anterior surface, and press it down into the hollow of the sacrum. Thus is brought about, if the uterus is rigid enough, an increased retroversion; but if it is soft, the retroversion is converted into retroflexion. Retroflexion may also be produced by the weight of a small fibroid tumour in the posterior uterine wall.

Results and Symptoms.—Versions of the uterus, unless of extreme degree, produce comparatively little effect upon the uterus itself, the symptoms being chiefly those due to dragging of ligaments or pressure on neighbouring structures, and those which belong to associated hyperæmia or inflammation. A flexion, when primary, may have little or no effect of any kind, for the uterine canal is then adapted to the curve of the uterus, and its calibre need not be diminished. When, however, a uterus, originally straight, or nearly so, becomes flexed, there is necessarily a tendency for its canal to become flattened, just as would happen with any tube of soft material. This will be more marked if, as is often the case in acquired flexion, the curve is not uniform throughout, but sharper near the junction of body and cervix, where the muscular walls are less thick than those of the body. The flexion may then have a double effect upon the uterus. First, the veins are compressed by the bending of the organ at the level where the vessels enter it ; and, secondly, the exit of the menstrual and other secretions is hindered. From the first cause may arise passive hyperæmia, menorrhagia, or metrorrhagia, and vulnerability to slight exciting causes of inflammation ; from the second, hypertrophy of muscular tissue to overcome the obstruction, and all the effects which have been enumerated as due to an obstructed canal, such as dysmenorrhœa, sterility, and especially endometritis from the irritation of retained secretions. All these effects are greatly enhanced if there be a tendency to hyperæmia or endometritis, independent of the displacement, and may be entirely absent in some cases in which the uterus is quite free from congestion, and the vessels have had time to accommodate themselves to the flexion. The passive hyperæmia is far greater in retroflexion than in anteflexion, since the enlarged fundus, pressed down into the hollow of the sacrum, rests between, and may be compressed by, the utero-sacral ligaments at either side. Menorrhagia, or metrorrhagia, is therefore a much more prominent symptom of retroflexion than of anteflexion, while sterility is not so general, the displacement more commonly occurring in parous women, so that there is no stenosis of the cervix to add to the effect of flexion. Repeated abortion at the third or fourth month is, however, a frequent result, the uterus being unable to rise out of the pelvis. Strangulation by the pressure of the utero-sacral ligaments is not, however, the sole cause of symptoms in retroflexion, for symptoms may exist, and be relieved by the use of a pessary, when the fundus can, without difficulty, be pushed up by the finger in the vagina. Since some degree of descent of the fundus below its normal level is hardly ever absent in retroversion or retroflexion, a dragging pain, from tension of the uterine attachments, is one of the most frequent symptoms. Adhesions are occasionally produced by a partial peritonitis, and the fundus then becomes tethered in a backward direction.

Many of the other symptoms of retroversion and retroflexion are common to most uterine maladies, being due to associated hyperæmia, endometritis, or metritis. Among such are pains extending down the thighs, digestive disturbances, hysterical manifestations, or, in hysterical subjects, functional paralysis. The most marked form of pain, however, in retroversion, and still more in retroflexion, is pain over the sacrum, increased in defecation. The pain in defecation is due to pressure on the tender fundus, which, when the displacement is considerable, encroaches on the calibre of the rectum (Fig. 27, p. 74). It is often associated with rectal tenesmus and excessive secretion of slimy mucus from the rectal mucous membrane. Frequently, also, there is obstinate constipation, partly due to the degree of mechanical obstruction existing, partly to the pain in defecation. In both forms of displacement coitus becomes a mechanical cause of inflammation, especially in retroversion of about 90°, when the cervix lies almost directly in the line of the vagina, and is usually, also, too low down (Figs. 26, p. 73, 40, p. 113), and so becomes exposed to a direct impact, to which it is not normally subject. The bladder is affected less in retroflexion than in retroversion, when the pressure of the cervix may cause irritability, or, if the uterus be enlarged by tumour or by early pregnancy, may lead to retention of urine, which is the most characteristic symptom of retroversion of the gravid uterus.

Diagnosis.—In retroversion the os is found on vaginal touch to be tilted forwards, often so much so as to look in the direction of the vagina, or still more anteriorly. By bimanual examination the absence of the fundus from its normal position is ascertained, the external hand coming close down upon the finger in the vagina. More or less of the body of the uterus is felt by the finger behind the os, but without any concavity or angle between it and the cervix. It may be made to move in conjunction with the cervix, unless fixed by adhesions, and, if necessary, the diagnosis may be confirmed by the sound, introduced with its concavity looking backward.

In retroflexion the os may look in the normal direction, or even too much backward, but is more frequently more or less tilted forward. The fundus is absent from its normal situation, and is felt behind the os as a rounded tumour with a concavity between it and the cervix. If rigidity of muscles, distension of the abdomen, or the presence of inflammatory or other swellings, makes it impossible to ascertain the presence or absence of the fundus in front on bimanual examination, the diagnosis becomes more difficult. It may often be effected with the finger alone, by tracing the continuity between fundus and cervix, and their conjoint mobility, but the sound here affords decisive information. Its use, however, should be avoided, as a rule, if active inflammation be present.

If the os looks in a normal direction, the sound, which has been previously bent to a curve nearly as great as that which the uterine axis is supposed to have, is introduced with its concavity at first forward, and, when it has reached the internal os, is reversed by a *tour de maitre*, the converse of that previously described (*see* p. 9 *and* Fig. 31, p. 88). It is then passed on to the fundus by carrying the handle far forward if necessary, and at the same time pushing up the fundus by the finger in the vagina. If the os is tilted forward, however, the concavity of the sound should be directed backward from the first. If the fundus be restored by the sound in the mode described under the heading of treatment, the swelling will disappear from behind the fundus. The most difficult cases for diagnosis are those in which the fundus is involved in, or adherent to, fibroid or other tumours or inflammatory swellings, and in such case the sound alone can usually afford certain results. A small fibroid in the posterior uterine wall is apt to be very misleading, especially since it generally produces more or less retroflexion. The diagnosis must then be made by completely restoring the uterus with the sound, and then observing whether the swelling previously felt behind the cervix has entirely disappeared.

Treatment.—In the slighter degrees of backward displacement, not accompanied by much descent, especially those which occur in single women, the uterus not being enlarged, or after the menopause, when the uterus is atrophied, mechanical treatment may be unnecessary. The need for keeping the uterus in place is greater the more the fundus is enlarged, and the lower it descends into the posterior cul-de-sac. Hence, in the majority of cases of retroverion or retroflexion of any notable extent, excepting those in which the displacement is secondary to periuterine inflammation, it is desirable to commence the treatment by replacing the uterus, and maintaining it, as far as possible, in position, after which remedies for the relief of any coincident hyperæmia or inflammation are likely to be much more effectual. This depends upon the fact that

Fig. 28.—HODGE'S Pessary.

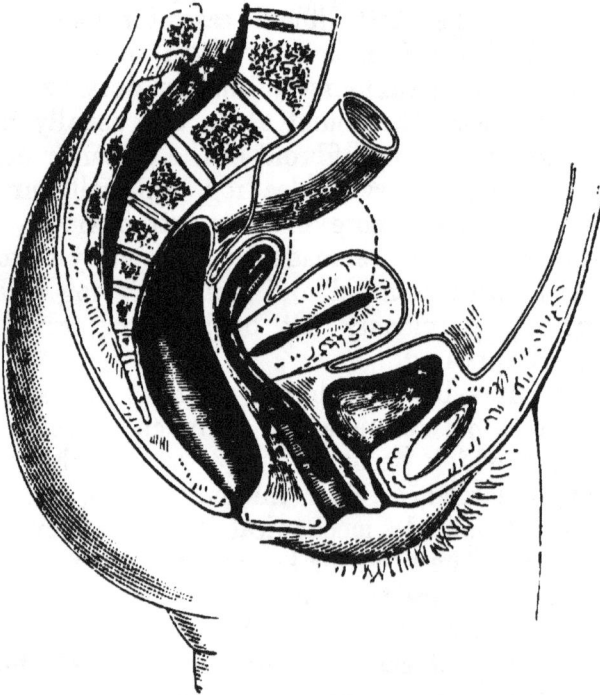

Fig. 29.—HODGE'S Pessary in position.

the displacement can generally be rectified in a more or less complete manner by some form of Hodge's pessary (Fig. 28), which can usually be tolerated even when the uterus is tender. The mechanical action of this pessary, which is sometimes termed the lever pessary, is two-fold. In the first place, its posterior limb stretches the posterior vaginal cul-de-sac backwards and upwards (Fig. 29, p. 80), and thereby draws the cervix backward, and tilts the fundus forward. The uterus itself may here be regarded as a lever, the fulcrum being at its centre, and the power applied to the cervix. This mechanism therefore tends to remedy retroversion, but has no direct effect upon retroflexion. The second action is that by which the posterior limb, when sufficiently long and curved upwards, directly pushes up the displaced fundus, or, more frequently, prevents its return when it has been restored by other means (see dotted outline in Fig. 29). The pessary is here the lever : the fulcrum is a transverse axis, nearly through its centre, upon which it is capable of oscillating as it is grasped by the vaginal walls : the power is the pressure of the anterior vaginal wall upon its anterior limb, greatly increased during any expulsive efforts : the weight, or resistance, is the fundus uteri, which is pushed up by the posterior limb. Some authorities have denied the latter action, and have maintained that the pessary is useful only in retroversion and not in retroflexion. If this were the case, a rather flat pessary would be the best, as most efficacious in drawing the cervix backward. Experience, however, shows that a pessary with a long and strongly-marked sacral curve often succeeds in retroflexion when a flatter one has failed.

It is impossible, however, for the pessary directly to push up a retroflexed uterus completely into its normal position, and, when it is acting in the most successful manner, the fundus will be found no longer in contact with the posterior limb of the pessary (Fig. 29). This depends upon two causes : first, that the fundus can be pushed up to such an extent that the weight of the uterus itself, in the standing position, will tend to remedy instead of to aggravate the displacement (*see* dotted outline in Fig. 29) ; secondly, that, when the coils of intestine have once been allowed to come down into the retro-uterine fossa of the peritoneum, they resume their normal function of pressing chiefly upon the posterior surface of the uterus, and so tend gradually to reduce any retroflexion. Short of this result, however, the pessary may do good by directly supporting the fundus, especially in cases of fibroid in the posterior uterine wall.

Hodge's pessary has been made in many different shapes, and various names have been applied to these. That most generally useful is shown in Figs. 28 and 29, p. 80. The upper or sacral curve is considerable, the lower or pubic curve is slight, and only just sufficient to distribute the pressure equally over the anterior vaginal wall. The lower extremity is square in the centre, but well rounded at the corners. The whole instrument should be made thick, the bar being nearly $\frac{3}{10}$in. in diameter, that its pressure may be more easily borne. There is then no risk of ulceration being produced, even if the pessary is neglected, provided that the fit is suitable originally. When in place it should not rest against the pubic rami, or any bony support, but be held by the elastic vaginal walls.

It will be convenient here to speak of the materials used in the construction of pessaries in general. The best of all is vulcanite, since it is light, smooth, and non-absorbent, and can readily be bent to any shape. The bending may be effected by placing it in hot water, not far short of the boiling-point, and afterwards plunging it in cold water after the desired shape has been given. Another method is to oil the surface, and then to move the instrument rapidly backward and forward through the flame of a spirit-lamp, till it is sufficiently softened. The latter mode is more convenient for bending one part of a pessary at a time, but a little practice is required to avoid burning the surface, and so spoiling its polish. Celluloid or xylonite is also a good material, but has one disadvantage, namely, that it cannot be moulded by the spirit-lamp, but only in hot water. Hodge's pessaries are also made of pewter tubing, which can be bent by the hand. These answer very well, but they are rather heavier than vulcanite, are apt to separate at the point where the tubing is joined, and are not quite so perfect in cleanliness. A pliable form of celluloid has also been introduced. Pessaries may be made hollow in platinum or aluminium, when the exact shape required is known, but these cannot be moulded to suit altering require-ments. Of all materials gutta-percha is the worst, since it rapidly becomes roughened, and sets up irritation. India-rubber is far preferable to gutta-percha, but, being somewhat absorbent, it retains the secretions, and so is apt, before long, to become offensive, and often to produce some vaginal irritation.

Fig. 30. – THOMAS'S Retroflexion Pessary.

The form of pessary recommended by Dr. Thomas is shown in Fig. 30. The upper part is made very thick, so as, by its actual bulk, to prevent the return of the fundus, while, at the same time, its pressure is distributed. This is an excellent device, the only drawback to it being that it greatly increases the price of the instrument. The other peculiarity is that the lower end is bent much downward, to avoid pressure on the urethra, and is nearly pointed, so as to rest between the rami of the pubes, and prevent rotation. A somewhat similar shape is preferred by Dr. Barnes, and a pessary much used in America, under the name of Albert Smith's pessary, has also a comparatively narrow anterior limb and a strong pubic curve. The objection to this is that the pointed end forms a wedge, to facilitate the escape of the pessary, and also, owing to its strong pubic curve, forms an obstruction in the vulva, very inconvenient to married women, while the pessary shown in Figs. 28 and 29 rests completely behind and above the apex of the pubic arch. The great advantage of Hodge's pessary is that it does not prevent coitus, but may lead to conception where sterility had previously existed. It is therefore important, in married women, to see that the lower limb of the pessary lies close against the anterior vaginal wall, and, at the same time, high up, and sheltered behind the pubes. In Dr. Greenhalgh's pessary, in order to fulfil this end, the lower limb is made of soft rubber tubing, the whole instrument being of somewhat elastic wire, covered with india-rubber, so that it can be pressed together during its introduction. It has, therefore, a disadvantage in point of cleanliness, and, the india-rubber becoming very soft in the vagina, the unsupported corners are apt to press injuriously. I have met with several instances in which they had ulcerated very deeply into the vaginal walls.

In some instances, when there is considerable hyperæmia, swelling, and tenderness of the fundus, as is the case more frequently in retroflexion than in retroversion, it is desirable, before attempting to use a pessary, to treat these conditions by rest for a few days in bed, with saline aperients and sedatives, and sometimes local depletion. Recourse should always be had to the same plan when a pessary has been tried, but cannot be tolerated on account of the pressure which it exerts upon the fundus.

If the Hodge's pessary, after such treatment, still cannot be tolerated, the elastic ring pessary of watch-spring covered with india-rubber, which will be described as one of the best pessaries for prolapse, often proves useful, at least as a temporary resource. It counteracts the descent and retroversion, though it rarely completely remedies retroflexion.

Before a pessary is inserted, the uterus should be replaced, if possible, by the finger. This may be done in the lateral or, what is better, the semi-prone position. By one or two fingers, the perineum is retracted, air allowed to enter the vagina, the fundus pushed upwards, and the cervix, if it looks too much forward, is afterwards drawn backward. Sometimes the external hand above the pubes may assist in bringing the fundus completely forward. For this, the patient is placed in the dorsal position; the fingers in the vagina first raise the fundus as far as possible, then push the cervix very far back, while the external hand, pressed deeply in a little below the umbilicus, endeavours to get behind the fundus and bring it forward into anteversion. The introduction of a Hodge's pessary itself will often effect or complete the restoration of the uterus. For its adjustment, the patient is placed in the left lateral or semi-prone position, and the pessary is turned edgewise, until it has more than half passed through the vulva, the perineum being meanwhile retracted by a finger of the left hand, and the pessary directed rather backwards, to avoid pressure on the symphysis. It is then rotated into the direction which it is to occupy, having the concavity of its upper or sacral curve looking forwards. The index finger of the right hand is then introduced behind the lower limb, and passing through the pessary, hooks the upper limb backward over the cervix, and, at the same time, pushes it upward into the posterior cul-de-sac. The upper limb always tends to run up in front of the cervix, and when the pessary has a strong sacral curve, it may be difficult to overcome this tendency. It is then often useful not to rotate the pessary completely into its destined direction, but to hold it somewhat diagonally, until the upper limb has passed behind the cervix. The pessary should cause no pain when once in position. If it does so, it is a sign that it is too large, too angular, or improperly adjusted, and it should at once be removed.

If in a case of retroflexion the fundus can be restored to a considerable degree by these means, the pessary may be left gradually to bring about a more complete reduction, and its leverage action is more effective when the patient is up and about than when she is confined to bed. In retroversion also, such treatment will rarely fail. It sometimes happens, however, in retroflexion, either that the pessary fails to raise the uterus at all, and only exercises painful pressure upon it, or that its upper limb fits it to the concavity in its posterior surface, and merely elevates the whole organ, while the fundus remains flexed over the pessary. It is then necessary, in the first place, to restore the uterus by other means if the bimanual method, mentioned in the preceding page, does not succeed. Of these the chief are : (1) pressure *per rectum* ; (2) the postural method ; (3) the use of the sound as a repositor.

(1.) Pressure by the finger on the fundus from the rectum is more effectual than by the vagina, since the leverage is greater, and it may sometimes be conveniently applied when the uterus is found incompletely restored after adjustment of a pessary.

(2.) In the postural method the object is to place the patient in such a position that the inlet of the pelvis looks vertically downward, the abdominal muscles are relaxed, and the weight of the abdominal contents tends to produce a negative pressure in the pelvis. If air be at the same time allowed to enter the vagina, by separating the labia, if necessary, the vagina becomes distended into an actual cavity, the uterus recedes, and the fundus may be restored by this means alone, its own gravity assisting in some small measure. The recession of the fundus may also be assisted by pressure from one or two fingers in the vagina or rectum. Sometimes, however, the fundus merely recedes out of reach, the retroflexion or retroversion remaining unrectified, and the third method is then the only effectual one. In carrying out the postural method, the patient may be placed in the knee-elbow or genu-cubital position on a hard bed or sofa. Care must be taken that the thighs are exactly vertical, and the chest low. In this way the axis of the trunk may be inclined as much as 35° to the horizon, the hips being higher than the shoulders—a position which will give the best result, taking the normal pelvic inclination as 55°. Some authorities reject the knee-elbow, and are content only with the genu-pectoral or knee-chest position. Most persons, however, will find, if they try the experiment of kneeling in this position on the floor, that it is impossible to touch the same plane with the chest. Moreover, to increase the inclination of the trunk beyond 35° would diminish the efficacy of the position. It is often of use, in retroflexion or retroversion, to instruct patients themselves to adopt this position once or twice a day, as well as to lie, as far as possible, in the prone or semi-prone position when in bed.

(3.) The use of the sound as a repositor is the most effectual of all methods, but is not so safe as the two already mentioned, unless both caution and dexterity be employed. Those, however, who possess the necessary skill will generally be able to restore the uterus in this way with much less discomfort to the patient than by either of the other means. The sound has its intrauterine portion made nearly straight, and is introduced in the mode already described, the handle being necessarily carried far forward, and the point directed backward. If it can only be introduced by giving it an increased curve, it should be withdrawn, and introduced a second or third time with a gradually diminished curve, so rendering the axis of the uterus nearly straight, and converting the retroflexion into a retroversion. The first stage of replacement is then to carry the handle of the sound backward toward the perineum, thereby partially elevating the fundus. The operator should do this with great gentleness, remembering the powerful leverage he is exercising, and any excessive resistance, as from adhesions, will then be discovered at this stage, and the attempt abandoned. It is not very usual, however, for the fundus to be tethered by adhesions without the existence of some periuterine thickening which may be detected by a skilful observer. It is more common to find a fixation which is only apparent, and due to the swollen fundus having become gripped between the utero-sacral ligaments at either

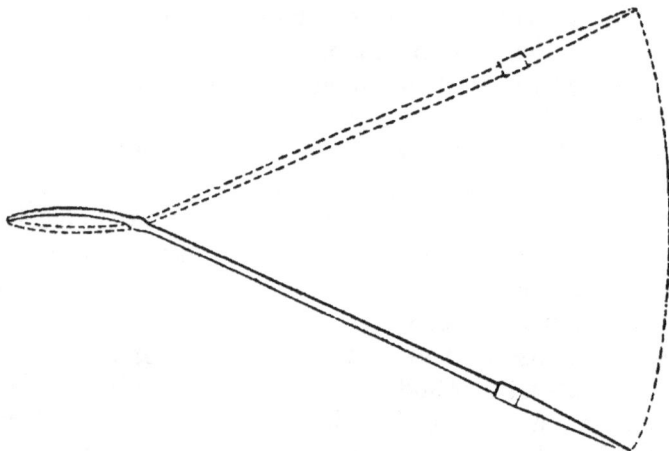

Fig. 31.—Diagram to Illustrate the Mode of Replacing the Uterus by the Sound.

side. The second stage in reduction is to sweep round the handle of the sound through a rather wide semi-circle, so that the handle and stem describe a semi-cone, and the intra-uterine portion rotates nearly on its own axis (Fig. 31). The third stage is to carry the handle again backward toward the perineum, and so bring the fundus completely forward. If the handle of the sound were simply rotated, its point would necessarily describe a circle, and press injuriously upon the fundus.

A uterine repositor has been invented, in which the

handle is not rotated, but the direction of the intra-uterine portion is changed by means of a screw. The sound, however, if used in the way described, is more convenient and quite as safe. After withdrawal of the sound, the pessary may often be adjusted before the displacement has had time to recur. If, however, the fundus drops back at the moment of its withdrawal, one of two methods may be used. With great caution, either the pessary may be passed into position over the handle of the sound, or restoration with the sound may be effected while the pessary is in the vagina, the uterus, in both cases, being held in perfect position until the pessary is fully adjusted. If the first method is adopted, it is desirable to have an assistant, who holds the handle of the sound steady while the pessary is passed through the vulva, the left hand being employed at that moment in separating the labia and retracting the perineum.

There is another method of keeping a retroflexed uterus in place, which is mechanically the most perfect, although for other reasons undesirable, and which may be tried in some very exceptional cases if all other means fail. This is the use of an intra-uterine stem, in conjunction, either with a simple Hodge's pessary, or with some vaginal support, which must not be rigidly connected with the stem. If, however, a Hodge's pessary be chosen with a posterior limb long enough and curved enough, and if sufficient perseverance be shown in the use of the sound and postural treatment as adjuncts, a stem pessary will rarely, if ever, be required. The Hodge's pessary may fail, however, when the posterior cul-de-sac is too ill-developed, or too atrophied, to admit an instrument of sufficient length, or when the vaginal walls are so excessively relaxed as to take no grasp of even a large pessary ; but in such a case, Cutter's retroflexion pessary, shortly to be described, may be tried, and is free from the danger which attends the use of an intra-uterine stem. Also, in the rare cases of congenital or primary retroflexion not dependent upon an antecedent retroversion, the mechanism of Hodge's pessary is less effective, since it can only act by direct pressure upon the fundus, and by this means can only partially elevate it. The choice of a stem pessary, and the precautions which must be observed if its use is ventured upon, will be considered under the head of anteflexion (*see* pp. 108 —112).

When there is difficulty in keeping the uterus in place, it is sometimes necessary to use at first a Hodge's pessary which is rather a tight fit for the vagina, but which after a while may be exchanged for a smaller one. In very obstinate cases of retroflexion, when there is great difficulty either in getting the fundus into position or in keeping it there, the following plan may succeed when other means fail. The uterine canal is fully dilated with laminaria tents, thus straightening the uterus for the time being; then an anæsthetic is given; the finger is then passed well up into the uterine cavity and used as a repositor, until the external hand can be got behind the fundus, and bring it forward into a position of anteversion. A full-sized pessary is then adjusted, while the external hand still holds the fundus in anteversion.

The action of a pessary should be observed every week or two for the first few weeks. Afterwards, patients should be enjoined to come for observation about every two months, and to use a vaginal injection twice a day.

In cases of retroversion or retroflexion, in which the uterus is too tender to allow any pessary to be tolerated, it may be supported temporarily by tampons of cotton wool soaked in carbolized or iodized glycerine. The uterus is first restored by one of the methods already described. A small tampon is then placed behind the cervix, to prevent the fundus again dropping backward; a second and larger tampon is adjusted in front of the cervix in such a way as to press it backward. The tampons should be changed every second day.

In the case of retroflexion, however, this method rarely effects more than a very partial restoration.

The use of Hodge's pessary may be rendered impossible by the presence of one or both prolapsed and tender ovaries, pressure on which cannot be tolerated. It is then often necessary to commence with treatment directed to the ovaries, but it is very desirable in such cases to restore the uterus, since the ovaries are then elevated at the same time, and a form of pessary may sometimes be found by trial, especially a very thick

Fig. 32.—CUTTER's Pessary for Retroflexion modified by THOMAS.

pessary, or one with an expansion at the upper end like that of Thomas's pessary, which elevates the ovary, and does not press painfully. If this is not tolerated, an elastic ring pessary may be tried.

A useful form of pessary for cases when the vagina is too lax to keep a Hodge's pessary in place, or when it is desired to effect a gradual stretching of a short posterior cul-de-sac, is Cutter's pessary for retroflexion (Fig. 32). From its having an external support, this pessary is more likely to communicate shocks to the fundus ; it has also the drawback that it must be introduced by the patient herself daily, and removed at night, and hence it is liable, either to be pushed up in front of the cervix in introduction, unless some dexterity be used, or to slip into that position afterwards. To diminish the chance of this the vaginal portion of the instrument should have but a slight curvature. The single band of the instrument is carried backward over the perineum and attached to a waist-belt, as shown in Fig. 48, p. 128.

ANTEVERSION OF THE UTERUS.

Pathological Anatomy.—The normal mean position of the axis of the uterus, being nearly that of the axis of the pelvic brim, is one of anteversion in reference to the axis of the vagina. Moreover, in the standing position, when the bladder is empty, it is a normal condition for the uterus to be anteverted even in reference to the axis of the brim. A pathological anteversion, therefore, only exists when, in all positions of the body, there is a notable and persistent anterior inclination of the uterus in relation to the axis of the brim, its shape remaining unaltered. It follows from this that the angle of possible deviation from even the theoretical mean position of the uterus cannot exceed 90° at the utmost (*see* Fig. 26, p. 73), while the deviation from the limiting normal position can scarcely reach 45°. Moreover, the deviation must be more or less rectified whenever the bladder is full. Hence, in comparison with retroversion, anteversion is of very little importance.

Causation.—All the causes before enumerated for displacements in general (*see* p. 71) which, when the uterus is low in the pelvis, produce retroversion or retroflexion, tend, so long as the centre of that organ remains at its proper level, to produce rather anteversion or anteflexion. Anteversion is therefore especially associated with increased weight of the body of the uterus or excessive intra-abdominal pressure, as from tight-lacing or weight of clothing, without a proportionate relaxation of the supports which maintain the centre of the uterus in its position in the pelvis. While anteflexion is frequently primary, anteversion (like retroversion and retroflexion) is usually secondary; and the commonest of all its causes is hyperplasia of the body of the uterus. Anteversion may also be produced by fibroid tumours in the uterine wall, or adhesions the result of periuterine inflammation.

Results and Symptoms.—Anteversion in itself generally produces little or no symptoms, and symptoms associated with it are most frequently due rather to the hyperplasia, hyperæmia, or inflammation which was anterior to the displacement. If, however, the displacement is considerable, and the uterus is also large and hard, especially if the enlargement is due to the presence of a fibroid tumour, signs of pressure upon neighbouring organs may appear. These may closely resemble those produced by a high degree of retroversion, the fundus and cervix lying in exactly the reverse direction (*see* Fig. 26, p. 73). When the cervix especially is enlarged and indurated, and particularly when the whole uterus is at the same time far back in the pelvis, pain in defecation and rectal tenesmus may be produced. Patients may suffer, also, from frequent calls to micturition, and dysuria in addition. The intensity, and indeed the existence, of all these symptoms depends to a great extent, first, upon the nervous susceptibility of the individual; and, secondly, upon the degree of congestion and consequent tenderness of the uterus. In some persons both bladder and rectum will tolerate a great deal of merely mechanical interference with very little complaint, as is seen in certain cases of prolapse.

Diagnosis.—By vaginal touch the os is found to be directed too much backward and high up in the hollow of the sacrum, so that in extreme cases it can with difficulty be reached. The anterior vaginal wall is tense, from the traction exercised by the cervix, and more than usual of the body of the uterus is felt resting low down upon it. On bimanual examination, the body of the uterus is readily defined in this position, the fundus being close behind the pubes when displacement is considerable. No concavity or angle is detected by the finger in the vagina between cervix and body. If no adhesions or tumour be present, the fundus may be pushed up by the finger through the anterior cul-de-sac until the external hand, pressed immediately above the pubes, is able to get below it, and carry it still further back, while the finger in the vagina draws the cervix forward. In simple anteversion the sound is hardly ever necessary either for diagnosis or replacement, though it may be of use to determine the degree of enlargement, or to decide as to the presence or absence of a fibroid tumour. It is to be remembered that when the examination is made in the dorsal position, the degree of anteversion will generally be less than that which exists in the erect posture.

Treatment.—Anteversion is generally rather the indication of increased weight of the fundus uteri than the cause of symptoms in itself. Moreover, there is no possible pessary for anterior displacements which is either so effective or so free from any injurious influence as the Hodge's or ring pessary in backward and downward displacements. Hence it is only very exceptionally that mechanical treatment is found useful in anteversion, and generally it is preferable to direct the treatment rather to the cause of enlargement. Any acute symptoms of hyperæmia should be relieved by rest in bed, aperients and sedatives, with local depletion if required. In this, as in other displacements, all tight clothing should be forbidden, and the skirts should be suspended from the shoulders. When any abdominal laxity exists, and even when this is not the case if the uterus is much enlarged, an elastic abdominal belt, with a pad to press above the pubes, often gives great relief. It acts partly by directly pushing the fundus backward and partly by diminishing the mobility of the viscera, and keeping up a gentle pressure upon the pelvic organs. It is thus often of value in cases of hyperæmia, even without any displacement.

Most vaginal pessaries for anteversion or anteflexion have a double action, like those for retroflexion. By pressure on the anterior vaginal wall they directly push up the fundus, and at the same time they render that wall arched, and so shorten it by bringing its extremities nearer together, and thereby draw the cervix forward (Fig. 34, p. 96, and Fig. 36, p. 98). The base of the bladder, however, appears to be more vulnerable to injury from any considerable pressure than the posterior cul-de-sac, and all these pessaries not only adapt themselves less naturally to the vagina than Hodge's pessary, but by occupying mainly the lower part of that canal, form an ob-

Fig. 33.—Cradle Pessary.

stacle to coitus, and are liable to extraordinary displacements in married women. It is, therefore, generally desirable to try first the effect of treatment adapted to any concomitant hyperæmia or hyperplasia, combined with the general measures already mentioned, and to use a pessary only if such treatment fails to relieve symptoms, and the displacement is considerable in degree. The cases in which a pessary is most likely to prove useful are those in which irritability of the bladder is associated with marked anteversion.

The condition of drawing the cervix forward is fulfilled best by an instrument which lies almost entirely in front of the cervix. Of these, perhaps the most convenient is the cradle pessary of Dr. Graily Hewitt (Fig. 33, p. 95), which should be made of vulcanite, and not, as originally constructed, of wire covered with gutta-percha. Its author now recommends that it should be made with the rings more unequal than those shown in the figure, and that the smaller ring should be placed foremost. In the pessaries sold, the rings are often made too thin. In another form of cradle pessary,

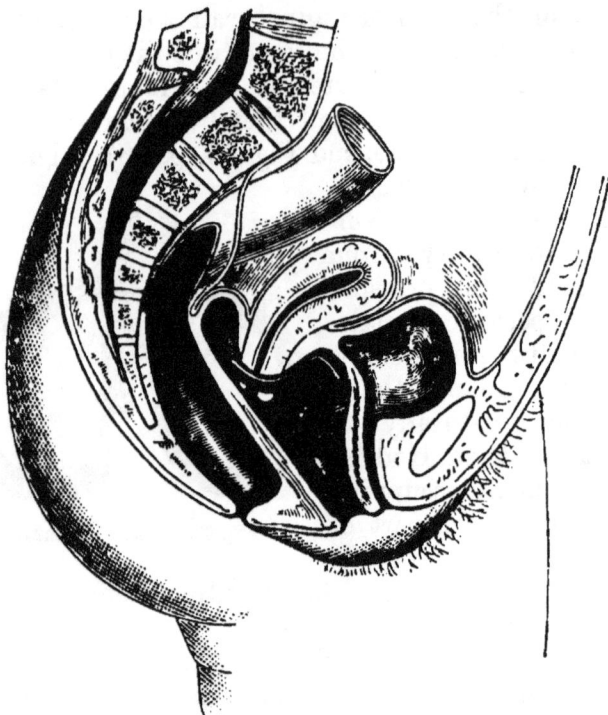

Fig. 34.—Cradle Pessary in position.

the apex of each side, instead of being united to the other by a transverse bar, ends separately in a rounded bend or "crutch," with the view of preventing lateral displacement of the fundus. The surfaces of the crutch part, which is in front of the uterus, are to be opened out so as to present a concave surface, against which the uterine body rests. The position in which

the cradle pessary usually rests is shown in Fig. 34.
The anterior, as well as the posterior end, takes its
purchase from the floor of the vagina, while the bridge
pushes up the anterior cul-de-sac, and the vaginal
entrance is thus blocked. An intelligent patient, however, may sometimes be taught to remove and replace
the pessary herself, and this objection is then obviated.

Numerous anteversion and anteflexion pessaries
have been devised by Dr. Thomas and others, on the
principle of attaching a moveable or elastic bow to a
Hodge's pessary as basis.

All these instruments form a great obstacle to
coitus, unless re-
moved for the time,
since they take their
purchase upon the
posterior vaginal wall
at its lower part, to
make pressure upon
the anterior vaginal
wall. For use in
married women, I
have devised the in-
strument shown in

Fig. 35.
The Author's Anteversion Pessary.

Fig. 35, with the object of extending to the treatment
of anteversion and of corporeal anteflexion the principle
of leverage, which is so useful in posterior displacements.
In the case of Hodge's pessary, the leverage is really of
an alternating kind, the power and the resistance being
interchanged. In an expulsive effort, the anterior
vaginal wall is depressed, and the upper limb of the
pessary tilted upwards ; when the effort is relaxed,
the upper limb is depressed again to some extent by
the elasticity of the posterior cul-de-sac, and the
anterior vaginal wall elevated, the power in the latter
case acting at the upper end of the pessary. In the
ordinary form of lever pessary, the shape is such as
to make the fulcrum nearer to the upper than the
lower end, and then the first kind of leverage has

H

the mechanical advantage. If, however, the pessary be so shaped that the fulcrum is nearer to its lower end, the opposite will be the case, and a moderate tension of the posterior cul-de-sac will produce a force greater than itself, pressing upwards the anterior cul-de-sac. The pessary shown in Figs. 35 and 36 is made of vulcanite, and resembles a thick short Hodge's

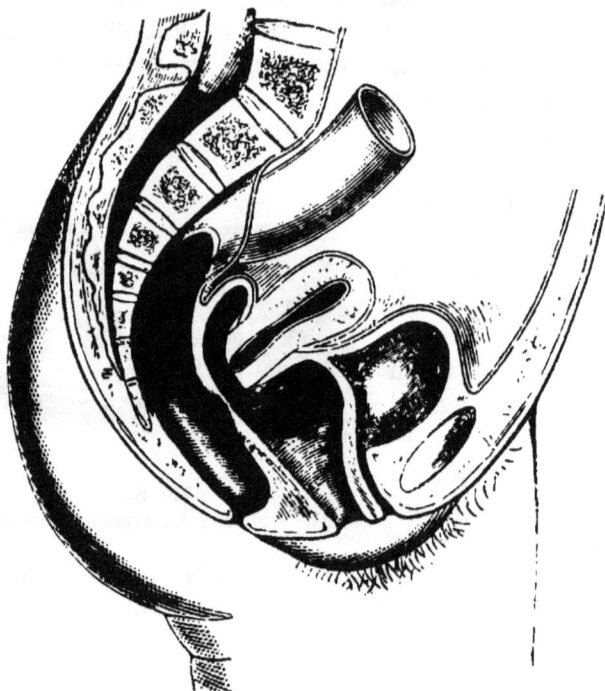

Fig. 36.—The Author's Anteversion Pessary, in position.

pessary, with its anterior limb replaced by a broad arch directed upwards, and nearly square at its summit. The position in which it lies is shown in Fig. 36. By its shape alone, without any leverage, it elevates the anterior vaginal wall in considerable degree, but it will be found in practice that the lower corners do not lie against the posterior vaginal wall, but the whole of the anterior extremity is tilted somewhat upwards, in consequence of the tension of the posterior cul-de-sac. In introducing the instrument, it is first passed entirely within the vulva, with the upper limb in front of the cervix ; the index finger is then passed through it, and hooks the upper limb backward over the cervix and into the posterior cul-de-sac. It is withdrawn by hooking the index finger over one of the lower angles, and making traction upon that. Since it occupies a higher position in the vagina than even a Hodge's pessary, it can be worn, without discomfort, by married women, and I have found, in some instances, that it has rendered coitus free from pain, when it could not otherwise be tolerated. Care must be taken that the posterior limb is not too long, since otherwise it would tend to draw the cervix backward through the medium of its vaginal attachment. As compared with the cradle pessary, it has the disadvantage that a patient cannot remove it herself. When in position, however, it is very easily tolerated, and there is this safety in its use, that it would be difficult to introduce a pessary which, when in place, would prove too large. As being more difficult to introduce and withdraw than the cradle pessary, it is not suitable for virgins, or cases in which the vaginal outlet is narrow. A small instrument, of the pattern shown in Fig. 44, p. 124, may also be used in anteversion, especially if the uterus is low, and the vagina lax.

In some cases of anteversion associated with congestion, benefit is found from the use of a Hodge's or elastic ring pessary, notwithstanding that the pessary tends rather to increase the anteversion. In such a case the pessary probably does good by somewhat elevating the whole uterus, and limiting its mobility.

As in the case of posterior displacements, it is impossible, by direct pressure through the vaginal wall, to restore the fundus completely to its normal position, but it can only be elevated to a certain extent, while there is, in this case, no mechanism tending to complete the restoration. All anteversion or anteflexion pessaries require more careful watching than is necessary with the ordinary Hodge's pessary. They should be used only tentatively in the first instance, and continued only if they are found actually to give relief.

Pathological Anatomy.—A slight anterior curvature of the uterine axis is normal, or at any rate very common, in the nulliparous uterus, and, when the uterus is soft, an increased anteflexion, instead of merely an anteversion, of the uterus is produced when the bladder is empty. A pathological anteflexion, therefore, only exists when the curve is very considerable as a whole, or very sharp at one part of the uterine axis. Acquired anteflexion may be combined with anteversion, so that the os is tilted too much backward, the uterus having partly yielded as a whole to the displacing force, and partly undergone bending. In primary anteflexion the os is most frequently directed too much forward. The classification of Thomas into corporeal, cervical, or cervico-corporeal anteflexion, according as the body alone, the cervix alone, or both cervix and body, are flexed forward, is a useful one, since each condition calls for a distinct mode of treatment (Fig. 37). Anteflexion combined with retroversion may also exist, the axis of the uterus being concave forwards, but its body displaced backwards (Fig. 37, E F). In acquired anteflexion the curvature is generally sharper at one part, usually near the internal os, and the canal is apt to be here flattened, and so obstructed. In primary anteflexion the curve is generally more uniform, and the uterine tissue harder, so that there is not necessarily any flattening of the canal, though its diameter is usually less than normal. Statistics vary very widely as to the relative frequency of anterior and posterior displacements of the uterus. The general opinion is, however, that anterior displacements, especially primary

anteflexions, are much commoner, but that of displacements of any consequence, calling for mechanical
treatment, descent and retroflexion are the most
common. The discrepancy is explained, in great
measure, by the difference between various authors as
to the degree of anteflexion or anteversion which they
would regard as pathological.

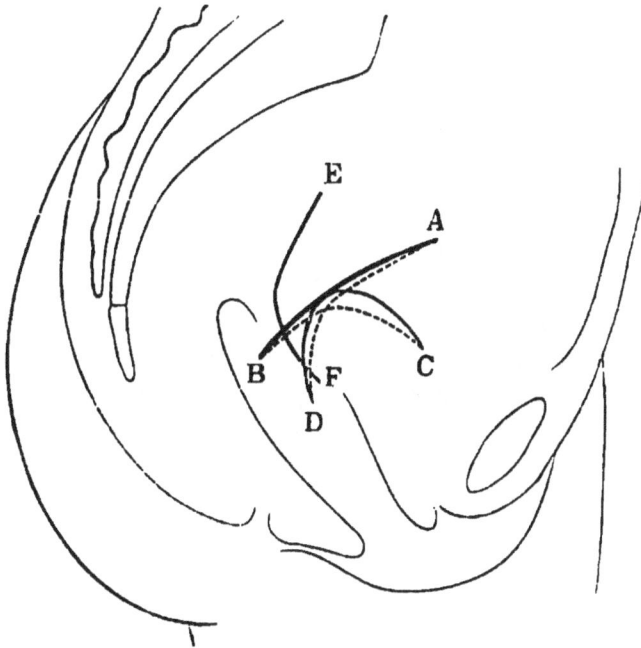

Fig. 37.—Diagram to illustrate the Varieties of Anteflexion.

A B, normal direction of uterine axis, the bladder being in a medium
condition of fulness; A D, cervical anteflexion; C B, corporeal
anteflexion; C D, cervico-corporeal anteflexion; E F, anteflexion
with retroversion.

Causation.—Acquired anteflexion is produced by
the same causes as anteversion, with the addition of
softness of the uterine tissue, or it may result from
morbid softness alone, such as is not uncommon in
ill-nourished girls, the uterus yielding from its own
weight, or from the effect of forces which are normally in action. Before the age of puberty it is

normal, or at any rate very common, for the uterine axis to have a greater physiological anteflexion than it has in the adult, while it straightens itself when the uterus reaches its full development and becomes firmer. Hence anteflexion is very frequently primary, and consists in an abnormal persistence or exaggeration of a state which, in childhood, is hardly to be considered pathological. In this case the anterior uterine wall is often thinner and less developed than the posterior ; the anterior lip of the cervix is usually too short, and frequently the anterior vaginal wall is itself short. Such a condition is often associated with a conical cervix and small external os, and sometimes with general smallness of the whole uterus. Frequently there are also signs of imperfect ovarian activity ; the vagina may itself be small, and even the bony pelvis may share in the want of full development. An acquired cervico-corporeal anteflexion has been ascribed to contraction of the utero-sacral ligaments, the result of localized cellulitis or *parametritis posterior*, dragging the centre of the uterus backwards. Acquired cervical anteflexion may also result from pressure against the posterior vaginal wall, when the whole uterus is displaced downward and backward, as from tight-lacing. Cervical anteflexion is also frequently the result simply of an undue length, generally associated with a conical form, of the vaginal portion of the cervix, which then most readily accommodates itself to the vagina in this flexed position. In extreme cases of cervico-corporeal ante-flexion the total angle of flexion may approximate to 180°, fundus and os looking nearly in the same direction.

Results and Symptoms.—The importance of ante-flexion has been one of the most controverted points in gynæcology, and the matter is still far from settled. By some this condition is regarded as a very common and important source of trouble ; by others it is held that a flexion causes no impediment whatever to the uterine canal, and that a large proportion of unmarried women in perfect health have an anteflexed uterus. It appears to be evident that, while a naturally bent canal may be quite patent, any sharp bending of an originally straight or nearly straight canal of soft tissue, like that of the uterus, must tend to flatten it, and so diminish its calibre. The smaller the canal is originally, the more likely is such flattening to produce the effects of obstruction. Even when the canal is small as well as flattened, the obstruction, like that produced by extreme stenosis, may not necessarily produce dysmenorrhœa, provided that the menstrual flow is moderate and uniform in quantity, and perfectly fluid, but only when shreds of membrane or clots have to pass.

The difference of opinion as to the frequency of anteflexion, apart from symptoms, may be in part accounted for by differences in the mode of estimating the anteflexion, or in the degree of curvature regarded as pathological. General experience appears to show that acute anteflexion is more frequently met with among women suffering from dysmenorrhœa, or dysmenorrhœa and sterility, than it is at post-mortem examinations amongst nulliparous women in general, and therefore that there is probably a causal relation between these conditions. It is true that theory does not receive the same confirmation by a therapeutic test from the successful removal of symptoms by the cure of the anteflexion as it does in the case of retroflexion from the successful results of treatment by Hodge's pessary. This may, however, be explained by the fact that it is impossible in anteflexion so effectually to straighten the uterus by any pessary except an intra-uterine stem, which, by the irritation of its presence, is apt to cause more evils than it can remove. It has been argued that flexions cannot produce any obstruction because the uterine cavity is not, post-mortem, found to be dilated. Such dilatation, however, is not to be expected. Just as, in stricture of the urethra, the bladder is found to be small, with thick walls, so in the case of the uterus, unless the obstruction is complete, the muscular walls become hypertrophied to overcome it, as is found to be the case in extreme stenosis of the cervix. The only post-mortem evidence, therefore, which obstruction of the canal would leave, is hypertrophy of the body of the uterus, and this is not unfrequently found in cases of anteflexion associated with dysmenorrhœa. An anteflexed uterus, if examined soon after puberty, is usually found to be small, and the anteflexion is probably often part of a general want of full development. But, in later years, the cavity even of a nulliparous anteflexed uterus is not unfrequently found to be elongated, though the sound is very apt to be arrested by the flexion short of the full length, and so to give an erroneous impression of smallness.

The general conclusion is that anteflexion, especially primary anteflexion, frequently exists without any symptom, and that the importance of this condition has been much exaggerated by some authors, but that an acute anteflexion, especially if acquired, may diminish the calibre of the canal, and so tend to produce, or assist in producing, dysmenorrhœa, endometritis, and probably also sterility. Practically an acquired cannot always be distinguished from a primary anteflexion, but, if the flexion is found to be specially acute at one point, it is more likely to be acquired. If a uterus have a primary anteflexion, and the flexion be afterwards much increased, it is obvious that there may be the same tendency to flattening of the canal as in the case of a uterus originally nearly straight.

Diagnosis.—The direction of the os and cervix is readily discovered by vaginal touch. In cervical anteflexion, a considerable length of the cervix may sometimes be traced behind the os, and the use of the sound may be necessary to distinguish between cervical anteflexion and partial retroversion. In corporeal anteflexion the fundus is felt resting low upon the anterior vaginal wall, and may be defined on bimanual examination as described in the case of anteversion. It is found to move in conjunction with the cervix, and a concavity or angle is felt between the two. On passing the sound it is generally arrested near the internal os, and can only be carried on to the fundus either by taking the handle far back toward the perineum, by pushing up the fundus by the finger in the vagina, or by withdrawing the instrument and re-introducing it with an increased curve. The slighter degrees of corporeal anteflexion may be difficult to detect, especially in virgins when the vaginal walls are tense, and the exact curve of the uterine axis can then only be determined by means of the sound. The conditions chiefly to be distinguished from anteflexion are a fibroid in the anterior uterine wall, thickenings due to cellulitie or peritoneal inflammation, or to blood effusion, or tumours, or calculi in the bladder. All of these conditions, except calculus, are distinguished by their fixity and ill-defined outline, but the most perfect evidence is that derived from the sound, especially in distinguishing the case of a fibroid in the anterior uterine wall. If the swelling felt anteriorly disappears when the uterus is straightened by the sound, or turned into a position of slight retro-flexion, it is proved to have consisted of the fundus alone.

Treatment.—Treatment is, of course, only necessary when symptoms exist which are referable to the ante-flexion. Reduction of the flexion is generally easy, but does not have a permanent effect. In the case of corporeal anteflexion, when the body of the uterus is bulky and soft, as in acquired flexions, it may some-times be replaced with the fingers alone, by pushing up the fundus from the vagina, and then pressing it backwards by the outside hand above the pubes. Generally the sound is necessary for replacement, but it should not be employed if there is any sign of peri-uterine inflammation or acute hyperæmia. If it can only be passed with an extra curve, it is withdrawn and introduced a second or third time, till it can be passed with its intra-uterine portion nearly straight. If treatment by rest is indicated for the relief of any concomitant hyperæmia or inflammation, and if at the same time the uterus is soft and the fundus large, the use of the dorsal position has the advantage that the effect of gravity on the fundus tends to reduce the flexion. It must, however, be remembered that in all chronic uterine maladies, it is undesirable to keep a patient too much to her bed. If undue softness of the uterus appears to be the result of malnutrition, it is important to see that a sufficiently nutritious diet is taken, and to adopt a generally tonic treatment.

Since the most important mechanical effect of ante-flexion is that of causing an impediment in the cervical canal, and anteflexion is often associated with a canal rather smaller than the average, it is generally desirable, especially in anteflexion of the nulliparous uterus, when dysmenorrhœa or sterility exists, to take care that the canal is of full calibre. Dilatation may be effected by metallic bougies, by Priestley's dilating sound, or by the occasional use of a tent, if repeated manipulation be considered undesirable (*see* p. 60). This treatment tends at the same time to straighten the canal in some measure. In the great majority of cases of corporeal anteflexion, according to the author's experience, no further direct mechanical treatment than this is found beneficial.

In pure cervical anteflexion pessaries are useless. In minor degrees dilatation of the os and cervix may be sufficient. In higher degrees, when symptoms are present, such as those of obstructive dysmenorrhœa, and especially when the os is also small, the best treatment is to incise the cervix backward, so as to convert the os into an elongated opening, more nearly in a line with the upper part of the cervical canal (*see* Fig. 20, p. 58). The mode of making the incision and the after-treatment are described under the head of stenosis of the external os (pp. 53—57). For cases in which the fundus is flexed forwards as well as the cervix, Marion Sims recommends the incision of the anterior uterine wall near the internal os in addition to that of the posterior wall of the cervix. It is impossible, however, when there is considerable corporeal flexion, to straighten in this way the cervical canal without a dangerously deep incision. Such an incision,. moreover, generally tends to close again. When the uterus is soft in corporeal anteflexion, it is possible to straighten it in some measure by a vaginal pessary,. which pushes up the fundus. Such straightening, however, can hardly ever be so complete as that which Hodge's pessary often produces in retroflexion, since there are not, as in that case, any natural forces. called into play to complete the restoration, and the result of such mechanical treatment is usually much less satisfactory. It is generally preferable, as in ante-version, first to direct the treatment to any cause for enlargement of the fundus which may be discovered. If other means fail, and the anteflexion is believed to be the cause of symptoms, one of the vaginal pessaries described under the head of anteversion may be tried (Fig. 33, p. 95; Fig. 35, p. 97), but should not be per-severed with unless actual relief is found from its use.

When the uterus is rigid as well as flexed, a vaginal pessary will only tilt the fundus upwards and cervix forwards without altering the shape of the uterus. The occasional use of a laminaria tent, reaching nearly up to the fundus, is then sometimes of service. It acts by softening the walls of the uterus, as well as by straighten-ing it for the time being, and dilating the canal. The most effectual means of straightening the uterus, how-ever, is one which has been the subject of much con-troversy, namely the use of intra-uterine stem pessaries. While some distinguished authors have denounced them entirely, others have expressed apparently over-sanguine views of their efficacy. But, while the advocates of this treatment have described its results as brilliant, none of them has ever adopted the only method which, in such a doubtful matter, could establish it on a. satisfactory basis, namely, to give a complete record of the results of every one of a series of consecutive cases. The objection to stem pessaries is, not that they are mechanically ineffectual, but that they always excite a certain degree of irritation and hyperplasia, while they have not unfrequently produced severe and even fatal metritis and peritonitis. It might be inferred *à priori* that a mucous surface covered by a cylindrical epithelium is not likely to tolerate pressure and friction altogether with impunity, and this conclusion is confirmed by experience. Even with the most modern form of stems, and in the practice of careful physicians, one or two fatal results after the use of these pessaries have been recorded. It is comparatively common to find that pain, or a rise of temperature, necessitates the removal of a stem before it has been long retained. Special precautions are therefore required in applying stems, and they should be used, if used at all, not as permanent supports, but as a temporary treatment, to be continued for a limited number of months, while a patient is kept under strict observation. No one should venture on the use of an intra-uterine stem who has not complete confidence both in his own power to judge of the suitability of the case, and also in the implicit obedience of his patient to directions. Their use is not to be recommended to those who have not given special study to the diseases peculiar to women.

The first precaution necessary is never to use an intra-uterine stem in a case in which periuterine inflammation has at any time existed, since this is always liable to be rekindled by any slight exciting cause. The second is not to use one when any considerable tenderness or hyperæmia of the uterus is present, until this condition has first been subdued by suitable treatment, especially rest in bed and local depletion. The cases in which stem pessaries can be used with least risk are those in which menstruation is scanty, in which the uterus is small, and has never been the subject of any considerable inflammation, and in which it is desired to stimulate the uterus, and dilate as well as straighten its canal. When the flexion is secondary to chronic hyperæmia or metritis, it is more frequently found that the stem pessary cannot be tolerated. Before applying a stem, the sound should be introduced occasionally, to ascertain the tolerance of the uterus. It is better to keep a patient in bed for two or three days on the first introduction of the pessary, and if it cause at first any great pain, it should be withdrawn for a while and again introduced, being worn a little longer on each occasion, until the uterus is gradually brought to tolerate its presence. A patient who is wearing an intra-uterine stem should always either be within reach of immediate medical assistance, or should have the means of removing the pessary herself by a thread attached to its lower end, in case any great pain, or any rigor or feverishness, should supervene. The stem should be withdrawn, for the purpose of cleansing it, as often as once a month, and at first it is generally better to remove it during menstruation, although afterwards, if the uterus is sufficiently tolerant, it may be left during the periods. The mere presence of a stem always tends to produce some hyperæmia of the uterus, with increased secretion, and to increase the menstrual flow ; and its use is thus specially to be avoided in cases of profuse menstruation. Generally the irritation caused by the stem preponderates over any advantage gained by straightening the uterus, and the flexion usually recurs when the stem is removed.

The varieties of intra-uterine stem which are available under different circumstances may be divided into three classes—simple straight stems, stems with expanding branches, and straight stems combined with a vaginal support. Stems attached to external supports are to be altogether condemned, since they destroy the natural mobility of the uterus, and expose it to dangerous shocks. A straight solid stem of vulcanite or glass, ending in a disc or sphere, is the least irritating of all these pessaries, but it is generally impossible to keep it in place, except as a temporary measure by placing a tampon soaked in carbolized or iodized glycerine beneath it, the patient remaining in bed.

The stem must be a quarter of an inch shorter than the uterine cavity, that it may not touch the fundus.

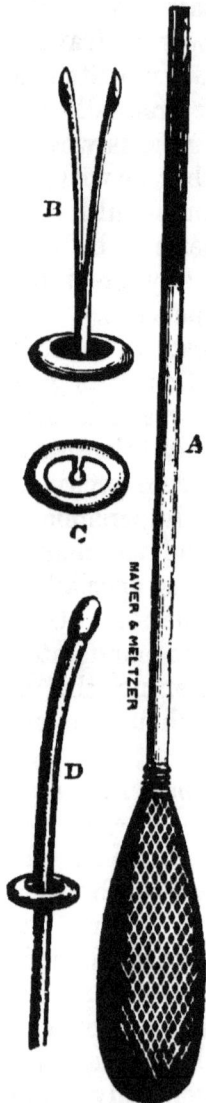

Fig. 38.—CHAMBERS' Intra-Uterine Stem Pessary.

A, introducer; B, stem; C, transverse section at base of stem; D, Introducer holding stem in position for Introduction.

MAYER & MELTZER

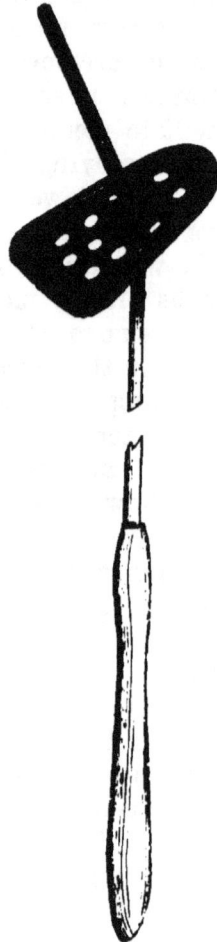

Fig. 39.—WYNN WILLIAMS' Intra-Uterine Stem Pessary.

Elastic stems of india-rubber have been recommended, but they do not effectually straighten the uterus, and the india-rubber is apt to absorb the secretions, become offensive, and so cause endometritis.

Expanding stems are made with two elastic lateral branches, which are introduced closed, and then spring open so as to lie against the sides of the body of the uterus, and render the instrument self-retaining. It follows that the tendency to flexion must be resisted by the comparatively sharp edges of the diverging branches. On this account these pessaries are still more likely to set up irritation in cases of flexion than a simple straight stem, though they are convenient if an intra-uterine stem is used after incision of the cervix, a proceeding not recommended by the author. Perhaps the best form of this pessary is that of Dr. Thomas Chambers (Fig. 38). This is made of vulcanite, and the introducer is a hollow cylinder, which slides over the branches, and holds them together until it is withdrawn. Another form of expanding stem is hollow in its lower part, and the introducer is passed through the canal so formed, which afterwards affords a ready exit for the uterine secretions. Previous dilatation of the cervix is generally required before the introduction of this pessary, while that of Dr. Chambers is no larger than the ordinary sound.

Uterine stems which rest upon vaginal supports should not be rigidly connected with these, that the natural mobility of the uterus may be respected as far as possible. In the pessary of Dr. Wynn Williams (Fig. 39), the support consists of a somewhat oval ring, having an india-rubber diaphragm, near the centre of which is a perforated cup. The straight stem is first placed by the introducer, used like the uterine sound, then, by means of the perforation, the shield is passed over the handle of the introducer, so that the cup is guided into its position enclosing the bulb of the stem. In the pessary of Dr. Thomas * a simple straight stem rests, by its bulbous extremity, in

* This and the other pessaries of Dr. Thomas are made by Messrs. Krohne & Sesemann, 8, Duke-street, Manchester-square.

a cup fixed near the upper extremity of a Hodge's
pessary. The former of these is, perhaps, the most
likely of any to prove useful if a stem pessary is tried
for the relief of anteflexion; the latter keeps the
uterus more effectually in position in retroflexion.

PROLAPSE OF THE UTERUS AND VAGINA.

Pathological Anatomy.—From the close connection
of the uterus with the bladder and anterior vaginal
wall, these structures necessarily take part in all downward displacements, and it will therefore be convenient
to consider prolapse of the vagina in association
with prolapse of the uterus. Descent of the uterus
has commonly been termed prolapsus so long as the
cervix remains within the vulva, and procidentia when
it passes outside, although, from the derivation of the
words, an opposite usage would have been more appropriate. A better classification is that of Dr. Thomas
into three stages of prolapsus—the first stage, in
which the uterus remains entirely within the vulva
(Fig. 40, p. 113); the second, in which it passes partially outside (Fig. 41, p. 115); and the third, in
which the whole uterus is extruded externally (Fig.
42, p. 116). As the uterus descends, the cervix tends
to move in the direction of the vagina as being that
of least resistance, and thus the axis of the uterus
follows the curved axis of the pelvis, and becomes
more and more retroverted in proportion as it becomes
lower (Fig. 40). The two chief causes of retroflexion
then come into play (*see* p. 75), so that this displacement is commonly added to the retroversion, and the
fundus lies low in the hollow of the sacrum (Figs. 27,
p. 74, and 41, p. 115). When the uterus is finally
extruded, it is always in a position of combined
retroversion and retroflexion (Fig. 42, p. 116), unless
it has been previously fixed in a position of anteflexion by the presence of a fibroid tumour or other

such cause. An important distinction must be made between simple prolapse of the uterus and prolapse associated with elongation of the supra-vaginal cervix, a much commoner condition.

Causation.—In *simple prolapse of the uterus*, the uterus itself may be the prime factor in the descent, and may overcome the resistance of its supports, or prolapse may primarily affect the anterior vaginal wall, with the base of the bladder, and these may draw down the uterus. A third, and still commoner, condition is that in which the two influences are more or

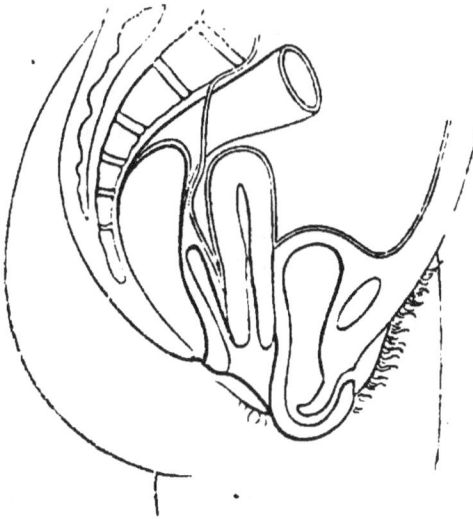

Fig. 40.—Prolapse of the First Degree.

less combined. The first causation, in its most pure form, is seen sometimes in the case of virgins when the prolapse is due to hyperplasia of the whole uterus or of the vaginal cervix, to the presence of a fibroid tumour, or simply to excessive muscular exertion, without any uterine enlargement. The resistance of the vagina, and even that of an intact hymen, are thus overcome. In most cases of this kind, however, there is some antecedent relaxation of the vaginal walls from chronic leucorrhœa or other cause. Sometimes

I

prolapse occurs in old women, even when the uterus is atrophied and considerably lighter than normal, the displacement then often arises from deficient support in the soft parts, owing to disappearance of fat. Among the exciting causes of prolapse, apart from the uterus itself, the most notable are laborious occupations, chronic cough or constipation, too early getting up after delivery, and rupture of the perineal body. In the last of these conditions the vagina not only loses its supporting power, but becomes an active factor in causing the prolapse. The lower third of the posterior wall of the flattened cylinder, which the nulliparous vagina normally forms (*see* Fig. 25, p. 69), being destroyed, its anterior wall is unsupported, and bulges through the vulva, carrying down with it the bladder, while at the same time it makes traction upon the cervix. Destruction of the perineal body also takes away the direct resistance to considerable descent which the pelvic floor affords, and renders the downward path of the uterus shorter. Even without any perineal rupture, subinvolution of the vagina after delivery or relaxation of its walls may transform it from a support into an agent of displacement. Another important cause contributing to the weakening of the anterior vaginal wall is habitual or occasional overdistension of the bladder.

Prolapse associated with · Elongation of the Supravaginal Cervix.—In the great majority of cases in which the cervix appears externally, the uterine cavity is found to be much elongated. An old doctrine was revived by Huguier, who separated this condition entirely from prolapse, and considered that the fundus usually remained at its normal level, the cervical hypertrophy being primary. Out of sixty reported cases of prolapse, in which the cervix was protruded externally, he found only two cases of true prolapse. Huguier has been followed in the main by Barnes and others. On measurement with the sound, however, it will be found that the elongated uterine cavity is frequently about 4½ inches long, and rarely much exceeds 5 inches, while the procident mass may protrude from 1 to 3 or more inches outside the vulva. A line drawn along the pelvic curve from the normal position of the fundus to a point 2 inches outside the vulva measures more than 6½ inches, and it is, therefore, clear from measurement, that the fundus is almost always in these cases depressed more or less below its normal level. In the majority of cases also the sound will show the fundus to be more or less retroflexed,

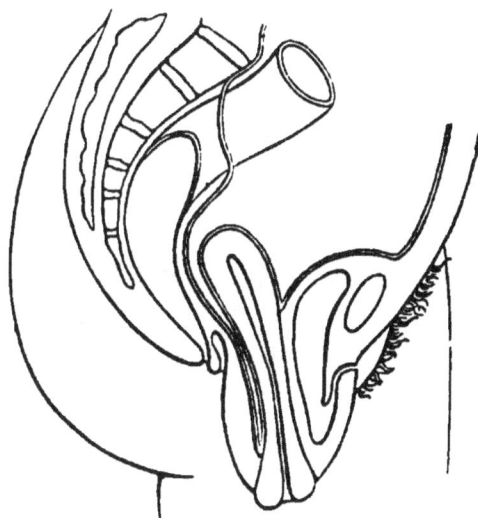

Fig. 41.—Prolapse of the Second Degree (or Procidentia).

lying in the hollow of the sacrum (Fig. 41). Moreover, the elongated cervix is invariably increased in length out of proportion to its breadth, and often it is actually attenuated, and has become elastic instead of being a firm muscular structure. This is a proof that the change is, in the main, the result of tension, although the hyperplasia may have been due in part to a state of hyperæmia, or subacute inflammation, with subinvolution,· taking its departure from labour.

There are two ways in which tension, tending to

elongate the cervix, may arise. The first is due to primary prolapse of the anterior vaginal wall, with the base of the bladder, which drag the uterus downward at the point of vaginal attachment, while the uterine ligaments, attached near the centre of the organ, tend to keep it in place. In most cases, the yielding takes place partly in the ligaments, leading to prolapse, and partly by stretching of the intervening portion of cervix. The next mode is one which is not noticed in text-books, but is capable of evoking a much greater

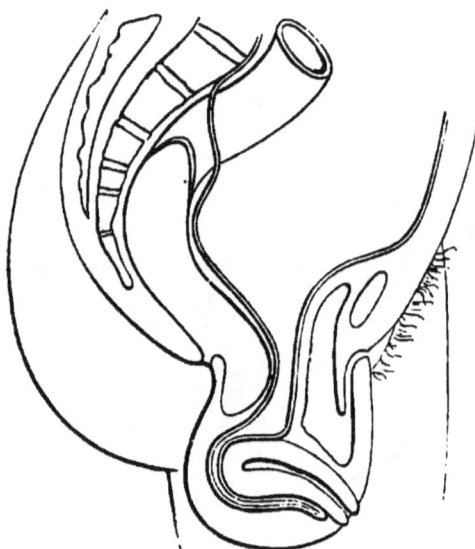

Fig. 42.—Prolapse of the Third Degree (also called Procidentia).

force. It arises when the cervix, already partially prolapsed, is extruded through the vulva by any sudden effort, and is there gripped and partially strangulated, and its return prevented for a greater or less time. For it is a common experience that, notwithstanding the relaxation of the vulval outlet in these cases, some force is required to reduce the swollen cervix and prolapsed portion of vagina through this outlet. The elastic attachments, stretched for the moment, thus tend to restore the uterus and exert a tensile force on the cervix

which may approximate in magnitude to the primary
expulsive force of which it is the recoil, and is likely
to be much greater than the mere weight of the
anterior vaginal wall with the base of the bladder.
This view of causation agrees with the fact that the
great elongation with attenuation of the cervix is only
met with in those cases in which the os uteri is
generally or frequently external to the body; and
that, if a case of prolapse afterwards reach the third
stage (Fig. 42) and the uterus remains entirely external, it may again be reduced in length, and its
measurement be as low as, or even lower than, normal.
Some reduction of length may even take place immediately upon the uterus being restored, from the
shrinking of the elastic cervix. This abnormal
elasticity is due to the muscular fibres having become
atrophied in conjunction with hyperplasia of the other
elements of the tissue.

Primary prolapse of the *posterior* vaginal wall
rarely exists, except as the sequel of destruction of
the perineal body. It may occur to a considerable
degree without affecting the position of the uterus.
The swelling so formed may or may not carry down
with it a pouch of the rectum, forming a rectocele.
If the whole posterior vaginal cul-de-sac is carried
down, the pouch of Douglas generally descends with
it, and in rare cases the small intestines descend into
the procident mass, and form a large bulging tumour,
much exceeding the size to which a rectocele usually
attains (Fig. 43, p. 118).

Results and Symptoms.—The chief symptoms of
prolapse of the first degree are dragging pain in the
back, hypogastrium, and groins from the strain upon
the uterine ligaments. The anterior vaginal wall, with
the base of the bladder, almost always descends first,
even when excessive weight of the uterus is the
primary cause of displacement. As this begins to
form a swelling, bulging externally (Fig. 40, p. 113),
it is often mistaken by the patient for the womb

itself. Though the bladder is often tolerant of this displacement, some difficulty of micturition is usually produced, and sometimes tenesmus and even cystitis, from decomposition of the urine retained in the pouch. The presence of the uterus low down in the vagina causes a sensation as of a foreign body, and tends to excite expulsive efforts, which accelerate the progress of the prolapse towards the second stage. The posterior vaginal wall, as a rule, does not, like the anterior, descend in front of the cervix, but is

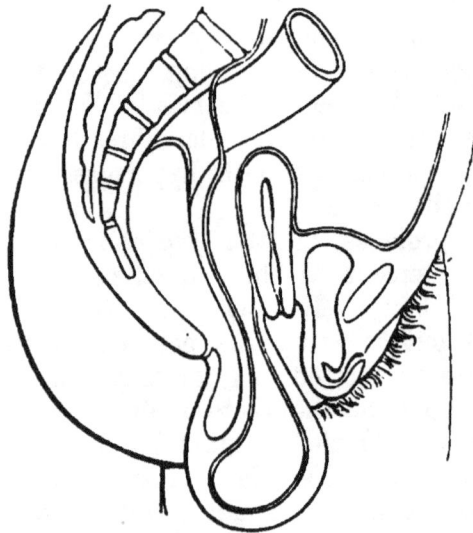

Fig. 43.—Prolapse of the Posterior Vaginal Wall with Rectocele and Enterocele.

invaginated by it from above (Fig. 40, p. 113). As the cervix protrudes externally, the anterior vaginal wall is first completely inverted, while the posterior wall for a long time maintains some duplicature posteriorly (Fig. 41, p. 115), although the inversion of this also may be complete at last. From the more loose attachment of the vagina to the rectum than to the bladder, the rectum is not necessarily carried down, and rectocele may or may not be associated with pro-

lapse of the second or third degree, while it rarely
attains to any great size.

When the procident cervix remains extruded through
the vulva, it becomes swollen from the interference
with venous circulation ; leucorrhœa, and sometimes
also menorrhagia or metrorrhagia are excited.   Frequently ulcers are formed from the effects of exposure
or friction on the cervix or vaginal walls, and these
may be the source of frequent slight hæmorrhage.   In
old standing cases the rugæ of the vagina are lost, and
the mucous membrane becomes hardened, like skin.
From the effect of œdema and tension the mucous
membrane also loses its close attachment to the cervix,
the vaginal reflection becomes more indefinite, and the
vaginal cervix may, in consequence, appear to have
disappeared.   In the third stage of prolapse one or
both ovaries may descend externally with the uterus.
When in this position, the uterus is commonly found
rather small, and the os contracted.   The pouch of
Douglas often descends externally even in prolapse of the
second degree, but rarely contains any intestines (Fig.
41, p. 115, Fig. 42, p. 116).   In recent prolapse, there
is ectropion of the cervix from tension of the inverted
vaginal walls, but in old cases the os may be found
small, and sometimes minute, even in prolapse only of
the second degree.   After the menopause the cervical
canal may become more or less completely occluded.
Displacement of the base of the bladder may, in rare
cases, produce such obstruction to the ureters as to
cause hydronephrosis or other kidney lesion, and
unreduced procidentia may lead to extensive and even
fatal sloughing.

Diagnosis.—When the cervix is external, it is impossible, if due care be taken, to make any error of
diagnosis.   The use of the sound is generally desirable,
to ascertain the length and direction of the uterine
cavity, and to learn how far the fundus is below its
normal level.   In prolapse of the first degree, the
extent of displacement can only be accurately estimated

by making the examination while the patient is standing, and by testing the effect of bearing-down efforts.
By the recumbent position a prolapse of the second
degree may be converted into one of the first, but it is
generally reproduced if the patient bears down. The
existence of rectocele is detected by rectal touch, and
the degree of cystocele may be estimated by passing a
curved catheter or sound into the bladder, and turning
its point downward into the prolapsed pouch.

Treatment.—The view here adopted as to the
essentially secondary character, in the majority of
cases, of cervical elongation associated with prolapse,
has an important bearing on treatment. Huguier and
others, accepting the logical result of their theory, have
considered the only curative treatment to be excision,
not only of the vaginal, but of a portion of the supra-
vaginal cervix. The dissection of this portion of the
cervix away from the bladder in front and the peritoneum posteriorly involves the probability of considerable hæmorrhage, with the consequent risk of septic
absorption, as well as some danger of opening the
bladder or peritoneal cavity. Hence few authorities in
Britain have thought it desirable, for the cure of an
affection not endangering life, to subject a patient to so
serious an operation. If, however, the elongation be,
in the main, secondary, it may be expected that, if
means be found to maintain the cervix for a sufficiently
long period at its normal level, and if, at the same
time, any chronic inflammation present be suitably
treated, the elongation will, in the end, diminish or
disappear. The necessity for amputation is then
limited, as a rule, to cases in which, not the supra-
vaginal, but the vaginal portion of the cervix is
elongated or enlarged. The treatment of these will be
considered under the head of hyperplasia of the cervix.

Replacement of the Procident Mass.—After evacuation, if necessary, of the bladder and rectum, the
patient should be placed in the semi-prone position, the
procident mass well lubricated, and compressed steadily

by both hands, so as to diminish its bulk. It may then be pressed gently upwards in the direction of the pelvic outlet, in such a way that the portion last prolapsed is returned first. The reduction is usually effected easily, but in some cases the bulk of the protruding mass may be so increased by œdema, that its return becomes very difficult. The patient should then be kept in bed for a time, in the first instance, while the swelling is supported, and treated by the application of cold by means of cooling lotions or ice, or gradual pressure by strapping or elastic bandages. In rare cases, in consequence of inflammatory adhesions in the pelvis, especially when these are associated with some tumour connected with the uterus, the procidentia may be irreducible, and it is therefore essential to use no excessive force in attempting its restoration. For an irreducible procidentia, the only available treatment is a suspensory bandage, which may support and, by gradual pressure, eventually diminish the displaced mass.

Methods of Retaining the Uterus.—The indications for effecting a radical cure are : (1) to diminish the size of the uterus, if excessive ; (2) to take away all sources of excessive downward pressure or traction ; and (3) to restore the uterine supports to their normal condition. In general, however, the first two conditions can only be attained by protracted treatment, and the third only by an operation for restoration of the damaged perineal body, or for artificially contracting the vagina. In a large proportion of cases, therefore, it is desirable to use palliative means, and support the uterus by a pessary, which should be so chosen as to help and not hinder the means which may be used for radical improvement.

Prolonged rest in bed is, in all cases, of the greatest use in diminishing the size of the uterus, if that organ can be kept in place meanwhile, and this treatment should always be adopted in severe cases, if it is possible for the patient to carry it out, especially when any ulceration exists. The ulcers will then usually heal readily, but the healing process may be accelerated, if necessary, by passing lightly the solid nitrate of silver over the surface, or applying a solution of the same salt (gr. x.—xxx. ad ℥j.). If relaxation of the vagina be only moderate, and the stage of prolapse early, the use of astringents may suffice to effect a cure. Alum, tannin, iron alum, or sulphate of zinc may be used in the form of vaginal injections (℥iij.—iv. ad Oj.), but a more effective plan is to insert into the vagina daily two or three teaspoonfuls of either of these substances in powder, in a muslin bag, or wrapped in cotton wool, while copious cold water injections are used from time to time. Benefit is also derived from the constitutional effect of cold hip-baths or sea-bathing, as well as tonic medicines, especially iron and strychnia. The administration of ergot may also give tone to the muscular walls of the vagina, as well as diminish hyperæmia of the uterus. Cough or chronic constipation is to be treated by suitable remedies. As a preliminary measure, to bring the uterus and vagina into a suitable condition for a pessary, and to allow ulcerations to heal, it is often useful to keep up the uterus by a large vaginal ·tampon, kept in place, if necessary, by a perineal band. This plan is especially indicated if the uterus will not remain in place even while the patient rests in bed, but it may sometimes obviate the necessity for confinement in bed. The congestive hypertrophy of the cervix may also derive benefit from the pressure exercised by the tampon. A sponge may be used for this purpose, but, from its tendency to promote decomposition, it requires frequent removal, and the utmost care in cleansing it. Unless these can be secured, it is better to use tow, oakum (the so-called "antiseptic marine lint"), or sheep's wool, which retains its elasticity in a state of moisture and pressure better than cotton-wool. If soaked in carbolized glycerine, to which alum or tannin may be added, such a tampon may be left in place two or three days.

Pessaries.—The pessary which of all others has the fewest drawbacks, and which will generally prove effectual in an early·stage of prolapse, if the perineal body has not been much damaged, is a Hodge's pessary of the ordinary sigmoid shape (Figs. 28, 29, p. 80). The action of this instrument is to stretch the posterior cul-de-sac backwards and upwards, and so hold the cervix at its normal level, and keep the uterus in a position somewhat of anteversion. If the lower limb be somewhat square (Fig. 28, p. 80), and have but a slight pubic curve, so that it rests behind and above the pubic arch, it will also support the base of the bladder, and so prevent what is often the first step in displacement (Fig. 29, p. 80). It is generally necessary that the instrument should be rather broad, but it should not be larger than is necessary to secure its retention. When the perineal body is very deficient, and the weight to be supported is considerable, a pessary of this form will usually be forced out. Another pessary, somewhat similar in its action, namely, the elastic ring pessary, may, however, be retained. The best form of this is that made of steel spring covered with india-rubber. Owing to its elasticity, a ring of considerable size can easily be introduced by compressing it laterally. The two largest sizes, $3\frac{1}{4}$ and $2\frac{7}{8}$ inches in diameter respectively, will be found most useful. The spring commonly used is often not stiff enough for the larger-sized ring, and the consequence is that the ring is compressed and forced out. In such a case, a pessary with stiffer spring may be retained. The diameter of the spring, with its rubber covering, should be at least $\frac{1}{2}$-inch, that its pressure may be readily tolerated. This pessary has the advantage that intelligent patients may be taught to introduce and·remove it themselves, since, from its flatness, it naturally passes in the right direction, namely, behind the cervix, while a Hodge's pessary usually passes in front of the cervix, if introduced by an unskilled hand. If it is frequently removed, the disadvantages it has in consequence of the more absorbent character, and, therefore, inferior cleanliness of its material, are in great measure obviated. If it is worn continuously, antiseptic vaginal injections should be used daily.

As an alternative to the elastic ring pessary may be used a form of pessary which will be frequently retained, when the sigmoid Hodge's pessary would fall out. This is one in which the anterior limb is bent upward, so that, viewed laterally, the instrument forms nearly an arc of a circle, about 110° in length (Figs. 44, 45). The essential point in the mechanism of this instrument is that its anterior limb rests high up above the pubic arch, distending the anterior vaginal wall, with the base of the bladder, into a pouch, and does not press against the rami of the pubes at all (Fig. 45). Its escape is thus prevented by the posterior surface of the pubes and the posterior vaginal wall, without any assistance from the vulval outlet or perineal body. I have had the anterior limb of the instrument made in the form of a cylinder ⅝-in. in diameter, so that its pressure may be as widely diffused as possible (Fig. 44). If, however, the posterior cul-de-sac is not capacious enough, the pressure of the pessary may not be tolerated ; and, if the vagina is so dilated that its width is nearly equal to its length, it is liable to turn round sideways. Again the instrument fails, if the vaginal cervix ƚ is so atrophied that it does not retain the posterior limb of the pessary behind it in the posterior vaginal cul-de-sac. Under such circum-

Fig. 44.
The Author's Lever Pessary for Prolapse.

stances the elastic ring is preferable ; otherwise the
lever pessary has the advantages that its material is
more cleanly, that it does not stretch the vagina so
much laterally, and that it tends to push the fundus
directly upwards if retroflexed, while the ring merely
draws the cervix backwards. It is therefore specially
indicated when any considerable retroflexion exists.
The mode of introduction of the instrument is the same

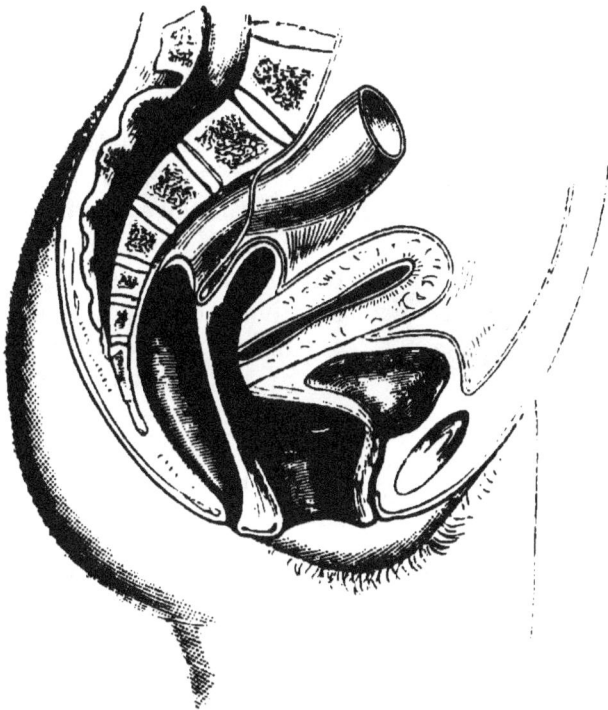

Fig. 45.—The Author's Lever Pessary for Prolapse, in position.

as that of the sigmoid pessary (*see* p. 85), but its shape
renders it rather more difficult to hook the posterior
limb backward over the cervix into the posterior cul-
de-sac. In proportion to this difficulty is the security
of its retention. With this, as with every other pessary, it is essential that it should be removed at regular

intervals, in order that it may be cleaned and the state of parts observed.

If the uterus be replaced, and the cervix maintained at its normal level by any of these pessaries, it is obvious that, if the organ be elongated, it must, for the time at least, either be more or less doubled upon itself, or the fundus must be elevated above its normal height. What most frequently occurs is a combination of the two effects. The uterus most readily tends to become doubled upon itself in a position of retroflexion, and it is often necessary to correct this tendency by replacement with the sound, so as to bring the axis into a position rather of anteflexion.

A form of instrument which will sometimes retain the uterus within the vulva when no other will do so except by external support is Zwancke's pessary. This consists of two plates, like butterflies' wings, hinged together, and capable of being expanded either by a screw, or by two arms which are secured by a simple catch as in Dr. Godson's modification, which is the best form of the instrument (Fig. 46). The pessary stretches the vagina laterally, and merely retains the cervix within the vulva, without tending to elevate it to its normal level, or remedy the retroversion or retroflexion.

Fig. 46.
ZWANCKE'S Pessary modified by GODSON.

It thus secures only a very partial alleviation, and most authorities justly denounce its principle as unsound. Some patients who have used it, however, derive from it so much relative comfort that they are not easily persuaded to change it for any other instrument. The cervix resting upon the hinge is apt to become inflamed or ulcerated, but this effect may be obviated in great measure if the pessary be so made that the hinge forms no projection. The chief advantage of the instrument is that the patient can easily remove it herself, and she should be stringently enjoined to do so every night. Through neglect of this precaution rectal and vesical fistulæ have not unfrequently been produced. On the same principle act Simpson's shelf pessary, and a modification of this praised by Dr. Matthews Duncan, namely, the disc and stem pessary, the stem of which projects through the vulva, after the disc has been introduced like a button through a buttonhole, but these are less easily removed by the patient herself than Zwancke's pessary.

The numerous old forms of pessary which kept up the uterus mainly by their bulk filling the vagina, such as globes or discs of wood or other materials, have properly fallen into disuse. One instrument, however, acting on this principle, is sometimes useful when, from the presence of a pelvic

Fig. 47.—Cup and Stem Pessary.

tumour or inflammatory swelling, neither a rigid pessary nor even the elastic ring can be tolerated, namely, the air-ball pessary. This consists of a hollow spherical ball of india-rubber. A tube is attached, through which it is inflated by means of a small air-pump, and the patient can easily introduce and remove it herself. Owing to the material used, frequent removal is necessary for the sake of cleanliness.

If the lever or elastic ring pessary cannot be retained, it is generally best, in default of a plastic operation, to resort to the cup and stem pessary (Fig. 47), supported from a waist-belt by four bands. Before it is used, any ulceration present should be cured by rest. This pessary has the advantage of not stretching the vagina, and the patient will scarcely fail to remove it every night, at which time astringents may be used. The bands should be made of india-rubber tubing, not of uncleanly webbing, as in the instruments commonly sold, and the pessary itself of vulcanite. Its lower end should be fixed in the centre of a square sheet of india-rubber of suitable

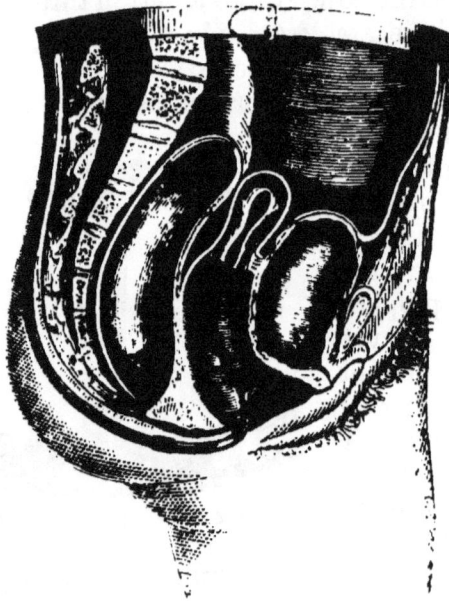

Fig. 48.—CUTTER'S Pessary for Prolapse, in position.
(After THOMAS.)

width, to the corners of which the bands are attached, so that they may not cross the labia, but lie in the groove at each side. They will then not produce chafing. The stem may have a pelvic curve if preferred, but the straight instrument better corrects the tendency to retroversion by pushing the cervix backward. Pessaries of the same shape, made in red rubber, are sometimes useful when much tenderness is present, but their material is less cleanly, and they will scarcely resist a very powerful displacing force.

As an alternative to the cup and stem pessary may be used the form of Cutter's pessary having a cup at its upper extremity to receive the cervix uteri. In this the stem curves backward over the perineum, and ends in a single band of india-rubber tubing, which passes backward and is attached to a belt (Fig. 48). This instrument is praised by Dr. Thomas as the most perfect of all those resting upon external support, but I have found it to be more liable to displacement, and more apt to cause chafing than the cup and stem pessary adjusted in the manner described.

When a patient declines operative treatment, and objects to more effectual forms of pessary, some degree of comfort may often be attained by one of the utero-abdominal supporters which were formerly more used than at present. These consist of a belt combined with a padded metallic plate, fitted above the pubes or over the sacrum. A strap passing between the legs supports a perineal pad, which may succeed in keeping the uterus within the vulva, while the pressure of the plate tends to relieve sympathetic pains.

Operative Treatment.—If there is hypertrophic elongation of the vaginal portion of the cervix, the elongation is probably the primary cause of the pro-lapse, and the only satisfactory treatment is amputation. Again, if there is general enlargement of the uterus and broadening of the vaginal cervix, even without any notable lengthening, it is often useful to amputate the vaginal portion by the plastic method, bringing the mucous membrane over the stump by sutures. The alterative effect of the operation, and the rest in bed, cause a diminution of the uterus, which is after-wards more easily retained by a ring or other pessary. The operation is described under the head of hyper-plasia of the cervix.

Since damage to the perineal body in parturition is generally the starting point of the conversion of the

K

vagina from a uterine support into a cause of displacement, the simplest operation for the restoration of
that support is the repair of the perineum, which may
be carried out without any need for quilled sutures.
Care must be taken not merely to make a thin
perineum, but to restore the triangular shape (in
longitudinal section) of the perineal body. If effectually performed, this will at least render the retention
of a lever or elastic ring pessary possible, although, by
itself, it usually fails to effect a cure, since the new
perineum dilates under the pressure of the descending
uterus.

Operation for Restoration of the Perineum.—The
cases of prolapse for which this operation is most suitable, are those in which more or less destruction of the
perineal body, as the result of parturition, is discovered,
and in which the hypertrophy of the upper part of the
vagina is not too extreme. The following is a convenient mode of performing the operation. The patient
is placed in the lithotomy position. The need for
assistants to support the thighs is avoided if "Clover's
crutch" is used. By this instrument the thighs, just
above the knees, are fixed by circlet straps at the end
of an iron bar, the length of which can be regulated
by aid of a screw which fixes it in any position. The
thighs are then flexed to any required degree by means
of a padded strap which passes from one end of the
bar round the neck, and is then attached to the other
end. Thus the knees can be kept widely apart while
the operation is performed, and brought closer together,
by altering the screw, when the time arrives for tightening the sutures.

The extent of surface to be freshened is indicated,
to some extent, by the cicatrix left by the rupture.
It is well, however, to go a little beyond the limits of
this in all directions, especially up the median line of
the vagina and toward the lower halves of the labia
majora, both in order to secure, if possible, a perineal
body somewhat longer and deeper than the original

one, and to allow some margin, in case the surfaces do not unite completely up to the edges. To put the mucous membrane on a stretch, an assistant at each side places one or two fingers on the skin of the thigh and draws the vulva outwards (Fig. 49). The skin

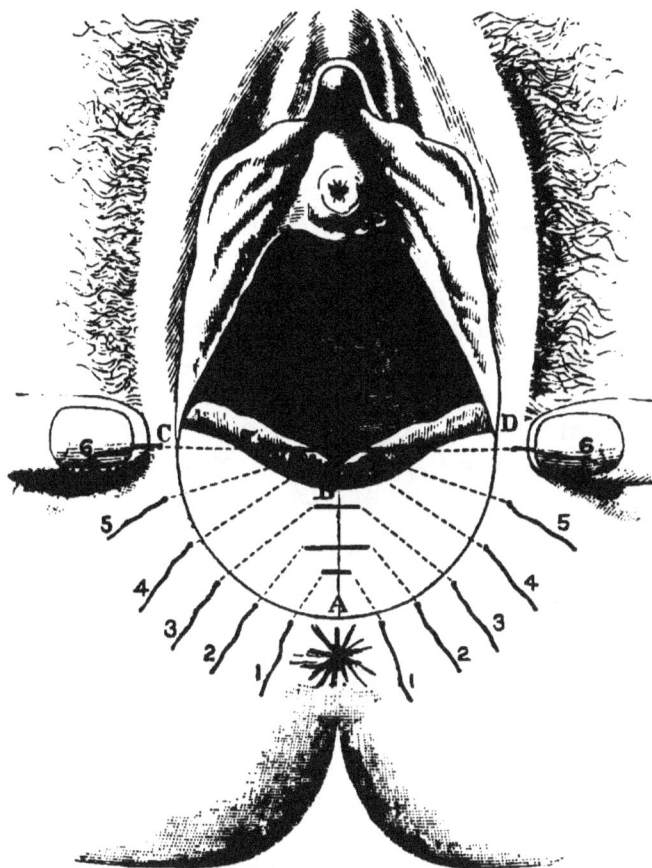

Fig. 49.—Operation for Restoration of Perineum.

just beneath A (Fig. 49), in front of the anus, may also be seized by a tenaculum and drawn downwards. If still the mucous membrane is not sufficiently on a stretch, from laxity of the vagina, the posterior vaginal wall, some distance above B, should be seized by long-

handled tenaculum forceps, such as those shown in
Fig. 4, p. 15, and pushed upward.   Incisions are then
made through the mucous membrane from B to A (Fig.
49), in the median line of the vagina, and from A to
C and D through the junction of mucous membrane
and skin.   These should not be extended in the direction of C and D further than the lower extremity of
the nymphæ at the utmost.   There are then two
triangular flaps, A B C and A B D.   These are to
be dissected up from the apex A toward the base B C
and B D, the corner of the mucous membrane at A
being seized with dissecting forceps.   The dissection
should not be deeper than necessary ; and, if it is done
with the knife, the surfaces are more ready to unite.
If, however, there is much tendency to bleed, scissors
may be used.   The apices of the flaps are then cut off
with scissors, leaving an upturned border along B C
and B D.   When the surfaces are drawn together,
these borders form a slightly elevated ridge toward the
vagina, and if there is any failure of union just along
the edge, they fall over and cover it.

The best material for sutures is the silkworm or
fishing gut, which should be stout, of the thickness
used for salmon flies.   It may be stained with ma-
genta or other dye, to render it more easily visible.
This has all the advantages of silver wire, as being
non-absorbent, while, at the same time, it is easier
to manipulate, and the exposed ends do not cause
discomfort after the operation, like those of wire.
The sutures are placed as shown in the figure.   The
most convenient needle is a slightly-curved one,
not too thick, mounted in a handle.   This is passed
in, unthreaded, rather close to the edge of skin,
brought out on the raw surface, there threaded with
one end of the suture, which is so drawn through.
By passing the needle in the same way on the other
side, the other end of the suture is drawn through.
Another mode is to use a more curved needle and to
bury the sutures 1, 2, 3 in the tissues throughout

their whole course. If, however, they are brought
out in the centre for spaces alternately short and
long, as shown in the figure, the surfaces are more
easily brought into contact at all levels without undue tension. In passing sutures 4, 5, 6, the needle
should be brought out precisely on the margin along
which the borders of mucous membrane B C, B D are
turned up from the vagina, not passing through the
mucous membrane itself. The sutures are then tied
in the order of the numbers from 1 to 6, care being
taken that the surfaces are brought just sufficiently
into apposition, and that no clots or blood is left
between them. The bleeding, if any continues, is
arrested by bringing the surfaces together, and if they
are properly united, there will be no secondary
hæmorrhage, unless the sutures begin to cut through
from excessive tension. The sutures may be left
from seven to ten days.

Operation of Posterior Colporrhaphy.—In those cases
in which the vaginal wall has become very voluminous
by hypertrophy, and those in which the prolapse does
not appear to have been due to defect in the perineal
body in the first instance, it is better to extend the
operation, so as to make it one of posterior colporrhaphy, or narrowing of the posterior vaginal wall
by suture. Some of the redundant mucous membrane
is thus removed, and the vagina contracted in its
lower part. I have usually performed this operation
according to the following method, which is a modification of that of Simon. If the vagina is not completely inverted posteriorly the mucous membrane is
put on the stretch, as in the former case, by seizing it
with long-handled tenaculum forceps above the point B
(Fig. 50, p. 134), and stretching it upward. If, however,
the inversion of the vagina is quite complete, the
freshening may be done, if preferred, quite outside the
vulva, by seizing the mucous membrane at the same
point, and drawing it upward in front of the pubes.
As before, the skin below A is drawn downward by

a tenaculum, and that near c and d outward by the fingers of assistants. Incisions are made through the mucous membrane from B to A, and through the skin near its junction with mucous membrane from A to c and A to D; but the point B is now taken far up the

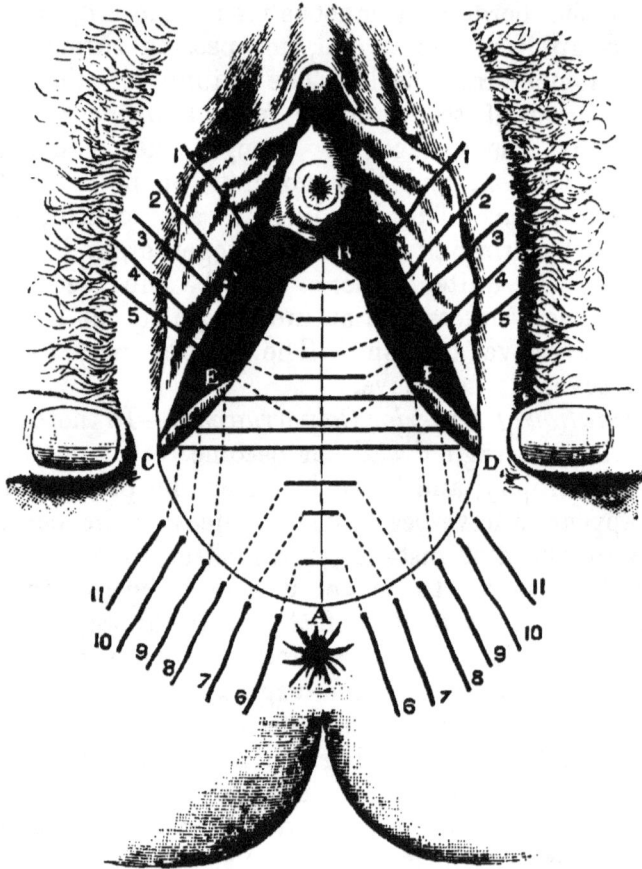

Fig. 50.—Operation of Posterior Colporrhaphy.

vagina, so that the line A B is twice or three times as long as in the former instance. The irregularly triangular flaps are then dissected up from the apex A toward the bases B c and B D, in such a way that the freshened surface is left with a very obtuse angle toward

B, as shown in the figure. In going far up the vagina it
is better to use scissors, on account of the vascularity of
the tissue. For the upper half of the surface, from B to
E and B to F, the flaps are completely cut away ; in the
lower half, from E to C and F to D, a small upturned
margin may be left, as in the former operation. The
vaginal sutures (1 to 5) are best introduced by a half-
curved needle held in a needle-holder. They are
alternately superficial and deep, and are made to enter
and emerge very near to the edge of mucous membrane, so that the mucous membrane is not turned in
when they are tightened. The remaining sutures (6 to
11) are placed exactly in the same way as those in the
simple operation for restoration of the perineum. The
vagina is then replaced (if not replaced previously),
and the sutures tightened and tied in the order of the
numbers in the figure, from 1 to 11. Most care is
required in making sure that the edges near E and F
come into exact apposition. The greater the resistance
to be overcome, the more numerous should be the
sutures, that the tension on each may be diminished.
By the plan of making the upper extremity of the
freshened surface end by a very obtuse angle, as shown
in Fig. 50, a sort of pouch is formed above the cicatrix,
in which the cervix may be retained, instead of gradually distending the contracted portion of the vagina.

In Simon's operation for posterior colporrhaphy, a
fenestrated speculum is used—a modification of that
of Sims. Through the fenestra the mucous membrane is freshened by scissors or knife over a surface
2¼ inches wide at the vaginal outlet, where the freshening is carried out upon the posterior halves of the
labia majora, and extending the same distance into the
vagina, narrowing slightly towards its further extremity, where it is completed by two incisions, meet-
ing above at a very obtuse angle. The freshened
surface thus forms a pentagon, and is similar to that
shown in Fig. 50. Alternately deep and superficial
sutures of silk are used in the vagina, and simple

sutures for the perineal border. The operation of Hegar, which he calls *perineauxesis*, is very similar, except that, to freshen the mucous membrane, he draws it down externally by a tenaculum, that he uses silver sutures, and makes the freshened surface triangular.

Other operators have preferred to narrow the anterior vaginal wall by removing a portion of mucous membrane near the cervix, an operation called *anterior colporrhaphy*. Marion Sims freshens by curved scissors, and brings together, by silver sutures, a V-shaped portion, the open arms of which, directed towards the cervix, are partly united by transverse portions, leaving a pouch of mucous membrane within. Emmet prefers to close the pouch by completing the transverse portion. Schroeder freshens the whole of an oval surface, and brings it together by alternately deep and superficial sutures. Some have proposed, in the case of old women, to close the vagina at its outlet, with the exception of a small aperture, by uniting the labia majora, or to close it completely at a higher level. Le Fort recommends that a longitudinal septum should be made by uniting the anterior with the posterior vaginal wall, so as to produce an artificially duplex vagina, which he supposes not to interfere with coitus or parturition.

After any of these operations, except complete occlusion of the vagina, however perfect the result may appear to be at first, an entire relapse is apt to take place after a considerable interval. It appears preferable, therefore, not to aim at absolute cure by the operation alone, but rather at rendering it possible for a convenient vaginal pessary, of moderate size, to be used effectually. The operation indicated for this purpose is the easier and less serious one, namely, that of narrowing the vagina at its lower portion posteriorly by one of the methods already described, either simple restoration of the perineum, or posterior colporrhaphy, according to the condition of the vagina. This will always enable some form of lever or elastic ring pessary to be retained, and generally a Hodge's pessary, of sigmoid shape and moderate size. If the cervix be kept by this means at its normal level for several years, any hyperplasia of the uterus may at length be removed, and the ligaments so recover their tone, that the pessary may be permanently discarded without fear of a relapse. Such a plastic operation is generally to be recommended in the case of women not much beyond middle age, when there is so much deficiency of perineum or relaxation of vagina that none of the forms of lever or ring pessary proves efficient, in order to save them the inconvenience of wearing for many years external supports. When even a pessary supported externally fails to retain the prolapse, or cannot be borne after judicious preparatory treatment, an operation becomes the only tolerable alternative.

INVERSION OF THE UTERUS.

Pathological Anatomy.—Inversion of the uterus may exist in three stages : the first, when the fundus is only partially invaginated and remains within the os ; the second, when the fundus has passed through the os ; the third, a stage very rarely attained, in which the whole of the cervix as well as the body is completely inverted, so that not even a groove remains between the inverted cervix and the vaginal vault. Either of the two latter stages may be complicated by extrusion of the inverted fundus outside the vulva. Acute inversion of the uterus belongs rather to obstetrics, since it is generally the result of parturition. We have here to deal with the chronic stage, which may be regarded as reached, in cases occurring after delivery, when the process of involution is complete.

Causation.—The conditions necessary for the production of inversion are laxity of the uterine wall, and a force of traction or pressure applied to some part of the fundus. In the puerperal state, the force is generally that of traction applied through the funis to an attached placenta, pressure by the external hand on a relaxed uterus, or simply the weight of the relaxed and prominent placental site, with or without the placenta itself in addition. Apart from parturition, it chiefly arises through the traction of a submucous fibroid, or fibroid polypus. In either case the uterus, grasping the invaginated portion as a foreign body, is stimulated to contract, and so increase the inversion.

Results and Symptoms.—The immediate symptoms of inversion are usually severe shock and collapse, frequently accompanied by sudden and severe hæmorrhage. The uterus is felt like a foreign body in the vagina or outside the vulva, and excites expulsive efforts. In the chronic stage, menorrhagia or irregular hæmorrhage is usually a prominent symptom, and often calls imperatively for relief. The presence of the tumour gives rise to bearing down, with rectal and vesical tenesmus, and frequently it becomes difficult or impossible to retain it within the vagina. The uterine mucous membrane becomes inflamed from the irritation to which it is exposed in its unnatural situation, and thus arises muco-purulent leucorrhœa. If the tumour be protruded externally, ulceration is likely to be produced, and even sloughing may occur. Death may result from sloughing, from hæmorrhage, or from gradual exhaustion. In some cases adhesion arises between the peritoneal surfaces of the uterus, but this is very rarely found, even when the difficulty of reduction has been extreme. Inversion may sometimes persist for years with but slight symptoms, but this is usually found only in those instances in which, after the menopause, toleration has become established, and the structure of the uterine mucous membrane has been profoundly altered. In very rare cases spontaneous replacement has occurred, even after a long interval, and it is not uncommon to find reduction spontaneously completed, after it has been commenced by art.

Diagnosis.—The tumour formed by the inverted uterus will be found externally, or felt by the finger in the vagina. The most essential point in diagnosis is to distinguish between inversion of the uterus and polypus, as well as to discriminate the case in which a polypus has produced by traction a partial inversion. In the case of polypus, the fundus may be made out on bimanual examination, the finger being introduced into either vagina or rectum, while in inversion it is absent from its normal place. The readiest mode of distinction is the use of the sound, which in inversion is arrested at less than the normal length, when passed up between the base of the tumour and the os, but in the case of polypus, passes to the full length, or generally to a greater distance. In rare cases the surface of a polypus may become adherent to the edge of the os at all points, but the sound can almost always be forced through at some part, without excessive pressure. The surface of the inverted uterus, unless modified by long exposure, is highly injected, velvety, and readily bleeds. It is also distinguished from a polypus by being painful, and sensitive to acupuncture, or the tightening of a ligature or écraseur wire. A crucial test is to pass a female sound into the bladder with the point directed backward. If a finger be introduced into the rectum, the point of the sound can then, in the case of inversion, be felt above the os, with only the walls of rectum and bladder intervening, at the point which would otherwise be occupied by the body of the uterus. In some cases the funnel-shaped depression formed by the inversion can be felt from the rectum.

Treatment.—In recent cases arising independently of parturition, as well as in those resulting from labour, reduction may be effected by taxis. The patient should be placed under an anæsthetic, and the hand passed into the vagina, so as to compress the tumour, and make steady and prolonged pressure upwards in the direction of the pelvic axis, while the other hand makes counter-pressure upon the abdomen. The effort should be to return first the part last inverted, not to indent the fundus, the effect of which would be to double the thickness of uterine wall to be passed through the constriction. If the inversion is chronic there is considerable risk of producing laceration by attempting to reduce it immediately by forcible taxis, and it is preferable to commence by the method of prolonged elastic pressure. If this fail to complete the reduction, taxis may then be tried, and repeated from time to time, elastic pressure being continued in the intervals. In easy cases an air-ball pessary in the vagina, supported by a perineal bandage, may effect reduction. In a more difficult one a repositor should be used, either straight, or having a compensating perineal as well as a pelvic curve, as in the instrument of Dr. Aveling. In the early stage of reduction of a complete or nearly complete inversion, when the uterine axis is likely to lie nearly in the line of the vagina, the straight repositor will answer well, but when the restoration is partially effected the doubly-curved repositor appears to afford a great advantage, allowing pressure to be made in the axis of the pelvic brim.

The repositor should be fitted with at least two terminals of different sizes, to be screwed on to its upper extremity; first, a cup-shaped disc large enough to receive the inverted fundus, for use in the earlier stage of reduction ; secondly, a smaller disc which may pass through the os, when the fundus has been once reduced inside it, and so complete the restoration. It is still better to have the smaller terminal in the form of a cylinder, cupped at the end, since the disc mounted on a small stem is apt, when reduction is complete, to be retained within the internal os, and to cause some difficulty in removal. To the lower end of the repositor are fixed four elastic bands, to be attached before and behind to a waist-belt. By the adjustment of these the direction and force of pressure may be regulated. The patient should remain in bed, and morphia injections should be given, if necessary, to enable the pressure to be tolerated. In the absence of a repositor made for the purpose, a fairly effective substitute for the earlier stage of reduction may be extemporized by cementing a small half-expanded air-ball, or a small india-rubber disc pessary on the summit either of a straight cup and stem pessary (Fig. 47, p. 127), or of Cutter's prolapse pessary (Fig. 48, p. 128), the latter of which gives the advantage of the double curve. The later stage of reduction may then be completed by taxis.

There are several stages of reduction at which impediments may arise, and special expedients be required to overcome it. Resistance may be offered at first by the constriction of the external os. This may be overcome by the plan of Dr. Hicks, namely, to distend the vaginal vault by a hollow annular elastic pessary, used in conjunction with direct pressure on the fundus, so as to stretch open the os; or by that of Dr. Barnes, namely, to incise the edges of the os. The former would generally be preferable. More frequently the point of constriction which causes most difficulty is near the internal os. If other means fail the expedient recommended by Barnes and Marion Sims may be employed, namely, to make two longitudinal incisions through the muscular fibres beneath the mucous membrane before applying taxis. This measure involves a risk of considerable hæmorrhage and of extension of laceration. It should scarcely be adopted until full trial has been given to the doubly-curved repositor. When the reduction has been more than half completed, but the actual fundus resists restoration, the mode of taxis proposed by Dr. Noeggerath, of New York, and found successful also by others, should be tried, namely, to make pressure with one finger upon one horn of the uterus, about the point of exit of the Fallopian tube.

If all means of reduction fail, there remains the

extreme resort of amputation of the inverted organ. In future this will very rarely be justifiable, and reduction will probably always be possible except in the very exceptional cases of firm adhesion between the peritoneal surfaces, in which case the risk in amputation is greatly lessened. Amputation should be performed by the galvanic écraseur, or, failing this, by the ordinary wire écraseur, an actual cautery being at hand to restrain hæmorrhage, if required. The tumour should first be drawn down, and a strong ligature passed through it above the line of amputation, in order to secure full command of the stump.

Dr. Thomas has boldly carried out the plan of opening the abdomen, in order to dilate the constricted cervix from above, and so effect reduction. Of two patients so treated, one died and the other had a narrow escape of death. The method thus appears at present to be too dangerous for imitation.

CHAPTER V.

HYPERPLASIA AND ATROPHY OF THE UTERUS.

SUBINVOLUTION, HYPERTROPHY, AND HYPERPLASIA OF THE UTERUS.

Causation.—Of all organs of the body, the uterus is that the tissue of which responds most readily by change of nutrition to any alteration in its vascular supply, or to any form of stimulus whatever. This quality is necessary to render it capable of growing during pregnancy from a weight of about $1\frac{1}{4}$ to one of about 28 ounces, and of being restored almost to its original size during about six weeks after delivery. Moreover, during the years of active sexual life it is never at rest, even apart from pregnancy, and its mucous membrane passes through alternations of growth, swelling, and exfoliation, in each menstrual cycle, accompanied by corresponding changes in the vascular conditions of the whole organ. Modifications in these changes are apt to be associated with hypertrophy, degeneration, or atrophy of its tissue.

Like any other hollow muscular organ, the uterus undergoes hypertrophy if any obstacle exists to the expulsion of its contents. Enlargement may thus be produced by stenosis or flexion, if either of these conditions produces actual obstruction to the canal. The most frequent cause, however, of enlargement of the uterus is subinvolution, or a failure to undergo a sufficient reduction in size after delivery or abortion.

For the proper performance of involution two conditions are necessary : first, a suitable diminution of
the quantity of blood in the uterine vessels and of the
blood-stream through them; secondly, a sufficient activity
in the process of absorption and nutrition. The former
is largely dependent both upon the periodical contraction of the uterus and upon the tone of its muscle
during the intervals of rhythmic contraction.   Subinvolution is thus promoted by muscular atony, and
also by a failure to perform the function of lactation,
since the suckling of the child, by reflex action, stimu-
lates the uterus to contract.   Other important causes
of active hyperæmia and consequent subinvolution are
retention of a portion of placenta, membrane, or clots,
and inflammatory conditions of the cervix or body of
the uterus or of neighbouring tissue, the commonest of
these being the effects of mechanical injury to the
cervix during labour. A too early return to the upright
posture, to muscular exertion, or to marital intercourse
has the same effect.   On the other hand, a too prolonged and absolute maintenance of the dorsal position
tends to passive hyperæmia, and so renders involution imperfect.   An important cause of passive
hyperæmia is the partial prolapse or other displacement
which often arises after parturition, especially in
consequence of a too early getting-up.   Any local or
general cause of venous obstruction tends to the same
effect.

After abortion subinvolution is still more frequent
than after delivery—first, because the uterine mucous
membrane, not being naturally prepared for the separation of the decidua, and in many cases having been
previously diseased, is more apt to be left in an
abnormal condition, or with a portion of placenta still
adhering ; secondly, because the stimulus of lactation
is wanting ; and, thirdly, because women, under-
estimating the importance of abortion, are more apt to
neglect the precaution of resting for a sufficient time,
and to return too soon to matrimonial intercourse.

After either delivery or abortion deficiency of absorption may arise from constitutional debility or malnutrition.

Apart from pregnancy, the main cause of uterine enlargement is active or arterial hyperæmia, either reflex or associated with inflammation ; but passive or venous hyperæmia also tends to cause the tissue to become infiltrated with serum, and, eventually, to produce overgrowth with degeneration. The causes of active and passive hyperæmia will be considered hereafter (*see* pp. 158, 165). According to Dr. Thomas, a uterus which has once been enlarged by the stimulus of pregnancy is afterwards far more prone than the nulliparous uterus to undergo a process of hyperplasia under the influence of any inflammatory or other exciting cause.

Pathological Anatomy.—In enlargement of the uterus, the result solely of obstructed outflow, the pathological condition is that of hypertrophy of the whole organ, but more especially of the muscular fibres. The muscular structure may also be hypertrophied equally with the cellular tissue in the earlier stages of a subinvolution which has arisen without the existence of any metritis, either as a cause or complication. In the great majority of cases, however, of subinvolution and other forms of enlargement, microscopic examination shows an undue proportion of fibrous tissue compared with muscular fibre. This leads at length to an induration of the tissue, which, in the early stage, was softer than normal from infiltration with serum. In the cervix especially this induration is manifest to the finger, and may lead to an erroneous diagnosis of scirrhous cancer, an error which formerly was probably often made. As might be expected from the well-known tendency of chronic inflammation to lead to induration by the production of fibrous tissue, this relative increase of cellular, as compared with muscular, elements, is greatest when chronic metritis, whether of body or cervix of the uterus, is the cause of enlargement, and, in mixed cases, it is more marked in proportion to the preponderance of the inflammatory

L

element. It may, however, be a degenerative change under the influence of venous stagnation, or due to constitutional causes of degeneration of tissue. The anatomical state finally reached, which has been variously called areolar hyperplasia, sclerosis of the uterus, or congestive hypertrophy, may thus be brought about by different causes, either by a chronic inflammatory process, or by conditions in which there is no proof of inflammation as commonly defined. Whichever name be chosen, therefore, should be reserved for the result produced, and not applied to the process leading up to it. The increase in thickness of the uterine walls is commonly greater in proportion than that of the length of its cavity, so that, apart from cases of descent and tensile elongation of the cervix, the organ assumes a more globular form. The cavity of the uterus, as measured by the sound, is usually not longer than 3½-in. or 4-in., though sometimes it is increased to as much as 5-in., or even more.

Varieties.—In most cases the hyperplasia affects both the body and neck of the uterus, though it commonly preponderates in one or other of these, and may be confined to one portion alone. Among special forms of hyperplasia is hypertrophic elongation of the supra-vaginal cervix. This is generally due in the main to the tension of the vaginal wall, or prolapsed cervix, associated or not with causes of hyperplasia, either primary or consequent upon the prolapse, and it has been discussed under the head of prolapse of the uterus and vagina. Another form is hyperplastic elongation of the vaginal portion of the cervix, to be distinguished from hyperplasia affecting the cervix in its breadth, or in all its dimensions. This is a comparatively rare affection, and is more usually found in virgins or nulliparous women, being apparently in some measure a congenital condition. It generally leads to prolapse of the first or second degree, and the prolapsed cervix then becomes congested or inflamed, and the hyperplasia is thereby increased.

Results and Symptoms.—Enlargement of the uterus is a very common cause of displacement, especially of prolapse, retroversion, or retroflexion, in consequence of the increase of weight, and the softness of tissue which generally exists in the early stage. Uncomplicated hyperplasia, especially when recent, is liable to cause dragging pain in the back, hypogastrium, and loins from the greater strain upon the ligaments. Increase of surface in the uterine cavity naturally leads to an augmented menstrual flow, except in the late stages of hyperplasia, when the tissue is degenerated and anæmic. In the early stage of subinvolution after delivery there is generally sufficient associated hyperæmia to lead to the recurrence of menstruation at an early period, notwithstanding the opposing influence of lactation. The remaining symptoms of hyperplasia are due, for the most part, to the hyperæmia, endometritis, metritis, or displacement, with one or more of which it is almost always associated.

Diagnosis.—Hyperplasia of the cervix is readily detected by vaginal touch, and its stage is indicated by the hardness or softness of the tissue. The differential diagnosis from cancerous degeneration is described under the head of cancer. In hyperplasia of the body of the uterus, the enlargement is detected by the bimanual examination, and the thickening of the walls is generally found to be greater than the increase in length. The length of the cavity is revealed by the sound, if its use is not contra-indicated. The most difficult point in diagnosis is to distinguish between simple hyperplasia and enlargement due to a fibroid tumour. In the latter case, the external surface of the uterus is often felt to be irregular. If, however, the tumour bulges internally, it may be necessary to dilate the cervix sufficiently to allow the index finger to be passed into the cavity of the uterus before the unequal enlargement can be detected. In cases in which the presence of a very small fibroid causes great hyperplasia of the whole uterus, the tumour is especially liable to escape recognition. From early pregnancy hyperplasia is usually distinguished by the greater sensitiveness of the uterus, and by the persistence of menstruation, whereas, in pregnancy, there has usually been amenorrhœa at some period, though hæmorrhage may be present when abortion is threatening. This distinction, however, may fail in the case of hyperplasia associated with the commencement of climacteric irregularities. The most valuable sign by which to distinguish the pregnant uterus is the more globular enlargement of its body as felt bimanually, and its greater softness and indistinctness, due to the chiefly fluid nature of its contents. Variation in the consistence of the uterus, due to the alternation of contraction and relaxation, may often be detected early in pregnancy, and is a very valuable sign if it exists, since it is far more marked in pregnancy than in any other uterine enlargement. In molar pregnancy, however, the uterus may never be soft or flaccid. Softening of the cervix is an important sign, if present, but it is often absent in the early stage of pregnancy in a multipara.

Treatment.—(1) *Prophylactic.*—The most important part of prophylactic treatment consists in the judicious management of the puerperal state and of abortion, in which should be included the utmost care to avoid causes of septic or traumatic inflammation. Rest for a due period, mainly in the horizontal position, should be observed, but a too continuous maintenance of the dorsal position, especially on a soft bed, should be avoided, as tending to cause venous stagnation and retain discharges. The child should be suckled, if possible, for at least from four to six weeks, even if the mother's milk requires to be supplemented. It would be of advantage if, in all cases, at the end of the puerperal period an examination could be made to ascertain that no displacement had arisen, or lesion of cervix remained. Whenever sanguineous or muco-purulent discharge continues too long after delivery, such an examination should not be omitted. After abortion in the early months, rest, more or less complete, and abstinence from coitus for as much as four weeks should be enjoined. After miscarriages between the third and sixth months much more prolonged rest and care are usually called for than after normal delivery at full term. It is also of essential importance in the immediate treatment of abortion to secure the complete evacuation of the uterus. After abortion, and also in the case of failure of lactation after delivery, it is desirable to administer a course of ergot for several weeks to promote uterine contraction.

(2) *Curative.*—In the earlier stages of hyperplasia a cure may result from the removal of its cause. Thus, in cases of subinvolution within a few months after delivery, great good is effected by remedying displacement, by curing cervical inflammation—one of the commonest causes which interfere with the normal process of involution—and by treating hyperæmia by suitable means. It is also of importance to remedy any constitutional debility or anæmia, which often accompanies lactation, and by which the activity of absorption may be impaired. For this purpose the most valuable drugs are iron, quinine, and strychnia.

In the case of enlargement of the uterus associated with hæmorrhage within a few months after abortion, the choice of treatment depends upon the question whether the subinvolution and hæmorrhage are common results of a piece of placenta remaining attached within the uterus. This will generally be found to be the case if the hæmorrhage has proved so severe as to produce marked anæmia ; and the probability is increased if the cervix is found to remain unduly open. In some cases even the internal os may remain so much open as to admit the finger and allow it to detect the foreign body within. If it does not fully admit the finger, it should first be dilated with one or more tents. For evacuating the uterus the finger is the best instrument. The necessity for giving an anæsthetic will depend upon the capacity of the vagina and the toleration of the patient. It is essential that the bladder should be completely emptied, for this allows the fundus to be brought much more easily under command of the external hand. The patient being in the dorsal position, the index finger alone is to be passed into the uterus, while the half hand is introduced, if necessary, into the vagina, the remaining four fingers being flexed upon the palm. The other hand is placed upon the abdomen, and its ulnar edge pressed in above the fundus so as to bring it close to the pubes. The uterus can then be pressed down so that, unless the uterine cavity is excessively lengthened, the finger can reach completely up to the fundus, without the necessity of passing the whole hand into the vagina, and very frequently this can be accomplished without passing more than one finger into the vagina, especially if an anæsthetic be given. If the uterus is at all retroverted, the finger, introduced into the cervix, may conveniently be used as a repositor, to bring it into a position of slight anteversion. It is very rarely that any other instrument than the finger is required. In some cases, however, a pair of forceps with flat serrated blades locking closely together may be useful to remove a piece of tissue which has been wholly or partially detached. Without an anæsthetic, in cases in which the abdominal muscles are at all rigid, there may be a difficulty in getting the uterus into the requisite position of moderate anteversion. Assistance may then be derived from the tenaculum forceps shown in Fig. 4, p. 15. The forceps are fixed into the anterior lip, the uterus drawn down sufficiently to allow the finger to penetrate well into the cervical canal, and the handles are then given to an assistant to hold, while the external hand obtains command of the fundus. If, however, an offensive discharge is present, it is better to avoid the puncturing of the cervix involved in this method. After the operation full doses of ergot should be administered for a time, and complete rest enjoined. Antiseptic vaginal injections should be used, and, if any offensive discharge should appear, or febrile symptoms arise, the uterus itself should be syringed out with a weak solution of carbolic acid or iodine.

When the cervix is closed, and the hæmorrhage not serious in quantity, there is a probability that the case may be one of simple subinvolution. The effect of treatment by rest and the administration of ergot and other uterine styptics should then first be tried. If hæmorrhage cannot be permanently checked by this means, the cervix should be dilated for exploration of the uterine cavity.

In the case of hæmorrhage persisting after delivery at full term, the treatment should be conducted on the same principles. The retention of some portion of placenta, membranes, or clot, must be regarded as a not improbable contingency, even though it may have appeared certain that the placenta and membranes came away intact.

In the earlier stages of hyperplasia, wholly or partially resulting from inflammation, absorbent remedies may be tried in the manner described under the heading of chronic endometritis and metritis. The later stage of fibroid induration is little susceptible to treatment, and is scarcely, if at all, affected by absorbents such as mercury or iodide of potassium. A degree of tolerable comfort may, however, frequently be obtained by treating the coincident hyperæmia, or displacement, though a tendency to relapse commonly remains. It is only in exceptional cases, and after fair trial of such means, that it is desirable to have recourse to the more surgical modes of treatment to be hereafter mentioned. The most powerful of all influences in diminishing the size of the uterus is the involution after delivery, during which even fibroid tumours may sometimes disappear. All means should therefore be taken (by curing any other morbid condition which may be discovered,

more especially endometritis) for removing the
sterility which commonly accompanies the late stage
of hyperplasia.

A process somewhat analogous to involution may
also be induced by the more powerful local remedies
sometimes used for endometritis, such as strong
carbolic or nitric acid, introduced into the uterine
cavity, and their occasional beneficial influence on the
size of the uterus may be thus in part explained. A
transient inflammatory hyperæmia is set up, on the
subsidence of which the blood supply is contracted, and
absorption becomes more active. A similar effect may
result from local applications to the cervix, and thus
hyperplasia associated with erosion of the cervix is
sometimes more amenable to treatment that when
erosion is absent. When there is no erosion a beneficial
effect may sometimes be produced by making an artificial
eschar upon the cervix either by heat or by caustics.
This acts most upon the cervix itself, but the body of
the uterus also partakes in the nutritive changes set up.

For the application of heat, Paquelin's benzoline
cautery is the most convenient means, but the galvanic
cautery, or cautery irons, of the size of a small button,
may also be used. It is best to employ a speculum of
wood or horn for this operation, but a large cylindrical
metal speculum may be used, if care be taken that it
does not become overheated. The application should
be made to the outer part of the cervix, to avoid subsequent contraction of the cervical canal. The poten-
tial cautery has the convenience that it does not
require the presence of an assistant. The best, in most
cases, is the potassa fusa cum calce. This is less superficial than nitric or chromic acid, while its action is
more easily limited than that of potassa fusa or chloride
of zinc. The cervix should be brought completely into
the field of a cylindrical speculum, and wiped dry. A
dossil of cotton-wool, soaked in vinegar and squeezed
nearly dry, should be tucked beneath the cervix at the
lower part of the speculum. A stick of the potassa

fusa cum calce* is then fixed in a long caustic holder, and rubbed several times, for not more than a minute at each application, over a surface about the size of a sixpence on one or both lips of the cervix, away from the cervical canal. A larger tampon, soaked in vinegar, should then be applied.

Potassa fusa causes a deeper destruction of tissue, and should be used with much caution, if employed, since a too vigorous application may cause serious inflammation. It is also more liable to run on to the vaginal walls. Chloride of zinc, made into sticks, is also sometimes used to cauterize the cervix. It is a powerful and rather painful form of caustic. In its use the vagina must be protected by an alkaline solution. Dr. Tilt recommends that some days before the application of potassa fusa cum calce or chloride of zinc, the spot should be rubbed repeatedly with solid nitrate of silver, to soften the epithelium. A second, or subsequent, repetition of the cauterization may be called for if the first has only a superficial effect. Such a prolonged treatment is specially applicable to the case of a localized induration of one lip of the cervix. When hyperplasia affects the vaginal cervix, as well as the rest of the uterus, and increases it in length as well as breadth, an efficacious treatment, in extreme cases, is to amputate a portion, either by the plastic method of Marion Sims (*see* pp. 155, 156), which is the best method, or by the galvanic écraseur, by which means not only is superficial tissue removed, but the alterative influence of the cautery is brought to bear upon the whole uterus. In the absence of the galvanic cautery, the cervix may be cut through with scissors, the actual cautery being used to stop bleeding, if necessary. After amputation of tissue or the application of cautery or caustic by any of these methods, the patient should be kept in bed for

---

* Potassa fusa cum calce may be made in the form of a stick by melting caustic potash with an equal part of quicklime.

at least a week, and tampons soaked in half an ounce
or more of glycerine kept applied to the cervix. Discharge is thereby promoted, and the influence of the
local inflammation and reparative action on the nutrition of the whole organ is increased.

Another mode of inducing an altered nutrition in
cases of chronic induration, especially as affecting the
cervix, is to introduce a succession of sponge tents at
intervals of some days. The effect of these is to soften
the tissue and produce a copious watery discharge for
the time.

Treatment of Hyperplasia of the Vaginal Cervix.—
Elongation of the vaginal, unlike that of the supra-
vaginal, cervix, is commonly, in the main, a primary
affection, the cause rather than the consequence of
prolapse, and hence the only satisfactory treatment is
the removal of the redundant portion. The only
difficulty of the operation is that of dealing with the
hæmorrhage, which is apt to be considerable, especially
when the tissue is soft, while the arteries cannot easily
be secured by ligature or torsion. The easiest and
quickest mode of amputation is that by the galvanic
écraseur. To avoid the risk of opening the bladder or
peritoneal cavity, the sound should be passed into the
bladder to ascertain the exact limit to which it extends,
and the uterus, if prolapsed, should be returned to its
place before the wire is adjusted. The stem of the
écraseur is then carried up in front of the cervix, and
the loop adjusted by the finger passed up into the
posterior cul-de-sac, without the use of a speculum. To
secure arrest of hæmorrhage, the wire must be tightened
slowly. After removal of the vaginal portion, a sound
should be passed into the cervical canal, to make sure
that its lips are not glued together, and a large bougie
should be passed at intervals for some time after the
operation, to counteract the gradual contraction which
is apt to occur after the use of the cautery.

In the absence of the galvanic cautery, the ordinary
écraseur (Fig. 68) may be employed, but in its use

there is a greater risk of lacerating bladder or peritoneum, in consequence of the extreme tension produced
when the tissues are tough. It is a good plan to make
an incision with the knife or scissors through the
mucous membrane around the cervix at the level
at which the wire is to be adjusted. A single steel
wire should be used, as in the case of removal by
écraseur of a fibroid tumour with thick base (see
section on *Fibroid Tumours*).

The cervix may also be amputated by knife or
scissors, and the hæmorrhage checked by actual cautery,
or by the plan introduced by Dr. Marion Sims, namely,
to unite the mucous membrane over the stump by
sutures, and so arrest the bleeding by pressure. By
this the advantage is gained that primary union may
be procured, and the patient is then saved from the
necessity of protracted suppuration and cicatrization of
the stump. The resulting cervix is therefore more nearly
of a normal character, and gradual contraction of the
cervical canal does not occur. The operation may be
carried out with the uterus in place, by means of Sims'
speculum, but it is then rather tedious and troublesome.
When the cervix can be drawn down externally without much force, as is generally the case under these
circumstances, it is a very easy one. I have usually
performed it in the following manner :—The cervix
having been drawn outside the vulva by tenaculum
forceps, and the lowest point of the bladder ascertained
by the sound, a strong hair-lip pin is passed through
the cervix from before backward, about a quarter of
an inch below this point, and another similar pin at
right angles to the first. A piece of thin india-rubber
tubing is then passed twice round above the pins, and
tied tight enough to prevent bleeding. The cervix is
then cut across transversely with scissors below the
pins. The incisions may be so made that the mucous
membrane is left longer anteriorly and posteriorly, but
it is unnecessary to dissect off flaps, since sufficient
mucous membrane to cover the stump can be obtained

by pulling it down. Scissors are preferable to the knife as causing less hæmorrhage. Sutures of silver wire are applied in the manner shown in Fig. 51 ; two at each side of the cervical canal, to unite the mucous membrane at front and back, and from one to three intermediate sutures, to unite the outer mucous membrane to that of the cervical canal. All these should be passed deeply enough to include somewhat more tissue than the mere mucous membrane. The elastic constrictor is then unfastened, to allow the mucous membrane to come together, and the sutures tightened. If the bleeding continues, or the flaps are not applied closely enough to the stump, one or more deep sutures

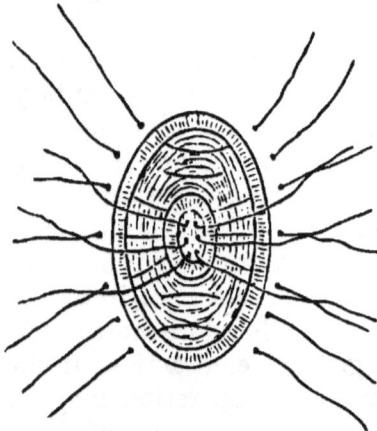

Fig. 51.—Mode of placing sutures after amputation of the vaginal cervix.
(After SCHROEDER.)

at each side may be passed through the whole thickness of the cervix at points intermediate to the more superficial sutures.

Of these modes of operating, the use of the galvanic cautery is the easiest, and may be adopted if the cervix cannot be drawn down. Dr. Marion Sims' mode of performing the plastic operation was merely to unite the mucous membrane at front and back, without passing any suture into the cervical canal, but the method above described has the advantage of keeping the cervical canal thoroughly open.

If the hyperplasia of the cervix is associated with eversion of its lips due to laceration in parturition, and is not too extreme in degree, the better mode of treatment is, in preference to amputating any portion, to perform the operation of trachelorrhaphy for repair of the laceration (*see section on Chronic Inflammation of the Cervix*). The hyperplasia will then afterwards generally gradually diminish. .

SUPERINVOLUTION AND ATROPHY OF THE UTERUS.

The process of involution may be excessive, although this fault is very far more rare than the opposite, and instances have even been recorded in which the uterus has been so reduced in size after parturition, that its presence could not be detected. Such atrophy may occur after normal delivery in ill-nourished women, who have a tendency to premature decay, or it may be the result of general or local puerperal disease. Atrophy of the uterus may also arise gradually in ill-nourished subjects, apart from parturition, and may lead to a premature menopause. There is a greater tendency to this in women in whom ovarian activity has throughout life been below par. Senile atrophy is a normal condition after the menopause, but does not usually proceed to a considerable extent till after the age of sixty. The vaginal portion and os then become especially small, and not unfrequently stenosis or even occlusion of the cervical canal occurs. The vagina shares in the atrophy, and becomes funnel-shaped, while the external generative organs also waste.

Results and Symptoms.—The symptoms of premature atrophy of the uterus are scanty menstruation or amenorrhœa and sterility.

Treatment.—In most cases it is preferable not to interfere, unless there are symptoms of unrelieved ovarian molimen. Otherwise, if the natural period of the menopause has not been nearly reached, and atrophy is not extreme, it may be desirable to try some of the means of local stimulus which will be described under the head of amenorrhœa. The same general and hygienic treatment as in the case of primary failure of development (p. 44) should be added.

CHAPTER VI.

HYPERÆMIA AND INFLAMMATION OF THE UTERUS.

ACTIVE HYPERÆMIA OF THE UTERUS.

As the uterine tissue is more prone than any other in the body to respond to stimuli by a change in its nutrition, so the uterus is, most of all organs, liable to physiological active hyperæmia, which readily passes into a morbid excess. Thus, hyperæmia occurs at each menstrual cycle, both during the period itself, and during the stage of growth and intumescence of the uterine mucous membrane which immediately precedes it. The tissue of the uterus and that around it are also, in a measure, erectile, and a more intense and transient hyperæmia thus arises through arterial dilatation under the influence of coitus or sexual excitement. To such forms of transient hyperæmia the term fluxion has been applied. While in the healthy uterus they are innocuous, they may become, in morbid conditions, serious sources of mischief.

The same susceptibility of the uterine vascular system to stimulus leads to a more chronic active hyperæmia, as the result of any source of reflex irritation. This may arise from any morbid condition or undue activity of the ovaries, from inflammation or other lesion of the cervix, or from more distant sources, as inflammation of the vagina or vulva, from sensitive caruncles of the urethra, or even from pruritus vulvæ. The same effect may be produced by any social conditions, or individual peculiarity, which may lead to undue, or especially to premature, development of the sexual emotions. Again, a fibroid or cancerous growth in one part of the uterus causes hyperæmia and consequent enlargement of the whole, and the same result may be produced by a neighbouring cellulitis or peritonitis. The persistent excess of vascular pressure leads to swelling of the tissue by effusion of serum, and eventually to hypertrophy. The pressure upon the nerves consequent upon swelling may lead, in persons of acute sensibility, to tenderness and pain, such as in them may be produced also by the fluxion of menstruation. Thus many of the conditions usually associated with inflammation may be present, while it cannot be proved that positive inflammation exists.

The fact of the hyperæmia being produced solely by reflex nervous influence does not, however, prove that it is not associated with a condition which partakes of the nature of inflammation. This is shown by the inflammation which may be produced over the field of distribution of a nerve by an injury to, or disease of, its trunk, as well as by the common phenomenon of catarrh or other forms of inflammation produced by the effect of cold. In the case of the uterus itself it is demonstrated by an occurrence which occasionally happens. There are some women so susceptible that the mere careful use of the uterine sound by a skilled hand may set up not only uterine but periuterine inflammation, the existence of which is made certain by the effusion and fixation produced. As there may be here no opportunity for septic absorption, and no perceptible injury to the mucous membrane, the case is clearly one of not merely hyperæmia but actual inflammation, produced by an impression upon the nerves. It can scarcely be doubted that, in such cases, some reflex influence is transmitted not only to vasomotor, but to trophic nerves. It must therefore be admitted that from reflex nerve stimulus an indefinite gradation may arise, from simple arterial dilatation up to undoubted inflammation.

The uterus has also an anatomical peculiarity which brings chronic hyperæmia of its tissue, when induced as the reflex effect of endometritis, into close relation with chronic parenchymatous inflammation. Most mucous membranes are separated from the structures lying beneath them by a layer of loose areolar tissue ; and, in such case, catarrhal inflammation of mucous membrane may exist without any perceptible implication of the muscular walls beneath, as is found to be the case in such mucous membranes as that of the intestines or of the air-passages. The mucous membrane of the uterus, however, is itself of a dense character, consisting mainly of closely-packed round or slightly elongated cells, and is intimately connected with the muscular wall, without any intervening loose layer. The extremities of the glands even dip more or less into the muscular layer; and it has been maintained, with much probability, by Dr. John Williams, that a considerable proportion of the thickness of the uterine wall really corresponds in development to the muscularis mucosæ, though in the human subject no line of demarcation can be traced. We may conclude, therefore, on anatomical grounds, that, if endometritis exists, the inflammation is not likely to be strictly limited to the mucous membrane, but will affect the uterine walls to some depth. The case may be compared to that of a sore and inflamed spot on the tongue, or on any sensitive surface, which gives rise to redness, swelling, throbbing, tenderness, and pain over a considerable region in its neighbourhood. The hyperæmia may be due mainly to reflex nerve irritation, but there is a zone of inflammation of lower degree, arising by continuity of tissue around the inflamed point, through which there is a gradation from simple hyperæmia up to the more acute inflammation.

The relation between mere engorgement and inflammation of the uterus has given rise to more divergence of opinion than any other in gynæcological pathology. Some distinguished authorities have omitted chronic metritis entirely from their nosology, while others, and the more numerous, have regarded it, as I believe, with greater accuracy, as one of the commonest of the special diseases of women. The difference is fortunately not so much with regard to the true nature of the condition present, or its treatment, but rather as to the question of definition—within what limits the word inflammation is applicable. The strongest argument for the view that, in most cases of hyperplasia of the body or cervix of the uterus, chronic parenchymatous inflammation plays some part, appears to be the fact, universally acknowledged, that in the later stages there is almost invariably an increase of areolar at the expense of muscular tissue, and eventually fibroid induration. In most cases we may find at some stage the old-fashioned surgical criteria of inflammation, namely, pain, redness, swelling, and, if not heat, at any rate the arterial hyperæmia which, on the surface of the body, produces local heat. To these is added a cell proliferation, not leading, as in acute inflammation, to unstable products, but to products of a lower grade than the normal tissue of the part. This occurs even when there is no cause of passive hyperæmia, and no constitutional tendency to degeneration or sclerosis of the organs ; whereas we might expect that, if the condition were solely one of active hyperæmia, the result would be true hypertrophy, such as occurs under the stimulus of pregnancy. Such a production of fibroid tissue is a characteristic result of chronic inflammation in other organs, as the lungs, liver, or kidneys, although in them also the distinction of inflammatory from degenerative changes has been the subject of much divergence of opinion.

Treatment.—The first effort should be to remove the inflammation which active hyperæmia reveals, or the cause which sets it up by reflex action. Of such causes the most common are ovarian irritation, and erosions, fissures, or other lesions of the cervix. O

M

internal remedies which have a direct influence upon hyperæmia, the most powerful is ergot, which acts in some measure by contracting the arteries, but in the main by its influence upon the muscular walls of the uterus. Half-drachm doses of the liquid extract of ergot, or of Richardson's liquor secalis ammoniatus may be given three times a day in chronic hyperæmia, especially if associated with menorrhagia or metror-rhagia. A similar influence, though in less degree, appears to be exerted by digitalis and strychnia, and these may often be usefully combined with ergot, while the general tonic effects of strychnia are at the same time valuable. Bromide of potassium, while acting as a general vascular and nervous sedative, has a special influence on the pelvic organs, which depends, in part at any rate, upon its effect as a sexual sedative, in which respect it is more trustworthy than any other drug. It may be given in doses of twenty or thirty grains combined with ergot or not, or, for long periods, in smaller doses, from ten to fifteen grains. Its supposed general depressant effect upon the system is not much to be dreaded, especially if a tonic be given in combination, but in susceptible subjects it is apt to produce the bromic acne. To avoid this inconvenience, the medicine may be discontinued for one week out of four ; or, if this is not sufficient, five-minim doses of liquor arsenicalis may be given in combination with it. The bromides tend to diminish the quantity of the menstrual flow and lengthen the intervals, and therefore do not act so well in hyperæmia from suppression of menses, or associated with scanty menstruation. In some cases of the latter, however, they may be beneficial when given in combination with iron. Bromide of ammonium and hydrobromic acid appear to have a similar effect to bromide of potassium, although Binz has maintained that the virtue lies in the potash and not in the bromine, and that chlorate of potash is equally efficacious. Iodide of potassium is especially useful in hyperæmia dependent upon ovarian enlargement or irritation, and in such cases it is desirable to combine it with the bromide. Quinine and the mineral acids are useful, especially in the more chronic stage, while quinine in large doses has a direct effect in promoting uterine contraction. In all cases of hyperæmia diet should be unstimulating, and alcohol should be avoided or taken sparingly.

Local Depletion.—Local depletion often gives great relief both in active and passive hyperæmia and also in a combination of the two, especially when the hyperæmia is a sign of inflammation, or much local pain and tenderness exist. It may be performed either by puncturing or scarification, or by leeches. The former is the most convenient, and is generally to be preferred in the earlier stages of hyperæmia or metritis, while the uterus is soft. It has the advantage that it is not liable, like the suction of a small quantity of blood by leeches, to set up a renewed fluxion to the part affected. A cylindrical speculum should be used, and two or three punctures made with a sharp-pointed bistoury, spear-headed scarificator, or triangular needle held in a pair of forceps. A sponge wrung out of hot water may afterwards be passed occasionally over the cervix to prevent clotting in the mouths of the vessels. One or two ounces of blood should be abstracted, and additional punctures may be made, if necessary, till this amount is obtained. Dr. Thomas recommends that if sufficient flow does not take place from one or two punctures, the cervix should be dry-cupped before puncturing by a cylindrical exhauster of vulcanite passed through the speculum.

The cases for which leeches are more applicable are those in which it is difficult to obtain sufficient blood by puncturing, as is usually the case in the later stages of hyperplasia of the cervix, and also those in which there is suppression of menstruation or too scanty a flow, so that it is advantageous to excite some temporary fluxion to the uterus. The cervix should be brought into the field of a large cylindrical speculum and the os plugged with a small piece of cotton wool to which a thread is attached for its removal. If this precaution be neglected, a leech may bite within the cervical canal, or crawl into the uterus and cause severe pain, although, in such a case, it is usually expelled after a time without very serious damage resulting. The cervix is first to be thoroughly cleaned, and may be slightly scarified to draw a few drops of blood. Three or four leeches should then be placed in the speculum, its lower extremity closed by a plug of cotton wool, and the speculum watched till the leeches have ceased sucking. The whole process is generally completed within half-an-hour. Single leeches may also be applied by a long glass tube provided with a piston, or by leech forceps. In case of excessive bleeding after removal of the leeches, a plug of cotton wool soaked in perchloride of iron may be applied to the cervix, or each leech-bite may be touched with the point of a heated knitting-needle.

Local depletion may be performed in the consulting-room or hospital out-patient room, but it is preferable that the patient should remain at rest in bed for some hours afterwards, especially if leeches are used. Several repetitions of puncturing or leeching are generally required at intervals of ten days or a fortnight. Neither should be performed within five or six days before a menstrual period is due; otherwise, its recurrence is apt to be interfered with. If, however, menstruation is scanty, and an increase of congestive pain occurs at its cessation, depletion immediately after the flow often gives relief.

A very convenient mode of causing a flow of copious secretion from the cervix and vagina, and so depleting their vessels, is the use of strong glycerine, by which the need for withdrawing blood may often be avoided. A tampon of cotton wool is to be thoroughly soaked in from half an ounce to an ounce of glycerine, and passed up to the cervix, a string or thread being tied round it to facilitate removal. It should be left twelve or

twenty-four hours. In the case of erosion or endometritis, an astringent may be dissolved in the glycerine,
but the pure glycerine produces the
most copious flow. It is often used
with advantage after puncturing or
leeching.

The patient may generally introduce
the tampon herself, by means of
Barnes' tampon introducer (Fig. 52),
and sometimes even without such
assistance. In the latter case, however, a good deal of the glycerine is
apt to be squeezed out in passing
through the vulva.

PASSIVE HYPERÆMIA OF THE UTERUS.

Passive hyperæmia may be by itself
a cause of subinvolution and hyperplasia, but is more frequently associated with active hyperæmia or
inflammation, and tends to aggravate
their effects. All grave displacements
of the uterus tend more or less to
interfere with the return of the venous
blood from that organ. Those which
have the most powerful influence
are prolapse of the second or third
degree (see p. 112) with strangulation, and acquired retroflexion, which
causes the veins to be compressed against the
utero-sacral ligaments, as well as from the effect
of the curvature of the uterus itself. Passive
hyperæmia is also produced by general causes of
venous obstruction in the heart, lungs, or liver, and
by any local pressure on the veins by ovarian or
other tumours, ascites, or fæcal accumulations, and
is promoted by want of exercise or constipation. Any

fixation of the uterus also tends to passive hyperæmia by interfering with the freedom of its motions, and, in most instances, leads to its enlargement. Such cases are frequently complicated by the effects of inflammation, but hyperplasia is brought about through fixation of the uterus even by a peritonitis which did not originate in the pelvis. Passive hyperæmia is apt to be promoted, in all classes of society, in the effect of posture : in the labouring classes by prolonged standing, with which is often associated a greater or less degree of prolapse of the pelvic viscera ; amongst the wealthy, by the excessive use of the dorsal reclining position in cushioned chairs or sofas, as opposed to the recumbent posture, and by the use of feather-beds instead of firm mattrasses. In the dorsal reclining position, the pelvic brim is rendered nearly horizontal, instead of being inclined about 55° to the horizon, as it should be in the upright position. The pelvis is thus exposed to the full weight of the abdominal viscera, and the return of venous blood from it is at the greatest disadvantage, while any tendency to retroversion or retroflexion is promoted by gravity. At the same time the use of soft cushions obviates the natural tendency which persons resting in a harder seat have to change their position frequently, and so assist, in an important degree, the venous circulation. In lying on a feather-bed also, the pelvis sinks in and becomes the lowest part of the body, whereas, upon a harder couch, in consequence of the greater width of the hips, the pelvis is somewhat higher than the shoulders.

The relation of passive hyperæmia to inflammation is that it does not, by itself, tend to produce inflammation, although it may lead to hypertrophy, and even to associated degeneration, but that it renders the tissue vulnerable to slight causes of inflammation, and makes the inflammation more obstinate when once excited, and repair more tardy. An example of this may be found in the case of ulceration of the legs, associated with varicose veins, and the same principle is largely exemplified in the case of the uterus.

Treatment.—The first indication is to remove,. if possible, all direct causes of venous obstruction, general or local, and especially to cure retroflexion or prolapse. Regulation of the bowels is of the utmost importance, and the practice of at least a daily evacuation at a regular time must be enforced, much trouble often arising from mere carelessness in this respect. The greatest relief is afforded by saline aperients, such as sulphate of magnesia and sulphate of soda, and a convenient mode of giving these drugs is in the form of one of the mineral waters, as Hunyadi Janos, Friedrichshall, or Pullna, to be taken the first thing in the morning with an equal quantity of hot water. When hyperæmia of uterus or ovaries is associated with much pelvic pain or tenderness, it is often desirable to secure a somewhat liquid motion at least twice a day, evening as well as morning, so as to diminish as much as possible the venous pressure during the hours of sleep. For this purpose drachm doses of sulphate of magnesia may be given two or three times a day. In all cases of passive hyperæmia postural treatment should receive due attention, since the blood pressure in the pelvis is necessarily increased in the upright position, and the ratio of increase, compared with the total pressure, is much greater in the veins than in the arteries. Long standing or sitting should be avoided, as well as the undue use of the dorsal reclining position on cushioned chairs, and the use of soft feather beds. Rest on a flat couch or bed in the lateral or semi-prone, rather than the dorsal, position should be frequently taken.

Passive hyperæmia receives benefit from all external agencies which act as stimulants to the general circulation, and especially to the heart. Of these some form of cold bath, the most generally useful being the hip-bath, taken on rising in the morning, is the most powerful, and in combination with this the cold vaginal douche, administered by Higginson's syringe, is a valuable adjuvant, provided that no active inflammatory state of pelvic organs be present. Failing the vaginal douche, the bath speculum, a small tube with perforations, may be used by the patient. In winter the water may be warmed to about 60° or 65° F. If neither cold douche, hip, nor sponging bath can be borne, alternate sponging with hot and cold water is a milder stimulant. If there is any weakness of the heart's action, the administration of digitalis helps to diminish general venous pressure, and a suitably nourishing diet and general tonics tend to the same effect. The veins of the uterus are emptied by the influence of ergot and other drugs causing contraction of the uterine walls, though these act more especially upon the arterial supply. The use of local depletion has been already mentioned (p. 163).

INFLAMMATION OF THE UTERUS.

Inflammation of the parenchyma of the uterus is termed metritis; catarrhal inflammation of its lining mucous membrane endometritis. In the most acute forms of inflammation all the tissues of the organ take part, and body and cervix are usually involved together, the affection of the body being the most important. Acute endometritis and acute metritis will therefore be considered together as a whole. Chronic endometritis, or metritis, may affect either the cervix alone, the cervix and body together—in which case the disease of the body is the more important—or, in rarer instances the body alone. It has already been described how, even in chronic affections, the inflammation is never entirely confined to the mucous membrane, but extends, in greater or less degree, to the adjoining parenchyma (*see* p. 160). Inflammation of the mucous membrane of the cervix, or cervical endometritis, will therefore be described in connection with inflammation of the substance of the cervix; that of the mucous membrane of the body, or corporeal endometritis, in connection with chronic metritis.

ACUTE METRITIS AND ACUTE ENDOMETRITIS.

Causation. — Acute inflammation of the whole uterus, in its most intense form, is very rare, except as the result of septic absorption after parturition or abortion, or after operations upon the uterus, the evacuation of retained menstrual fluid, or the use of tents. Next in intensity is that produced by a traumatic cause, such as intra-uterine injections, intra-uterine stem pessaries, cauterization of the cervix or cavity, of the uterus. In some of these cases absorption of septic material may also play some part. Acute endometritis, in which the whole thickness of the uterine walls also generally participates, but in a less extreme degree, is not unfrequently produced by exposure to cold, especially at a menstrual period, extension of gonorrhœal or other acute inflammation from the vagina, or excessive coitus. It may also arise in the course of specific fevers.

Pathological Anatomy.—Acute metritis is always complicated by endometritis, and in the more severe forms the inflammation extends to the peritoneal surface of the uterus, which becomes covered with lymph, and sometimes also, especially in the septic variety, to the neighbouring cellular tissue. The uterus becomes hyperæmic and enlarged by infiltration of serum, while, in the most acute form of inflammation, ecchymoses are scattered through its substance. In the septic variety, small collections of pus may be found between the muscular fibres, in the veins or lymphatics of the uterus, or still more frequently in those of the broad ligament adjoining. Purulent peritonitis may also be set up, and in cases dependent upon lymphatic absorption the affection of the peritoneum often preponderates over that of the uterus itself. Acute abscesses of notable size in the uterine wall have occasionally been recorded, but are very rare. Much more frequent are abscesses in the ovaries, or cellular tissue of the broad ligament. The disease may also end in acute or chronic pyæmia. In acute endometritis the mucous membrane is swollen, softened, and injected; that of the body of the uterus secretes at first thin serum, and afterwards muco-purulent fluid, often tinged with blood. The secretion of the cervix, normally clear and tenacious, becomes more copious, thin, and turbid. The inflammation is liable to extend along the Fallopian tubes and attack the peritoneum, even when the substance of the uterus is not involved in any great degree.

Results and Symptoms.—In most cases of severe septic or traumatic metritis, while the uterus itself is found to be swollen and excessively tender, the symptoms of periuterine inflammation, especially of that of the peritoneum, preponderate over those of the metritis proper. Both septic and traumatic forms are marked by rigors and considerable elevation of temperature. In the septic variety the increase in pulse-rate is often more marked than that of temperature, and as the disease advances the pulse, while becoming small, becomes at the same time compressible. In bad cases, in which the peritoneum is extensively affected, the abdomen quickly becomes tympanitic, and the breath acquires the peculiar sweetish odour of septi-cæmia.

In acute endometritis, with more or less participation of the uterine walls in the inflammation, but without any periuterine complication, the symptoms are pain, with a sense of weight and heat or throbbing in the pelvis, and pain also in the back, groins, and thighs. Considerable febrile action is present in the more severe cases. The pain is much aggravated by movement, or by any bearing-down effort; there is often much vesical tenesmus, and the urine is generally high coloured.

There may be paroxysmal aggravations of pain due to uterine contractions, and marked by their intermittent character. Occasionally there is active diarrhœa for a time, set up by reflex irritation, though, with the exception of these attacks, the bowels are generally constipated. When endometritis or metritis arises during menstruation, its immediate effect is usually the arrest of the flow. Septic metritis has a similar effect upon the lochial discharge, or that which follows abortion. In traumatic endometritis, however, especially when induced by caustic applications such as the insertion of the solid nitrate of silver into the uterine cavity, there may be profuse sanguineous discharge in the early stage. Ordinarily, at the outset of acute endometritis the discharge is scanty and serous ; after a few days it becomes profuse and muco-purulent, often offensive to the smell, and sometimes tinged with blood. Usually it has an irritating effect upon the vagina and vulva, and may cause excoriation of the thighs.

In septic metritis the prognosis is always grave, and bad cases pass rapidly into purulent peritonitis, and end fatally in spite of all treatment. Simple acute endometritis and metritis are apt to merge into the chronic form of the disease, and relapses are specially likely to occur at ensuing menstrual periods.

Diagnosis.—Endometritis and metritis uncomplicated by periuterine inflammation are distinguished by the mobility of the uterus, and the absence of any thickening round it. Constitutional disturbance is less than in pelvic peritonitis or cellulitis, but greater than in simple vaginal inflammation. On vaginal examination, the cervix is found swollen and sensitive, its arteries often pulsating strongly, and the os patulous. On bimanual examination, the body of the uterus is found to be very tender on pressure, and still more so if movement be imparted to it. It is often distinctly enlarged, and, if its previous size be known, the degree of swelling indicates the extent to which the uterine parenchyma has taken part in the inflammation. If the speculum be used, the cervix is seen to look red and œdematous, and to contain shreds of mucus, scanty serous fluid, or muco-pus. As a rule, the sound should not be used. If employed, it causes great pain, and generally some bleeding.

Treatment.—In *septic metritis* the first indication for treatment is to get rid of . the exciting cause. Any retained placenta, or clot, or decomposed polypus or other tumour, should be, if possible, evacuated at the very commencement of symptoms. When the inflammation is fully established, and the os does not admit the finger, it may be a difficult question whether artificial dilatation of the cervix is desirable. When, however, the discharge has any considerable fœtor, and it is suspected that there is something in the uterus, it is better to run the risk of interfering. It is preferable, if possible, to introduce the finger, with the aid of an anæsthetic, or to effect rapid dilatation of the cervical canal by a two-bladed, or three-bladed, dilator, such as that of Marion Sims (Fig. 15, p. 32), and avoid the use of tents. If tents are used, care should be taken to dilate the cervix by a single application, and not to leave them longer in place than necessary. The uterus being sufficiently evacuated, it should be washed out at intervals with antiseptic fluid. A solution of carbolic acid (1 in 40), or a weak solution of iodine (Tr. Iodi. ʒij. ad aq. Oj.), is preferable to one of permanganate of potash, since the latter rapidly loses its efficacy in contact with organic matter. The best apparatus to use is a funnel or other irrigator acting by hydrostatic pressure, attached by a flexible portion to a long silver tube with a rounded extremity, having openings on all sides. A simple syringe, large enough to hold ten ounces or more, is preferable to Higginson's syringe, since, with the latter, injections of air, together with the fluid, can hardly be avoided. It does not, however, give the security against undue pressure which is afforded by the hydrostatic method. In the absence of a metal tube, a large gum-elastic catheter may be used. To avoid the introduction of air, care should be taken first to fill the tube completely, and then a clip should be placed upon the elastic portion, until the terminal part is introduced up to the fundus.

Quinine in full doses is generally useful, and if the temperature is very high, it is well to begin with from 30 to 60 grains, given in two or three doses at short intervals, until the temperature is markedly influenced, or cinchonism produced. If vomiting interferes with the retention of the quinine, it may be given in the solid form, combined with a full dose of subnitrate of bismuth in a mucilaginous mixture, or a smaller dose may be given subcutaneously, the kinate of quinine * being the best form for this purpose. Opium, or morphia, must be given in sufficient quantity to allay the pain. Locally, fomentations or turpentine stupes assist towards this object. Other internal antiseptics, such as sulphite or sulpho-carbolate of soda, or salicylic acid, have scarcely shown themselves to be equal in value to quinine. In highly adynamic states, however, Warburg's tincture, containing quinine, with a great variety of other substances, among which are aromatic stimulants, has sometimes been found more serviceable than quinine alone. Two successive doses of half an ounce, undiluted, may be given at two or three hours' interval, brandy or beef-tea only being taken meanwhile. In a similar adynamic state, with much tympanites, turpentine, in doses of 15 or 20 minims, may be useful as a stimulant. If high temperature persist, it should be reduced by direct application of cold. For this purpose the most convenient means is Thornton's ice-water cap, or Leiter's temperature regulator, made in the form of a cap, whereby a continuous stream of ice-cold water is made to circulate round the head. Another method is to place the patient upon a water-

* See a Paper by Mr. Collier in the "Pharm. Journ." Sept. 1878.

bed, from which water is from time to time drawn off, and cold water added. It is of the highest importance to support the strength by administering such nourishment as milk, beef-tea, and eggs, at short intervals, as well as stimulants in ample quantities. If food is rejected by vomiting, recourse should be had at once to nutrient enemata. For this purpose "Derby and Gosden's fluid meat" (sold by Messrs. Savory and Moore, 143, New Bond Street), is of great value. A fluid which undergoes artificial digestion in the rectum may be made by mixing thick boiling gruel with an equal part of cold milk, and adding to half a pint of the mixture a drachm and a half of Savory and Moore's saline essence of pancreatine, or Benger's liquor pancreaticus, and ten grains of bicarbonate of soda.

In simple *acute endometritis* (with more or less implication of the parenchyma, but without periuterine inflammation), absolute rest in bed should be enjoined. If much fever and pain are present, from four to six leeches may be applied near the anus. This is better than applying them to the cervix, since too much disturbance of the patient is thereby involved, and increase of pain is sometimes produced. At the outset, minim doses of tincture of aconite every hour may be given to diminish the fever. Sedatives, with salines, especially the nitrate of potash, or acetate of ammonia, should afterwards be administered; or, when pain is acute, full doses of opium or morphia, either by rectum, subcutaneously, or by the mouth. Fomentations or linseed poultices, covered with oil-silk, should be kept applied to the hypogastrium. At a somewhat later stage, hot hip-baths, or copious warm vaginal injections of decoction of poppies, or of linseed or starch, with the addition of a drachm of laudanum to the pint, have a valuable sedative effect. Purgatives must be avoided in the acute stage, but the rectum should be unloaded, if necessary, by an enema. Later, saline laxatives are useful.

CHRONIC INFLAMMATION OF THE CERVIX, CHRONIC
CERVICAL ENDOMETRITIS, ECTROPION, EROSION, AND
FOLLICULAR DEGENERATION OF THE CERVIX.

Causation.—The majority of cases of inflammation
of the cervix may be divided into two great classes—
first, those in which the primary affection is catarrhal
inflammation of the lining mucous membrane, and in
which the parenchyma of the cervix becomes only
moderately swollen, and eventually indurated; secondly,
those in which the whole thickness of the cervix becomes inflamed from the injuries received in parturition, and eventually undergoes a process of extensive
hyperplasia and induration, while cervical endometritis
at the same time persists. The first class comprises by
far the greater part of the cases which occur in virgins
or nulliparous women, since in them it is rare for the
cervix to undergo any great degree of hyperplasia,
unless, either from congenital elongation or prolapse,
it becomes subject to mechanical irritation.

Of the first variety of endometritis, the predisposing
causes are similar to those of catarrhal inflammations of
other mucous membranes, such as general debility,
and the strumous, rheumatic, or gouty diathesis. Of
exciting causes, the most frequent are the effect of
cold, extension of inflammation, gonorrhœal or simple,
from the vagina or from the body of the uterus,
displacements of the uterus, excessive coitus, and
direct traumatic causes, such as the use of an intra-
uterine stem.

The second variety of inflammation arises from the
bruised condition in which the cervix is left after
labour, with numerous ecchymoses in its substance,
damage to its epithelium, which is soon afterwards
shed, and frequently more or less deep lacerations along
its edge. The failure in the healing of these lesions,
and their passing into a state of chronic inflammation,
may be due to the lacerations having been too deep to

heal spontaneously, or may be brought about by a too early getting up, by displacement of the uterus, or by any of the causes already enumerated which tend to produce subinvolution (p. 143), or hyperæmia (pp. 158, 165) of the whole organ.

Among the injuries produced by labour, the most important are lacerations of the edge of the cervix. If these are superficial, they may heal more or less completely; and this also happens more readily when the laceration is anterior or posterior. If, however, there is a deep laceration at each side, especially when the frequently associated complication of subinvolution and consequent partial descent of the uterus exists, the anterior and posterior lips of the cervix roll outwards, so as to evert the lining mucous membrane, and the condition termed *ectropion of the cervix* is thus produced. The delicate mucous membrane, turned outwards towards the vagina, is exposed to friction, and becomes inflamed. It then becomes swollen and deeply injected, and its surface granular from irregular proliferation, so that it closely resembles the surface of an erosion, or granular inflammation, at a spot originally covered by squamous epithelium. At the same time hyperplasia results in the portions of the cervix intervening between the clefts, and leads to distortion and induration. Similar results to those produced by labour may follow if the cervix is bilaterally incised to too great a depth, but generally to much less degree, since the effect is not then assisted by the enlargement of the uterus and bruising of the cervix.

A laceration on one side only may also lead to some degree of eversion of mucous membrane with granular inflammation. The deepest laceration, or single laceration, if there is only one, is more frequently on the left side, because the occiput of the child is generally directed that way. A lateral laceration, at the time of its production, often gives an opportunity for absorption, which leads to local cellulitis. From this a permanent band of thickening is apt to remain, which may be felt running from the angle of the laceration, generally on the left side.

Erosion or Granular Inflammation of the Cervix may originate simply from catarrhal endometritis. The inflammation of the cervical canal extends to the mucous membrane around the os. From the effect of irritation, the squamous epithelium proliferates and becomes softened, while it is, at the same time, macerated in the morbid cervical discharge. It is then shed, in the greater part of its thickness, either gradually or in bulk, and leaves behind a congested and slightly granular surface. Erosion, however, is found far more frequently in parous than in nulliparous women, and the more severe forms of the affection are very rarely seen in the latter. In the majority of cases it takes its origin from labour, commencing either with the shedding in bulk of the bruised and damaged epithelium, after parturition, or by its more gradual disintegration, in consequence of the inflammation which is a sequel of that event.

Pathological Anatomy.—In chronic cervical endometritis, the mucous membrane is swollen and hyperæmic, the glands more especially being enlarged. The secretion is increased in quantity, and becomes more opaque and stringy, often filling the cervix with a tenacious plug. In a later stage the mucous membrane becomes hypertrophied, filling the cervical canal, and protruding somewhat at the os, and considerable proliferation of the glands of the entire cervical canal may take place. The whole cervix is swollen and soft in the earlier period, but becomes indurated by areolar hyperplasia in the later stage. This change is much greater in those cases in which the disease commences with inflammation of the whole thickness of the cervix after labour, especially when its edge has been cleft by lacerations, in which case the diagnosis from carcinoma may become difficult.

When *simple erosion* arises by detachment of the squamous epithelium *en masse*, the slender papillæ,

N

which, in the normal state, reach nearly to the surface, are carried away at the same time. The surface left is only slightly granular. In more severe forms of the affection, to which the name of *villous or papillary erosion* has been applied, the inflammation proceeds further, and the mucous membrane becomes elevated into soft, deep-red papillæ, which readily bleed. It has generally been considered that the surface becomes entirely denuded of epithelium, more or less of the papillæ being left, and that the villous prominences are due to the overgrowth of these papillæ. According to the recent researches of Ruge and Veit,[*] however, the surface always remains covered with a single layer of cylinder-like epithelium, which is really derived from the deepest row of the original squamous epithelium. The normal papillæ are always thrown off; the cylinder-like epithelium grows inward, so as to form glandular crypts, and the villous prominences arise by growths of vascular connective tissue between these crypts. In the more severe forms the glandular crypts increase and proliferate. The very commencing stage of cancer, according to the same authors, differs from this condition only in the fact that the epithelium of the adventitious gland cavities proliferates, so as partially, or entirely, to fill up the acini.

I have found, in examining specimens excised during life, that in some instances the histological characters correspond to those described by Ruge and Veit, but in others there is actually complete loss of epithelium at some points, and the tissue near the surface is infiltrated with numerous inflammatory leucocytes, and contains many distended capillaries. The Malpighian layer of the adjacent squamous epithelium generally grows thinner as it is traced toward the inflamed surface, the horny layer being thrown off, and is destitute of the normal papillæ. Often over the apparently eroded surface, originally covered by

[*] "Zeitschrift für Geburtshülfe und Gynäkologie," Bd. ii. Hft. 2.

squamous epithelium, may be seen patches of cylindrical epithelium, adjoining the glandular crypts, and alternating with patches of ill-formed squamous epithelium, only one, two, or three cells deep.   In cases which were formerly described simply as villous erosion, it will often be found, if the experiment be tried of taking a tenaculum hook in each hand and drawing together the two lips of the cervix within a Sims' speculum, that the villi really belong to the cervical canal, and arise by hypertrophy of the prominences naturally covered by cylindrical epithelium, not of the normal papillæ beneath the level surface of squamous epithelium.   In old cases of laceration the epithelium of the exposed cervical mucous membrane may be found more or less completely converted into squamous epithelium, beneath which are often many of the glandular cysts described below.

Opinions have differed as to whether the so-called erosion deserves the name of "ulceration."   It is clear that, although in the initial stage there is a loss of substance of vascular papillæ as well as of epithelium, and the process must therefore be admitted to come, strictly speaking, within the definition of the word ulceration, yet there is no progressive ulceration, the condition being rather that of inflammation with glandular degeneration, and the term "ulceration" is therefore one which is needlessly alarming to patients.

In another and less important form of erosion, which has been called *aphthous or herpetic erosion,* inflammation of the mucous membrane leads to the formation of small vesicles, which burst, and leave an eroded spot.   These generally heal readily without treatment.

Cystic degeneration may arise from closure of the mucous glands by swelling of the mucous membrane, and adhesion of the edges of the orifice.   The glands then become distended into small cysts, known as *ovula Nabothi.*   Within the cervical canal, the swelling cysts force up the mucous membrane into an elevation, and often take the form of minute polypi. Similar small cysts are often found on the vaginal surface of the cervix, but these do not so easily elevate the denser mucous membrane. They may be seen, and more readily felt, as minute protuberances beneath it. According to Ruge and Veit these are not pre-existing glands, but are formed under the influence of irritation from the *rete Malpighii* of the squamous epithelium. The distended follicles may burst if the inflammation in them is more severe, and give rise to *follicular erosion.*

Results and Symptoms.—The cervix is, in general, but slightly sensitive, as is shown by the fact that nitric acid or other strong caustic may often be applied to its inflamed and eroded surface without causing any very great discomfort. Inflammation limited to the cervix, therefore, generally causes comparatively little pain, and pain may even be absent altogether. The pain most characteristic of cervical inflammation is situated over the sacrum, and not very severe in character. If the inflammation is not limited to the mucous membrane, but has affected the whole tissue of the cervix and led to hyperplasia, pain in the back and loins is generally more marked, and is often increased by walking, while pain may also be produced by coitus. A more constant symptom than pain is morbid secretion, and thus, in many cases, the presence of leucorrhœa may be the only indication for investigating the condition of the uterus. If, however, the discharge is thick in character, it may be retained in the vagina, and the patient may then not notice any leucorrhœa, although the altered secretion is manifest on the use of the speculum. In simple catarrhal inflammation of the cervix, the discharge is clear, glairy, like white of egg, and more tenacious than normal, often forming a plug in the cervical canal. When inflammation is more severe, and especially when it is combined with villous erosions, the discharge may be muco-purulent, or purulent, and is occasionally tinged with blood. By its irritating effect it may set up vaginal inflammation. If the discharge is profuse and long-continued, it may form a drain which tends to weaken the system. Hyperplasia of the cervix often leads to irritation of the bladder or rectum, as the result of pressure, especially if any anteversion or retroversion be present. The cervix is more richly supplied with sympathetic than with sensitive nerves, and thus its inflammation is apt to lead to reflex congestion of the body of uterus and ovaries. When this occurs there may be more severe pain, menorrhagia or metrorrhagia, and even more distinct reflex symptoms, such as nausea, vomiting, dyspepsia, vertical headache, intercostal neuralgia, and hysterical manifestations. Such reflex symptoms are generally not so marked in inflammation of the cervix as in that of the body of the uterus, and their existence generally implies that the body of uterus or ovaries are involved in congestion, if not in actual inflammation. The whole question of distant reflex symptoms dependent upon uterine conditions will be discussed more fully under the head of endometritis. In many cases of cervical inflammation there is more or less extension, not merely of congestion, but of actual endometritis, to the body of the uterus. Inflammation limited to the cervix, or erosion, may persist for a long time with but little affection of the general health, but it is often associated with dyspepsia and general failure of nutrition, which may be partly the cause and partly the consequence of the persistence of the uterine affection..

An erosion, while generally in the first instance the result of some other condition, as endometritis or inflammation of the whole substance of the cervix, itself often becomes a source of reflex irritation, and maintains a hyperæmia not only of the cervix but of the body of the uterus and the ovaries, all of which are frequently found to be enlarged and tender, in conjunction with such a condition. Under these circumstances menorrhagia is often a prominent symptom, and the first

thing necessary in its treatment is the cure of the
disease of the cervix. In the case of villous erosion,
coitus often gives rise to slight hæmorrhage, and this
may be the chief symptom which attracts the patient's
attention. In cervical inflammation sterility is often
produced by the obstruction to the spermatozoa formed
by the plug of mucus in the os, or by the deleterious
influence upon them of the cervical secretion. These
obstacles do not, however, always form a bar to conception ; and if pregnancy occurs the resulting hyperæmia tends to render worse any inflammation, and
especially any erosion, which exists. From this cause
may arise hæmorrhage during pregnancy, severe
vomiting, or other reflex symptoms, and abortion or
miscarriage.

The natural course, both of chronic cervical endometritis and of erosion, is a very tedious one, with but
little tendency to recovery, although a cure may result
by improvement of general health. They are fairly
amenable to treatment, but improvement is often
slow, and persistence in treatment for four or six
months is not unfrequently requisite. If there is
extensive hyperplasia of glands, cure can only be
effected by vigorous measures. Long-standing hyperplasia of cervix, with induration, is little amenable to
remedies. The granular inflammation of the mucous
membrane of the cervical canal everted in cases of
laceration of the cervix, is very similar in its effects,
as well as in its histological characters, to that of the
portion of mucous membrane originally covered with
squamous epithelium. In course of time, sometimes
after the lapse of years, cylindrical having been replaced by squamous epithelium, the leucorrhœa and
other symptoms may subside, although some degree
of hyperplasia of the everted cervix is apt to remain
permanently. Since cancer of the cervix is excessively rare in virgins, it appears certain that erosion
or other form of inflammation may be the starting
point of cancer in persons predisposed to that disease ;

and this view is confirmed by the close approximation found by Ruge and Veit in the histological characters of villous erosion toward those of commencing cancer. In one or two cases I have had the opportunity of observing epithelioma supervene upon chronic granular inflammation, and frequently, at a very early stage of epithelioma, I have found the disease to be situated just at the angle of a previously existing laceration of the cervix. Hence it is of great importance not to omit the due treatment of granular inflammation when the age has been reached at which cancer becomes probable; an age which, in the case of the cervix uteri, must not be reckoned as much beyond thirty years.

Diagnosis.—In simple cervical endometritis, vaginal touch may reveal only slight enlargement, or may fail to detect anything. The speculum will show the os to be congested, and generally either pouring forth copious clear viscid mucus, like white of egg, or filled with a more tenacious and opaque plug of similar mucus. The characteristic glairy mucus may sometimes be observed in the vagina when the os does not happen to contain any. If the plug be removed by twisting it round a Playfair's probe wrapped in cotton wool, the interior of the cervix is seen to be red, swollen, and granular—a condition which is more manifest if the bivalve speculum (Fig. 6, p. 21) is used, and expanded rather widely, so as to stretch open the os. The mucus is clear and alkaline as secreted by the cervix, but is rendered more opaque by contact with the acid vaginal secretion. In the mixed discharge, the acid usually preponderates. In most of the cases occurring after delivery, broadening of the cervix from hyperplasia will be detected by the finger, and frequently the clefts resulting from laceration will be felt. In ectropion arising from bilateral laceration of the cervix, the condition existing is often more manifest to the finger than to the speculum. If, however, a Sims' speculum is used, and the lips of the os are drawn together into their original position by two tenaculum hooks, the exact relation of parts will readily be seen. As seen in a speculum, especially if a bivalve is used, the antero-posterior diameter of the cervix appears increased, as shown in Fig. 53. When the two lips are drawn together through the Sims' speculum, by aid of the two hooks, the greater part, or the whole, of the apparently eroded surface, if granular inflammation exists, may be turned inward toward the

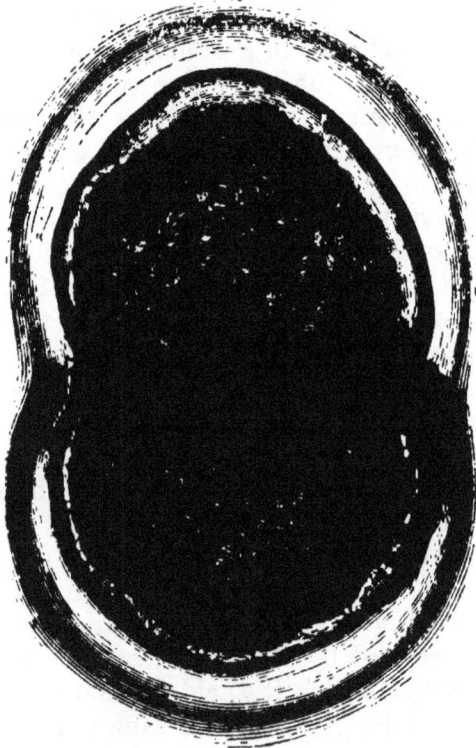

Fig. 53.—Bilateral Laceration and Ectropion of Cervix, with Severe Granular Inflammation of Exposed Mucous Membrane.

cervical canal, but this surface generally extends beyond the limits originally covered by cylindrical epithelium, which normally passes into squamous epithelium at a point about a quarter of an inch above the external os.

The more severe kind of erosion, or granular inflammation, is easily recognized by the touch as a soft, villous, velvety surface. A simple erosion feels softer and more granular than the normal mucous membrane, and is almost always associated with some broadening of the cervix. In the healthy cervix a *tactus eruditus* may always determine the negative as to erosion, but there may sometimes be an uncertainty in distinguishing by touch between a slight existing erosion and one that is healed, or an irregularity due to hyperplasia or degeneration of glands. The speculum will always resolve the doubt and show the erosion as a circumscribed, deeply red, granular, or villous surface, rather elevated above than depressed below the surrounding mucous membrane (Fig. 54). In the more severe form of erosion bleeding is readily produced by contact with the speculum; but a great proneness to bleed on a gentle touch with the finger should always raise the suspicion of the presence of commencing cancer.

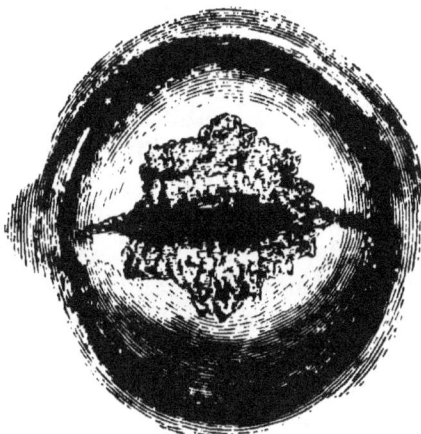

Fig. 54. — Simple Erosion, or Granular Inflammation, of the Cervix, with indication of slight laceration. (After BARNES.)

Difficulty is often found in introducing the sound in a case of cervical endometritis from its point catching in the folds of the swollen or hypertrophied mucous membrane, and when this is the case slight bleeding may be produced. Otherwise, if there is no complication with corporeal endometritis, the sound may be passed to the fundus without causing bleeding, or the pain which usually follows its introduction in that disease.

Treatment.—Constitutional remedies are of great

importance, though local treatment is usually required in addition. Nourishing, but not too stimulating, diet, with abundance of fresh air, and gentle exercise without fatigue, are to be enjoined. Causes of mental depression should be avoided as much as possible, and change of scene is often of great value. Any depressing influence, such as prolonged lactation, should be removed. The medicinal treatment should be of a tonic kind, with special reference to the impaired digestion which is a usual concomitant. Nitro-hydrochloric acid, with nux vomica, or strychnia, and a vegetable bitter, to be taken directly after meals, is a useful prescription.* If there is much stomach irritability, bismuth, with or without small doses of morphia, may be substituted for the bitter. When the digestive function is re-established, the liquor cinchonæ, tinctura cinchonæ flavæ, or quinine, may be given in place of a simple bitter. Iron is apt to disagree when there is any sign of liver inaction, or portal congestion, or when the case is complicated by metritis or hyperæmia with considerable tenderness of the uterus. But in the absence of these, especially in later stages, it may be usefully given in combination with a laxative.† Passive hyperæmia should be treated by the means enumerated under that heading, and any displacement of consequence rectified.

If acute pain or tenderness of the cervix is present, it is well to commence with local depletion (*see* p. 163), and if extensive degeneration of the cervical glands is detected, the depletion may be effected by scarification of the lining membrane of the cervical canal with a narrow-bladed knife. Any prominent glands on the

---

* ℞ Acid. Nitro-hydrochlor. dil. ℥x. ; Tinct. Nucis Vomieæ, ℥x.; Tinct. Gentian co. ℨj.; Aq. ad ℥j.—ter quotidie.

† ℞ Ferri et Ammon. Citrat. gr. v.; Magnes. Sulphat. gr. xxx. ; Sp. Chloroform. ℥xv. ; Glycerin. ℨj. ; Aq. ad ℥j.— ter quotidie. A formula for iron very readily tolerated, when a laxative is not required, is the following:—Ferri Tartarati, gr. v. ; Glycerini, ℨj. ; Sp. Vini Rectificat. ℨss. ; Aq. ad ℥j.— ter quotidie.

vaginal surface of the cervix should be punctured and touched with strong carbolic acid or solid nitrate of silver, since they keep up irritation by their presence. Coitus should be prohibited while any notable tenderness exists, and placed under strict limitation at all times.

Of local applications, the simplest are vaginal injections, which should always be used at least twice a day to wash away the secretion, if for no further object.

Injections of water, for removal of secretions, should generally be used moderately warm. For relief of congestion, as will hereafter be described, hot water at 100° to 105° F., and used in large quantities, is often very effective. Emollient or alterative lotions should be used warm. If an astringent effect only is desired, the lotion may be cold in summer, and in winter at a temperature of about 60°. Syringes of pewter and glass do not contain sufficient fluid for ablution, and the latter are dangerous from the risk of breakage. Higginson's syringe (Fig. 55), provided with a vaginal tube about six inches long, and having a central ball, by compressing which a steady stream is produced, can be used

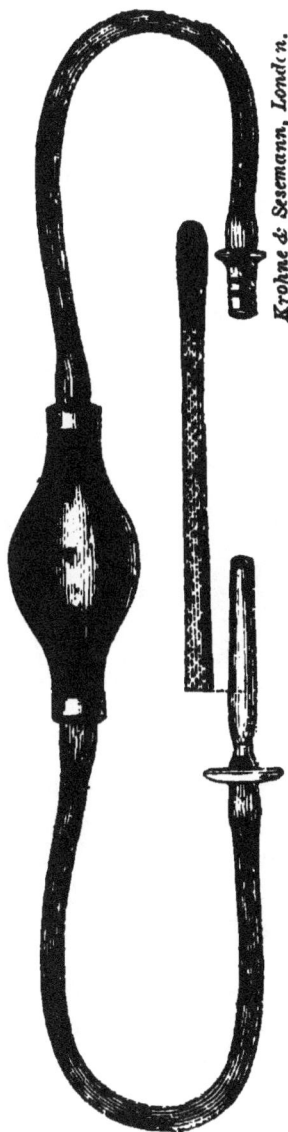

Krohne & Sesemann, London.

Fig. 55.—HIGGINSON'S Syringe.

by the patient herself more effectively. For the use of water for cleansing purposes, she may be in a sitting position over a bidet or ordinary chamber utensil. But water for the prolonged application of heat or cold, and lotions for any purpose, can only be used effectually when the patient is lying down in the dorsal position, so that the fluid does not flow away immediately, but distends, in some degree, the vagina. The best receptacle for her to lie upon is the "ladies' bed-bath" (Fig. 56), which may be provided with an elastic tube, to carry away the fluid to a vessel on the floor. In resting upon this the hips are elevated, the back being supported, if necessary, by a folded blanket

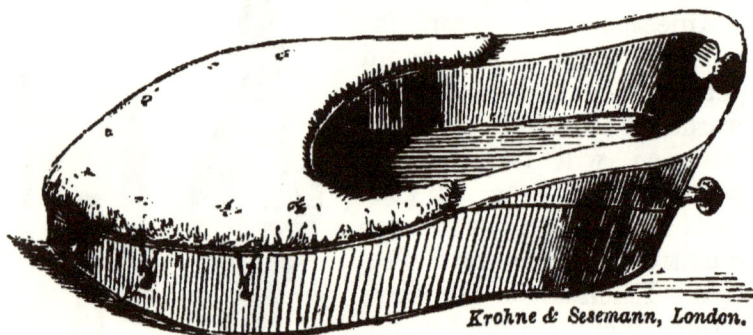

Krohne & Sesemann, London.

Fig. 56.—Ladies' Bed-bath.

or very thin pillow. In the absence of this, if a moderate quantity of fluid only be required, she may lie upon a round bed-pan, brought well under the hips, so that it is not too much tilted. A large quantity of fluid may also be used, as in irrigation by hot water, if the patient lies crosswise on a low bed or sofa, the feet on a couple of chairs, a mackintosh being so arranged as to carry down the fluid into a foot-pan.

In the dorsal position, the patient cannot generally work the Higginson's syringe conveniently herself. If the services of a nurse, or other attendant, are available, the best plan is for the nurse to use it. It is maintained by Emmet that, in the use of hot water, the discontinuous stream of the Higginson's syringe is more effective, in stimulating the vessels and absorbents, than the continuous stream of an irrigator. Generally, if the patient has to manage the injection herself, it is best to use some form of irrigator. The fountain irrigator (Fig. 57, p. 190) is a very portable form of apparatus, and may be made to fold up and be contained in a tin box, like a collar box. The bag to contain the water or lotion may be made to hold as much as three quarts, and is hung up on a nail two or three feet above the level of the patient. For prolonged use of hot water, it is better to have an ordinary hot water can fitted with a delivery tube attached to a tap near the bottom. In the use of either of these, the taps are first opened until the fluid begins to flow, so as to get rid of the air in the tube, and then the tap at the delivery tube is shut off, until the tube has been introduced into the vagina. By this tap, also, the rapidity of flow may be regulated. A cheaper and more portable form of irrigator than the can is the syphon irrigator (Fig. 58, p. 190), which has a weighted end, to keep it under the water, and is stiffened at the bend, to prevent its collapsing. In using this, the tap is first opened, and the whole tube gradually immersed, the delivery tube last, so as to fill it all with water. The tap is then turned off, and the tube taken out of the water, except the weighted end, and the syphon is by this means formed. Another mode of forming the syphon is to open the tap, place the weighted end in the water, and then draw the finger and thumb down the tube, until the water begins to flow. In using the can or syphon irrigator, the can or jug is placed on a shelf or chest of drawers, above the level of the patient.

For the employment of a small quantity of lotion a more convenient mode is to use a simple india-rubber enema syringe (Fig. 59, p. 191), containing about four or six ounces, and having a vaginal tube. If the effect of lotion only is desired, the patient may first use the Higginson's syringe herself with warm water, then lie down on

Fig. 57.—Fountain Irrigator.

Krohne & Seemann, London.

Fig. 58.—Syphon Irrigator.

Krohne & Seemann, London.

a bed-pan, and inject slowly one or two small syringefuls of the lotion, remaining five or ten minutes in the same position to allow the lotion to have its full effect. When she rises, the remainder of the lotion flows away. If a nurse administers the injection, before removing the bed-bath or bed-pan she should depress the perineum with one finger, and gently compress the hypogastrium, to get rid of the excess of fluid, and finally place a napkin against the vulva to absorb any that may remain.

For an emollient effect an ounce of glycerine, or a drachm of borax, carbonate of potash, carbonate of soda, chloride of ammonium, or chlorate of potash, to the pint of water may be used; while the salts tend also to diminish the cervical secretion. The glycerine may be added in combination with them, and if a more sedative effect is desired, from one to four drachms to the pint of tincture of opium. For a more astringent effect, alum, iron-alum, tannin, or sulphate of zinc may be used. The strength should be from twenty to sixty grains to the pint in the case of sulphate of zinc, and from one to three drachms or more to the pint for the rest. The liquor plumbi subacetatis dilutus is also a valuable remedy, and is less apt to irritate than most of the astringents, but it has the inconvenience of occasionally staining the linen brown, from formation of a sulphide. If the discharge is at all offensive, or if it is due to the extension of gonorrhœa to the cervix, antiseptic lotions are often of use. Carbolic acid (gr. xl. to gr. lxxx. ad Oj.), chloride of zinc (gr. xx. to gr. lx. ad Oj.), liquor carbonis detergens (3j. to 3ij. ad Oj.), perchloride of mercury (gr. iij. to gr. v. ad Oj.) may then be used. If the lotion is prescribed in

Fig. 50.—Vaginal Syringe for use with Lotion.

Krohne & S. sewann, London.

fluid form for use, it should be ordered of double
strength, and the patient should add an equal quantity
of hot water at the time of using. In rare cases severe
pain, uterine and peritoneal inflammation, and even
death have arisen from the use of a vaginal injection.
This has probably been due to the patient having
inserted the tube into the patulous cervix of a retroverted uterus.   Caution should therefore be used in
recommending injections while such a condition exists
unremedied; and whenever the os is at all patulous
it is a safeguard to have the vaginal tube with no
terminal, but only lateral openings. In some cases
the disastrous result may have been due simply to
the stimulus of the lotion causing contraction of the
uterus and Fallopian tubes, and so forcing purulent
fluid into the peritoneal cavity.   It is well, therefore, especially with neurotic patients, to begin with
a weak lotion, and gradually to increase the strength.

Astringent and alterative drugs may also be dissolved
in glycerine, and used in the mode described at page
165.   The most useful are borax, tannin, or acetate of
lead, in a strength of from thirty to sixty grains to the
ounce.   The last is especially serviceable in the case of
erosion, the astringent contracting the vessels, while the
glycerine depletes the congested surface.   For the same
cases, fifteen grains of iodoform in fine powder suspended
in three drachms of glycerine form also a very useful
application.   Astringents may also be employed in the
form of suppositories, of which the most serviceable are
those containing five grains of tannic acid or acetate of
lead.   As a basis for suppositories, a combination of one
part of powdered gelatine moistened and gently heated
with three parts of glycerine, is much preferable to
cocoa-butter.   The formula recommended by Dr. Tilt
of one part of pure paraffin to four of vaseline, may
also be used.

Local Applications to the Cervix.—In the cervical
leucorrhœa of virgins, a fair trial should be made of
the means already enumerated before resorting to the

speculum, which, for obvious reasons, should not be used in their case, if it can be avoided. In married women, however, the speculum may be used at once for diagnosis, and if severe erosion or glandular degeneration is detected, the necessity for stronger direct applications may be immediately recognized. The object should be to effect a cure, if possible, without leaving any cicatricial tissue, and, hence, the mildest remedy likely to prove effectual should be tried first. The solid nitrate of silver, which at one time was the favourite remedy in all cases, is now less generally preferred, since it may sometimes cause considerable irritation and hæmorrhage, if vigorously applied, and, in severe cases, is not so effective as other measures. It is most suitable for a case of simple erosion, the surface of which may be touched lightly over, so as rather to form a protecting film than to have any deep caustic effect. This may be repeated two or three times at intervals of a week, but not too often. A tapering pointed stick of nitrate of silver may also be passed, on one or two occasions, into the cervical canal, when there is granular inflammation of the cervix.

Liquid applications may conveniently be made to the vaginal surface of the cervix with a brush, and to the cervical canal by means of Playfair's probe (Fig. 60). If a little absorbent cotton wool is first spread out in a thin layer and then wrapped round it by rotating the probe, it becomes very firmly attached, and may then be dipped in the liquid to be used. If the probe is notched at the sides and bulbous at the extremity, as is the case with probes frequently sold, it is a very troublesome process to remove the cotton wool,

Fig. 60.—Playfair's Probe.

unless by burning it off. It is better to have no bulb at the end, and simply to roughen the probe slightly by rubbing it longitudinally with sand-paper. The wool is then held quite firmly enough for the application, if carefully wrapped, and yet can easily be pulled off. The terminal portion of the probe should be made of aluminium, that it may resist nitric acid, and should be as much as three inches long, so that it may be used for application to the body of the uterus if desired. For the application of nitric acid, in the absence of an aluminium probe, a vulcanite sound may be used. If the sound be first wetted, and a very thin layer of dry cotton wool be wrapped closely round it with some dexterity, the bulbous extremity prevents any risk of the cotton being drawn off and left behind in the uterus. For making the application, Sims' speculum and the semi-prone position are the best, but, in the absence of an assistant, the probe may be used with any other speculum, especially a short Ferguson's (see p. 19), short bivalve, or Neugebauer's speculum (Fig. 10, p. 25), which may be so manipulated as to bring the uterus into a position of slight retroversion. Before any application is made, the tenacious mucus should be removed from the cervical canal by entangling it in a swab of cotton wool, or, what is better, a small fragment of sponge, not to be used a second time. This is facilitated if a swab of glycerine, or white of egg, is first used.

Of the milder remedies, a solution of nitrate of silver, of thirty or sixty grains to the ounce, is by some preferred to all others, but it must be applied rather frequently, namely, at intervals of from five to seven days. A useful mild application to an erosion is Richardson's styptic colloid, consisting mainly of tannin dissolved in collodion. This forms a protecting film, as well as being astringent, and may be used at intervals after one or two applications of a stronger caustic. Dr. Atthill recommends the addition to it of fifty grains of carbolic acid to the ounce. The liquor or linimentum iodi may also be used, or a saturated tincture of iodine,* which Dr. Churchill recommends to be applied once a week to the whole cervix as an absorbent in hyperplasia, after a single application of nitric acid. For an erosion which very readily bleeds, the liquor ferri perchloridi fortior may be used.

Perhaps the most widely useful of all applications, both for the cervical canal and for erosions, is strong carbolic acid, a caustic of medium strength, which leaves little pain behind, since it has a somewhat anæsthetic effect upon the tissue, and is not likely to produce contraction or occlusion of the os. It may be used either simply liquefied by the addition of a sixteenth part of water, or an equal quantity of glycerine may be mixed with this. For erosions the stronger application is preferable. Care must be taken to protect the vagina and vulva. Two or three applications may be made at about a week's interval, and then about three weeks should be allowed for healing. Another good application is Dr. Battey's "iodized phenol."† For severe forms of villous erosion, and for extensive cystic degeneration of the cervical canal, strong nitric acid is the best application. Recourse should also be had to the same caustic, if an erosion resists all milder remedies. While it produces a superficial eschar, its action is not deep, if it is not left very long in contact, and it does not usually produce much pain when applied to the cervix, though in some susceptible persons it evokes hyperæmia of the uterus, with reflex nervous symptoms, lasting for some days. The vagina should be protected, the swab of nitric acid should be kept in contact not more than a minute or two, and a large swab, freely soaked in

* Iodine, 75 grains; iodide of potassium, 30 grains; rectified spirit, 1 ounce. Dr. Goodell recommends an application consisting of tannin 60 grains, and iodine 30 grains, or iodoform 120 grains, dissolved in an ounce of flexible collodion.

† Take of iodine, ℥ss.; crystallized carbolic acid, ℥ij.: water, ℥ij. Mix and combine by gentle heat. Use either pure or diluted with glycerine.

water, should be applied afterwards. One application of nitric acid is often sufficient, and it should not generally be used more than two or three times, at intervals of about four weeks. It may be followed by the milder astringents, as styptic colloid or a solution of nitrate of silver. The acid nitrate of mercury is used by some in place of nitric acid, but it does not appear to have any advantage over it, and has occasionally produced salivation in susceptible subjects. Dr. Marion Sims' favourite caustic for villous erosion is chromic acid, dissolved in an equal quantity of water. He applies a drop or two on a pointed glass rod to the granulations only. Unless its action is very carefully limited it is rather a painful caustic. For the treatment of the same affection Schroeder recommends the repeated application of acetic acid, poured into a cylindrical speculum.

Some cases of glandular degeneration round the edge of the os, and in the cervical canal, may resist the action even of nitric acid. The choice then lies between the use of deeper caustics, as potassa cum calce, potassa fusa, or the actual cautery, and the scraping away the diseased glands with a sharp steel curette (Fig. 61). The latter appears preferable, as less likely to cause contraction or occlusion of the cervix. After the use of any of the stronger caustics it is a good plan to apply a tampon soaked in glycerine. Care should also be taken that contraction of the cervix does not arise; and, if

Fig. 61.—Sims' Curette.

necessary, a large metallic bougie should be occasionally passed. Occlusion has been produced by the repeated use even of the solid nitrate of silver.

In the more chronic stages of cervical endometritis, the solid points of fused sulphate of zinc with alum, introduced by Dr. Braxton Hicks, are one of the most effective applications, but they are liable to cause a good deal of pain and irritation when any active hyperæmia is present. The zinc point is passed for its full length into the cervical canal, through a speculum, or by means of an applicator consisting of a tube provided with a piston, and left there to dissolve. Milder applications may be made in the form of crayons containing tannin or other astringents, but these are usually less convenient than liquid applications.*

In some cases of cervical endometritis, in nulliparous women, it is found that the os remains small, and the cervix has undergone little apparent change. Before local treatment can be satisfactorily used, the os must be dilated. This may be done by a sponge tent, or if the cervix is conical and the os congenitally small, by incision (*see* p. 53). The application of a sponge tent has a use apart from mere dilatation, if the disease is chiefly confined to the cervical canal, since by·its pressure it modifies the mucous membrane, and removes granulations or projecting glands.

The inflammation of the whole thickness of the cervix, tending to hyperplasia, is little affected by internal remedies. At the stage when it is beginning to pass into induration, absorbents used locally may be of some service. A convenient application is iodized cotton, containing 20 per cent. of iodine. A pledget

---

* Tannin may be made into a crayon with glycerine alone: tannin, gr. xxx.; glycerine, ℳ iij. For other drugs, fifteen parts of the drug may be used to fifteen parts of powdered gelatine and two of glycerine. The gelatine is first moistened with water and then mixed with the glycerine in a water bath, the drug being afterwards added. The mass is then rolled out into crayons like a pill-mass.

of this is placed in contact with the cervix, and kept
in position by a tampon soaked in glycerine. Iodide of
potassium and iodine may also be used dissolved in
glycerine or in the form of suppositories. The treatment of the resulting cervical enlargement has been
already considered (*see* p. 149).

Treatment of Ectropion of the Cervix.—After slight
laceration of the cervix, any exposed cervical mucous
membrane has its epithelium at length converted into
the squamous variety, a process which may be accelerated by the use of astringents or caustics. If,
however, there is deep bilateral laceration with eversion, this condition always remains a source of irritation and consequent hyperplasia, and the exposed
mucous membrane is always liable to granular inflammation. For these cases Dr. Emmet has introduced
the operation of *trachelorrhaphy*, or paring the edges
of the laceration, and uniting them by sutures. This
operation has not hitherto been much performed in
Britain, but deserves a more extensive use, though
the proportion of cases requiring it would seem to
have been much exaggerated by some.

Operation of Trachelorrhaphy.—If the uterus is high
up, and cannot be drawn down, the operation may be
somewhat troublesome and tedious, but if the cervix
can readily be drawn to the outlet of the vagina,
the perineum being retracted by a very short Sims'
speculum, it is a very easy one. Care must be taken,
however, not to use any undue traction, especially if
there is any trace of a past cellulitis, such as is often
associated with a deep laceration, for then there would
be a risk of rekindling the inflammation.

Dr. Emmet uses a double tenaculum with diverging
points, introduced within the cervical canal, in order
to steady the cervix. If, however, any considerable
traction is employed, it is better to make it by means
of two loops of wire or silk passed through the anterior
and posterior lips of the cervix.

It is convenient to place the patient on the left side

to freshen the right side of the cervix, and conversely. The mode of freshening the sides of the laceration and placing the sutures is shown in Fig. 62. The figure will be more readily understood if it is compared with Fig. 53, p. 184, representing the appearance of the laceration with granular inflammation. When the sutures are tightened, the two lips of the cervix, at the top and bottom of the figure, are brought into contact. In general, two sutures at each side are sufficient, but for a

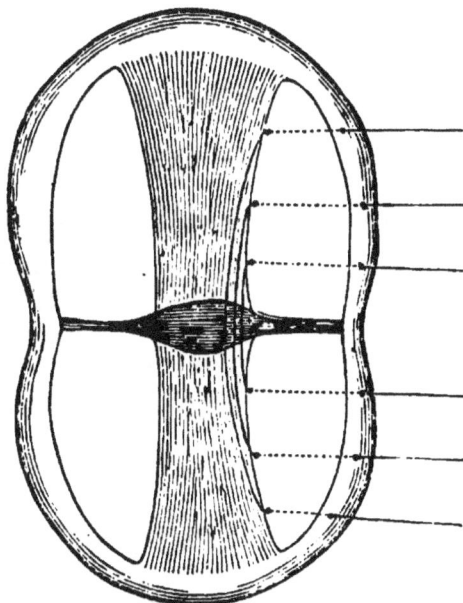

Fig. 62.—Lacerated Cervix, after denudation on both sides, and application of sutures on one side.

deep laceration three may be required, as shown in the figure. Dr. Emmet has introduced a "uterine tourniquet," in order to constrict the cervix during the operation. The hæmorrhage, however, is very rarely sufficient to call for any such expedient, and it is always arrested by tightening the sutures if the freshened surfaces are accurately adapted. If the cervix is drawn to the vulval outlet, it is preferable to freshen the edges with a narrow-bladed knife, such as that used for vesico-vaginal fistula. If the cervix remains in place, it is more convenient to use scissors. In either case the tissue should be removed from each angle in one piece if possible. The cervical canal should be left somewhat trumpet-shaped, as shown in the figure, to allow for some subsequent contraction of the os through diminution of the hyperplasia, and care must be taken not to leave any mucous membrane unfreshened at the extreme angles of the lacerations. For the sutures silver wire may be used, or silkworm gut of the thickness used for salmon flies. This has most of the advantages of silver wire, being perfectly unabsorbent, and is more convenient to manipulate. The sutures should be passed by short straight or slightly curved needles, held in a needle-holder. If silkworm gut is used, the sutures may be fastened by Aveling's coil and shot (Fig. 63). The coil is made by winding silver wire very closely round a metal rod or stylet. The two ends of the sutures are passed first through a bead, then through the coil, and then, when tightened, are fixed by a perforated shot. The suture is then removed with very great ease, for the coil being cut through with scissors just below the shot, and coil and bead being removed, the two ends of the suture are left projecting, and can be easily seized with forceps. The sutures should be removed after an interval of about ten days.

Fig. 63.—AVELING's Coil and Shot.

Experience has not yet fully decided the scope of trachelorrhaphy. It does not seem to be called for in all cases of laceration, or even in all cases of laceration with ectropion, for these conditions do not always produce symptoms. But if there is granular inflammation of the mucous membrane and hyperplasia, especially if there is also a tendency to descent of the uterus, the operation is likely to afford the most rapid and permanent cure. I have found, however, in some instances, that cervical endometritis has remained afterwards, and has required local treatment for its cure. In cases of laceration limited to one side, that one side may be united in a similar manner. In a case of very old standing, when the lips of the os are too much deformed by hyperplasia to allow them to be brought together, both lips may be amputated by the plastic method (*see* p. 155).

If, on account of fixation of the uterus by cellulitic thickening, or for any other reason, it is decided not to operate, and there is inflammation of the exposed cervical mucous membrane, the object will be to hasten the conversion of the epithelium into the squamous variety. For this purpose one or two applications may be made in the first instance of one of the stronger caustics, such as nitric acid, the benzoline cautery, or even potassa fusa cum calce.

SYPHILITIC ULCERATION OF THE CERVIX.—Primary chancre may occur on the cervix, but is very rare in this situation. The ulcer is marked by sharply-cut, indurated edges, depressed surface, and a tendency to become covered with false membrane. Mucous patches on the cervix are also rare, as also is tertiary syphilitic ulceration. The latter forms an excavated ulcer which readily bleeds, and is apt to be mistaken for an early stage of cancer. It is not generally accompanied by so much pain, or so great fœtor in the discharge, but the history of constitutional syphilis will guide much in the diagnosis. Syphilitic ulceration has occasionally even laid open the rectum or bladder. It is generally distinguished from simple erosion by its not being close to the os, in continuity with an inflamed cervical canal, but separated from the os by a bridge of intact tissue.

CHRONIC ENDOMETRITIS AND CHRONIC METRITIS.

Pathological Anatomy. — Chronic endometritis proper, or chronic corporeal endometritis, consists of inflammation of the mucous membrane of the body of the uterus. The inflammation is not absolutely limited to the mucous membrane, but extends to some extent, slight or considerable, into the substance of the organ (*see* p. 160), and is accompanied by more or less active hyperæmia of the whole uterus. Endometritis and metritis are therefore not separate affections, but the terms may be used respectively to indicate the preponderance of the disease of the mucous membrane, or that of the parenchyma, in different cases. Endometritis and metritis are frequently associated with subinvolution, of which they are often the cause, and with the effects of passive hyperæmia, which renders the tissue more vulnerable to irritating causes.

In milder and more recent cases of endometritis, the mucous membrane is swollen and hyperæmic. The inter-glandular stroma of this mucous membrane, which is constantly being renewed at each menstrual period, normally approximates towards the character of an embryonic tissue, and hence it shows much less marked histological changes under the influence of inflammation than the mucous membrane of the cervix. The microscope shows more marked changes in the glands than in the inter-glandular stroma. These glands become dilated irregularly, and are filled irregularly with more or less rounded cells, instead of being lined for the greater part of their course with uniform cylindrical epithelium. The cylindrical epithelium on the surface, if retained in the microscopic section, is also seen to have become deformed and irregular. The superficial layers of mucous membrane may eventually be thrown off, and by irregular proliferation villous or polypoid masses may sprout up. This constitutes the more severe disease of *fungoid or villous endometritis*, of which hæmorrhage is the prominent symptom. The secretion in milder forms of endometritis is an alkaline mucoid fluid, less tenacious than that of the cervix. When the inflammation is more severe it is muco-purulent, and may become rusty from slight admixture of blood, or more decidedly sanguineous. After long-continued endometritis, especially when the parenchyma is considerably affected, the mucous membrane becomes atrophied and thin, and its cells are infiltrated with an abnormal fibrillated tissue. The menstrual decidua is then imperfectly formed, and menstruation is generally scanty.

The parenchyma is most involved in those cases which originate in the more acute forms of inflammation, septic or otherwise, of the whole substance of the uterus, especially those which originate after labour or abortion. Even when the disease does not immediately follow upon parturition, but originates in catarrhal inflammation of the mucous membrane at a later period, after involution is complete, it tends to involve the parenchyma more in the parous than in the nulliparous uterus on account of the looser texture of the uterine walls. In the early stage of chronic metritis the tissue is soft, red, swollen, and succulent, from infiltration of serum, and therefore prone to flexion. The uterus becomes enlarged, even when not already large from the effect of subinvolution, but the enlargement is more in the thickness of its walls than in its length, especially in the nulliparous uterus. At a later stage the tissue is indurated by growth of connective tissue, and the state of hyperplasia, which has already been described (p. 145), is reached. Some degree of degeneration of tissue may arise from passive hyperæmia; but, in the absence of any cause of venous obstruction, the degree of fibroid induration may be taken as a measure of the degree to which inflammation has extended through the parenchyma. In the majority of cases, especially in the parous uterus, the cervix as well as the body is involved in chronic metritis.

Causation.—Among predisposing causes of chronic corporeal, as well as of cervical, endometritis, are general debility, mental depression, chlorosis, and a strumous, rheumatic, or gouty diathesis. A part of some importance is also played by syphilis, which specially affects the developing uterine mucous membrane in pregnancy, and so leads to abortion. After abortion the lining membrane of the uterus is apt to be left diseased. Apart from abortion, endometritis is common in syphilitic subjects, though it presents no distinctive signs. When the constitutional taint is active the leucorrhœal discharge may convey the contagion. The chief exciting causes are the results of acute endometritis and metritis, the retention of portions of placenta, clots, or decidua, extension of inflammation from the vagina and cervix, cold, especially at menstrual periods, sexual excess, obstruction to the escape of secretions from stenosis or flexion of the cervical canal, and direct mechanical irritations, such as intra-uterine pessaries, the use of the sound, or attempts to induce abortion.

Apart from the results of pregnancy, one of the most fertile causes of chronic endometritis is dysmenorrhœa, especially that due to obstruction of the cervix from stenosis or flexion; and, in this case, the mucous membrane of the body of the uterus may be inflamed, without that of the cervix participating. The menstrual blood is normally prevented from clotting by admixture with the acid vaginal mucus, but if retained in any quantity, or for any long time, within the uterus, clots are formed, and these have an irritating effect. A similar influence is exercised by any shreds of mucous membrane which may be detached, if the menstrual decidua does not become completely disintegrated. The retained fluid, whether blood or mucus, also undergoes, if not any noticeable decomposition, yet sufficient change to give it an irritant effect. When the tissue near the internal os is compressed in consequence of flexion, the resulting passive hyperæmia also predisposes to inflammation.

Inflammation of the cervix, extending from the vagina, is more likely to affect also the body of the uterus, the more acute in its character. This is especially likely, therefore, to take place in the case of gonorrhœa, although a non-specific inflammation may occur, so acute as to be indistinguishable from it. The foundation of chronic endometritis and metritis is often laid at the commencement of married life, and though this may result simply from marital impru-dence, yet gonorrhœal contagion is not an unfrequent cause. Dr. Noeggerath, of New York, has main-tained that gonorrhœa, in both sexes, persists for life in certain sections of the organs of generation, not-withstanding its apparent cure ; that this "latent gonorrhœa" may affect a healthy person either with an acute attack, or with a similar chronic inflammation, which, in women, is apt to lead not only to chronic endometritis but to ovaritis, pelvic peritonitis, or even puerperal septicæmia. He also regards this infection from latent gonorrhœa as the commonest cause of sterility. It can scarcely be doubted that this view as to the incurability of gonorrhœa is greatly exaggerated. But it appears to be the fact that a latent gonorrhœa or gleet in the husband very frequently infects the newly-married wife with an inflammation which is not acute enough for its nature to be obvious, but is yet the starting point of chronic endometritis and conse-quent sterility.

Results and Symptoms.—The most constant symp-tom of corporeal endometritis is leucorrhœa. The discharge is of a less clear and tenacious character than that secreted by the cervix, and is more frequently muco-purulent. Very often it has an irritating effect upon the vagina and vulva. The discharge may collect for a time in the uterus, and be expelled occasionally, leading the patient to imagine that an internal abscess has burst. When endometritis is not limited to the cervix, but affects the body of the uterus, some menstrual disturbance is almost invariable. In the early stages the flow is usually profuse, painful, and often irregular, and is followed for some days by an excess of leucorrhœal discharge, which is often rusty, from slight admixture of blood. Hæmorrhagic discharges in the intervals are not uncommon. Of fungoid endometritis the prominent symptom is profuse and intractable menorrhagia or metrorrhagia. In the later stages of endometritis, when, with general induration of the whole uterus, there is degeneration of the mucous membrane, menstruation becomes scanty, and generally painful. Sterility is a usual result at all stages, from the destructive effect of the altered secretion upon the spermatozoa, or from the mucous membrane having ceased to form a suitable nidus for the ovum.

The more prominent general symptoms of endometritis depend upon the whole parenchyma of the uterus being affected by reflex hyperæmia, or more or less involved by extension of inflammation to the deeper tissues. They vary greatly in intensity, according to the degree of such extension and the susceptibility of the patient to reflex nervous disturbance. Dragging pain is felt in the hypogastrium and groins, often extending down the thighs, and also in the back, generally at a somewhat higher level than in affections of the cervix—that is to say, over the upper part of the sacrum or last lumbar vertebra. The pain is frequently most acute in one groin, generally the left, a circumstance sometimes, but by no means always, explained by the participation of the ovary on that side in hyperæmia or inflammation. There is usually tenderness on pressure over the situation of the uterus. Pain is greatly increased by locomotion or coitus, and the latter often leads to an aggravation of distress of considerable duration. More or less disturbance of the functions of bladder and rectum is generally produced. There is pain in micturition and defecation from the pressure upon the tender uterus produced in any bearing down effort, and frequently also irritability of bladder. Sometimes there is diarrhœa, from a similarly irritable condition of rectum, but more frequently, constipation, arising, in great measure, from the reluctance of the patient to make any effort.

Numerous other general symptoms are more or less directly connected with uterine disease, and gynæcologists have been accused of claiming too much importance for the local condition as a cause of such disturbances. It would obviously be erroneous hastily to assume that any nervous affection or digestive disturbance in a woman was dependent upon the uterus or ovaries, in the absence of local symptoms pointing to such a cause. On the other hand, a reflex symptom may be the subject of much more complaint than symptoms directly connected with the primary cause, especially when the latter are of a nature which women often do not mention until cross-examined on the subject. Reflex symptoms are not necessarily proportional to the intensity either of the local disease or of the local symptoms, but depend much more upon the susceptibility of the nervous centres. This is proved by such familiar instances as the vomiting of pregnancy, which may occur when there is no local distress and pregnancy is quite unsuspected, or as an attack of asthma produced by a late supper, or a headache due to irritating material in the alimentary canal, and relieved at once by a purgative. In an individual case it may be a matter of extreme difficulty to decide whether a given symptom is reflex; and the only hope of solution will sometimes lie in the therapeutic test of treating the uterine or ovarian disorder, if evidence is found of the existence of one.

No one can doubt that chronic endometritis or metritis is capable of producing such reflex symptoms as pain in the dorsal, lumbar, or sacral regions, or extending down the thighs and legs; pain in the course of the ilio-hypogastric or ilio-inguinal nerves, not necessarily dependent upon any inflammatory change in the ovary; also pain and irritability in the bladder or rectum. To these may be added pruritus of the vagina or vulva, vaginismus, and pain and tenderness in the coccyx, apart from any inflammatory lesion of that structure, all which symptoms may arise through reflex hyperæsthesia. These pains are liable to exacerbations at or near menstrual periods, or from the effects of coitus, as well as from general causes, such as exertion or the effect of cold. As might be expected from the close sympathetic connection between the breasts and the uterus, the breasts may be affected by uterine disturbance. Neuralgic pain in the nipples or glands is not uncommon about the time of the maximum development and congestion of the uterine mucous membrane, shortly before the onset of the menstrual flow, especially in conjunction with dysmenorrhœa. In cases of chronic endometritis or metritis, the glands sometimes become enlarged, containing a mucoid secretion, and their areolæ darkened, so that a patient may often imagine herself to be pregnant, especially when tympanitic distension of the abdomen is present. Histological researches have shown that the type of scirrhous cancer of the breast corresponds to a condition which is normal at the very earliest stage of the natural evolution of the breast in pregnancy, at which stage it is the normal destiny of the epithelial cells of the acini to pass into the connective tissue stroma.* Hence it appears probable that a similar abnormal stimulus to gland activity, very protracted, but of a still lower intensity than that which evokes a mucoid secretion, may be a common cause of cancer of the breast, not having a traumatic origin, and clinical experience appears to give some support to this view.

The close nervous connection of the stomach and other digestive organs with the uterus is equally undeniable, and is demonstrated by the vomiting and other digestive disturbances of pregnancy. Also in chronic congestion or inflammation of the uterus, flatulent distension of the abdomen, eructation, nausea, and vomiting are frequent results, especially in hysterical subjects, and more or less dyspepsia is almost

* See Dr. C. Creighton's work, "Contributions to the Physiology and Pathology of the Breast."

invariably produced. The effects of ovarian irritation are very similar, and both are explained, on anatomical grounds, by the connection of the sympathetic nerve supply of the upper part of the uterus with the ovarian plexus, and through this with the upper aortic and renal, and so with the solar plexus. In accordance with physiological doctrine, irritation of the sympathetic system inhibits the secretion of gastric juice and other digestive fluids; and hence arises failure of digestion and fermentation of the food, by which catarrhal gastritis and enteritis may be subsequently set up. The failure of nutrition thus brought about is a prominent symptom of uterine or ovarian disorder, which, in this way, may be, in predisposed subjects, the starting point of phthisis.

Other general results of uterine disease are of a more indirect kind. The local malady may affect the general health, not only by its adverse influence on the digestion, but by interfering with locomotion. Owing to the pain produced by movement, the patient is deterred from obtaining sufficient air and exercise to fulfil the hygienic requirements for healthy life. It is often doubtful whether distant effects are produced in this way, or directly by reflex nervous influence, and probably, in many cases, both modes of causation are more or less combined. Among the symptoms common in patients suffering from chronic uterine inflammation, are neuralgic pains in various localities, as along the edges of the false ribs, but more especially at the top of the head or under the left breast, the last form of pain being often accompanied with palpitation. This infra-mammary pain would seem to have a direct nervous association with the tendency to pain in the left groin, rather than the right, and with the greater frequency of congestion and swelling in the left ovary. Other reflex neuroses are occasionally produced, such as asthma, or catarrh of the fauces or air-passages; but in individual cases of this kind the link of causation is very difficult to establish, unless it is proved either by

P

a relation to menstruation, or by the effect of local treatment directed to the uterus, or of the intervention of pregnancy.

Various changes of nutrition, somewhat resembling those of pregnancy, may result from chronic uterine or ovarian disorder, such as dark rings under the eyes or general darkening of the skin by pigmentation. Eczema is not uncommon, and acne still more frequent; while the time of outbreak of these eruptions often has a relation to menstrual periods. With the general break-down of health to which uterine and ovarian disorder often leads, are frequently seen loss of hair and failure of sight, especially a form of amaurosis depending upon chronic optic neuritis. In patients predisposed to hysteria the multiform manifestations of this disorder are an early result, and are generally aggravated at menstrual periods. In these cases, however, hyperæmia and tenderness, if not inflammation, of one or both ovaries are generally found to exist, in addition to the uterine affection. A vaginal examination, or pressure upon the uterus or ovary, will often excite a nervous paroxysm, noisy eructations, or a feeling of nausea or faintness. In patients having a different predisposition, the nervous disturbance may take the form of epilepsy, hystero-epilepsy, or positive mental aberration, generally of the melancholic type.

Some evidence has been adduced that in cases of uterine disturbance attacks of articular rheumatism may be repeatedly associated with menstruation. A form of rheumatism has also been described, analogous to gonorrhœal rheumatism, but distinct from it, in that it affects chiefly the smaller joints, and arising in connection with leucorrhœa or pregnancy.* In favour of the possibility of such a connection may be quoted the prevalent theory which regards ordinary gonorrhœal rheumatism in the male sex as produced rather by reflex

* See Papers by Dr. Ord, "Brit. Med. Journ." Jan. 31 1880; by Mr. Davies-Colley. "Obstet. Journ." June, 1878.

nervous influence than by an absorption of poisonous material approximating toward the pyæmic character. Chronic rheumatoid arthritis is much commoner in the female than in the male sex, and its subjects have often suffered from uterine disturbance, but the connection here may be merely the deteriorating effect of uterine disease upon the health.

Partial or complete paraplegia (such as occasionally arises after delivery, without any difficulty in parturition likely to cause direct lesion of nerve trunks) is another symptom which has been considered a reflex effect of uterine disease, but whose causation is difficult to trace. In most such cases the paralysis is functional, and more or less allied to hysteria. But to call a symptom hysterical is not completely to explain it. Indeed, the hysterical temperament chiefly implies that reflex, as well as emotional, sensibility is exalted, while the control of volition is diminished. Hence, in these cases of paraplegia, as in that of hysterical vomiting, it may be desirable to attempt to remove any uterine cause of irritation, although, especially in the unmarried, over-much local interference is undesirable, and moral treatment is of chief importance. Whether a reflex paraplegia, dependent upon actual chronic myelitis, can result from uterine inflammation, as the so-called reflex paralysis has been supposed to result from positive and severe inflammation (rather than mere irritation) of the bladder, prostate, or kidneys, there is, as yet, no evidence to decide.

A recent endometritis, in which the parenchyma is not much involved, will generally yield to treatment; but chronic metritis, when it has reached the stage of induration, is one of the most obstinate of diseases. Untreated, it is commonly limited only by senile atrophy; and even under the most judicious treatment, only a relative degree of cure is usually attained, and relapses frequently occur.

Diagnosis.—Corporeal is distinguished from cervical endometritis by the nature of the discharge, which has not the tenacious glairy quality distinctive of the cervical secretion, but is either thin and mucoid, muco-purulent, or, what is more characteristic, has a slight rusty tint. There are also greater tenderness and enlargement of the body of the uterus, as detected by bimanual examination, and disturbance of menstruation is a more constant symptom. The sound shows lengthening of the uterus, not accounted for by cervical hyperplasia. On reaching the fundus it usually causes considerable pain, and frequently nervous disturbance. Slight bleeding often follows upon its withdrawal. The cervical canal is generally more dilated than normal, but in cases of endometritis of the nulliparous uterus without affection of the cervix, it may be the opposite. The cervix may be normal in nulliparous women, but in other cases it is usually involved in the hyperplasia. In cases of doubt, whether inflammation affects the body of the uterus or the cervix only, the doubt may be resolved by the persistence of the symptoms after treatment of the cervical disease. The diagnosis of *fungoid endometritis* may be established in two ways. (1.) The cervix may be dilated by a tent, and the finger passed up to the fundus in the manner described with reference to evacuation of the uterus after abortion (p. 149). The villous surface will then be felt. (2.) The blunt wire curette of Dr. Thomas, an instrument shaped like Sims' curette (Fig. 61, p. 196), but having a loop a quarter of an inch in diameter, made simply of copper wire $\frac{1}{12}$-in. to $\frac{1}{16}$-in. thick, may be used as well for diagnosis as for treatment. This may be introduced without dilatation of the cervix if the canal be moderately patulous, and the cavity of the uterus gently scraped. The fungoid prominences will be brought away, and their character may be recognized by the microscope. If a sharp curette were used a fallacy might arise from the scraping away of the mucous membrane itself.

Treatment.—All exciting causes, such as inflammation of the cervix, or serious displacement of the uterus, should be removed if possible. Thus, when erosion or cervical endometritis is present, local treatment should be directed to the cervix first, since it is more readily accessible to such medication. When displacements are present the most important point to decide is whether to resort at once to a pessary, or first to treat directly the inflammation. As a general rule, in retroversion or retroflexion of any important degree, as well as in prolapse, a vaginal pessary may be employed with advantage either immediately or after a short course of rest and local depletion, and the successful use of other means will then be facilitated. In anterior displacements, however, in which the deviation from the normal is not generally so extreme, while vaginal pessaries act at less advantage, it is preferable in the first place to try the effect of treatment directed to the inflammatory condition. Then, if this proves insufficient, careful trial may, in some cases, be made of a pessary. Even when no displacement can be detected, relief is sometimes afforded by the use of a Hodge's or elastic ring pessary, which limits the mobility of the tender uterus, keeps it in a position of slight anteversion, and resists any tendency to partial prolapse. The hypogastric belt (*see* p. 95) may also give relief.

Active and passive hyperæmia are to be treated by the means enumerated under those headings (pp. 161, 167), especially frequent, but not too prolonged, rest in a completely horizontal position, the use of hot water injections, the administration of saline laxatives, bromide and iodide of potassium, ergot, strychnia, or digitalis, and, when much tenderness is present, local depletion either by blood-letting or the application of glycerine. Coitus must be strictly limited, though it is not always desirable to enforce an absolute prohibition, especially when ovarian hyperæmia exists. Internal remedies have little influence upon the inflammation of the parenchyma; but when the enlargement is passing into the stage of induration, the liquor hydrargyri perchloridi may be given in doses of from thirty to eighty minims three times a day.* This may often be usefully combined with small doses of quinine, or, in the absence of general or local hyperæmia, of the tincture of perchloride of iron. Iodide of potassium may also be tried as an absorbent, or if menstruation is scanty, tincture of iodine, in doses of from five to ten minims. The latter often acts as an emmenagogue in chronic metritis.

Every possible hygienic means should be taken to promote the general health, especially by fresh air, cold or sea bathing, sufficient mental occupation, and change of scene. A stay at a pleasant watering place or hydropathic establishment, or a sea voyage, is thus of great service. These advantages may be combined with the effects of bromine and iodine in mineral waters, both in the form of baths and internally, at certain celebrated watering places, especially Kreuznach, the virtue of whose water depends chiefly upon bromide of magnesium. The water of Woodhall Spa, in Lincolnshire, has a similar effect, but contains a greater proportion of iodine. The imported Kreuznach salt may also be used for hip baths, or in concentrated solution, for abdominal compresses. Diet should be nourishing but very simple, in view of the so constantly attending dyspepsia. Alcohol should be much restricted, since by relaxing the arteries it promotes active hyperæmia, and, moreover, chronic uterine disease is one of the commonest causes of intemperance in women, who are led to take spirits for the temporary relief of pain, or of the feeling of sinking or depression from which they so often suffer. Taken with meals, however, a moderate allowance of alcohol may be useful as a stimulus to digestion, and a good claret or Burgundy is generally the best form to recommend. If any

---

* For prolonged use, the mercury is most readily taken after meals in the following mixture :—Liq. Hydrarg. Perchlor. ℳlx.—lxxx. ; Acid. Hydrochlor. dil. ℳx. ; Syrupi. ℨj. ; Aq. ad ℥j.—ter quotidie.

tendency to excess be suspected, it is better to enjoin total abstinence.

In so protracted a disorder as chronic metritis, it is desirable to avoid as far as possible the administration of opium or morphia, lest the patient become dependent upon the drug. If required during exacerbations or at menstrual periods, a morphia suppository may be given *per rectum.* For soothing pain, warm hip-baths, or what are still more effective, whole baths, are ·a valuable resource. Vaginal douches at a temperature of not less than 110° F. have the further effect of stimulating absorption and diminishing the size of the uterus by the contraction of the uterine muscular fibres, and of the arterial walls, which they produce. It is well, however, to commence the injections at a more moderate temperature, and gradually to increase the heat. When sedatives are required, others than opium may be tried in the first place. Bromide of potassium, belladonna, hyoscyamus, cannabis indica, or camphor may be given internally, and chloral if required to procure sleep. Of these belladonna acts most upon the sympathetic system and is specially valuable in vesical tenesmus, while it is often a useful addition to opium : hyoscy·amus has a greater soporific effect, and both of these are useful for their laxative tendency. Cannabis indica has a special influence in neuralgia and headache, besides being a general sedative. Camphor is an anaphrodisiac as well as sedative if given in doses of as much as from five to ten grains. Bromide of camphor has been recently introduced, and may be given in Dr. Clin's capsules. Iodoform may be given with advantage in a suppository containing five grains, introduced *per rectum.* Belladonna suppositories are also more effectual *per rectum* than *per vaginam.* Sedatives are also useful when introduced *per vaginam,* but must then be employed in larger doses. Absorbents, such as iodine, iodide of potassium, or iodoform, may be used in the same way (*see* p. 192).

Counter-irritants are not only valuable for relief of

pain, but exercise an alterative effect upon chronic inflammation. They also appear to relieve reflex neuroses, such as vomiting, by a kind of inhibitory effect upon the nervous system. Flying blisters may be produced on the hypogastrium or groins by blistering fluid, and repeated at intervals. If, however, there is any tendency to vesical tenesmus it is well to avoid the use of cantharides, and employ some other counter-irritant, as the linimentum or strong tincture of iodine (*see* p. 195), repeatedly applied. Reflex pains may be relieved by applications to the nerve-terminations at the seat of pain, such as mustard poultices, turpentine fomentations, or a small quantity of the linimentum sinapis co. sprinkled on spongio-piline and kept applied six or eight hours. As a counter-irritant the linimentum crotonis may be applied with a sponge, or, as a sedative, equal parts of linimentum aconiti and linimentum belladonnæ may be used. In the same way comfort is afforded by plasters applied to the back, for which purpose emplastrum calefaciens or emplastrum belladonnæ may be used. Strong caustic applications to the cervix, which have been considered under the head of hyperplasia (p. 152), may prove more effectual counter-irritants than those applied externally. A vesicating effect on the cervix, followed by a flow of serum, may be produced by vesicating collodion, painted with a brush over the whole cervix, and followed by the application of a tampon soaked in glycerine. Care must be taken to protect the vagina and vulva in making the application.

Tonic treatment, such as that described under the head of inflammation of the cervix (p. 185), is generally useful in the course of chronic endometritis and metritis. Iron should not be given when there is a furred tongue, or any sign of portal congestion, till this condition has been relieved by occasional mercurial purgatives or other means. It is likely, also, to prove injurious while there is any marked hyperæmia, or tenderness of uterus. In the later stage,

however, when there is much general debility, and a
flabby tongue impressed by the teeth, it is often of
service, especially if combined with a laxative (see
p. 186). In the stage of hyperplasia with induration,
when menstruation is scanty, iron may be added to
bromide of potassium, and, if necessary, aloes also.
Bromide of potassium alone in such cases is apt to
diminish the menstrual flow and prolong the intervals.

Intra-uterine Medication.—When endometritis, not
complicated by any considerable degree of metritis,
fails to yield to general remedies and local treatment
to the cervix, the most efficacious method is to apply
remedies directly to the cavity of the uterus. If there
is considerable inflammation of the parenchyma, but
yet endometritis is the starting point of, or forms a
prominent feature in, the inflammation, it may similarly become desirable to treat the mucous membrane
directly. In this case, however, more caution in the use
of local remedies is necessary, and hyperæmia should in
the first place be relieved as far as possible. The cases
most urgently calling for intra-uterine medication are
those of villous endometritis with severe hæmorrhage.

Remedies are generally most conveniently applied in
a liquid form by means of Playfair's probe (Fig. 60,
p. 193), or other similar instrument, wrapped very carefully and closely in a thin layer of cotton wool, so that
the cotton is not liable to slip off, or become wrinkled
up. Unless it is desired to extend the application to the
cervix also, the cervix should be protected by an intra-
uterine canula (Fig. 64, p. 218), which also tends to
prevent so much of the fluid being wiped off before it
reaches the cavity of the uterus, while it renders the
use of a tent unnecessary if the cervix is somewhat
patulous. Dr. Atthill recommends a short platinum
canula, which is introduced by a guide, and held in
position by long forceps after withdrawal of the guide.
It will be found more convenient, however, to have
the canula fitted with a long handle as well as with a
guide to facilitate its introduction. If made of vul-

canite, it answers every purpose, while it is much less costly than if platinum be used. Sims' speculum and the semi-prone position should be employed for the operation. The most generally useful fluids for the application are the liquor iodi, linimentum iodi, or saturated tincture of iodine (*see* p. 195), a solution of nitrate of silver of from twenty to forty grains to the ounce, the liquor ferri subsulphatis or liquor ferri perchloridi, strong carbolic acid, or carbolic acid with an equal quantity of glycerine, iodized phenol (*see* p. 195), and strong nitric acid. Of these, carbolic acid and the strong tincture of iodine are the most generally useful in ordinary cases of endometritis, while the solution of iron may be used when hæmor-

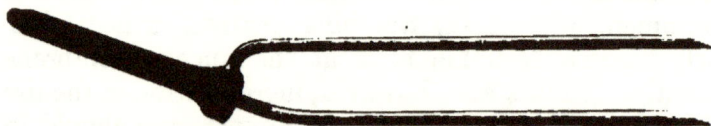

Fig. 64.—Canula for Intra-uterine Medication.

rhage is a prominent symptom. Strong nitric acid is the most efficacious when a profound degeneration of the mucous membrane is indicated by profuse hæmorrhage, or by the failure of milder measures to cure. The very free application of this remedy has been especially lauded, but though ordinarily it is well tolerated, if used with care, it may sometimes in British patients, if not in Irish, excite considerable inflammation. A probe of aluminium, vulcanite, or platinum, wrapped in cotton wool, as already described, and charged with the acid, is passed once up to the fundus uteri.

A convenient mode of making a mild application of the solid nitrate of silver to the interior of the uterus is to coat with it the point of a uterine probe. The bulbous end of the probe is slightly roughened, and then, after being warmed, dipped repeatedly in the nitrate of silver, fused in a platinum or porcelain capsule, until it is sufficiently coated. It is then passed

up to the fundus. A small piece of the solid nitrate
of silver, or one of the zinc points, is sometimes passed
into the uterus by Simpson's porte caustique—a tube
provided with a piston—and left there to dissolve.
The medicated crayons described at page 197 may be
used in the same way. The nitrate of silver thus used
is apt to cause violent uterine tenesmus, and even
inflammation ; and all solids excite irritation as foreign
bodies, while from their becoming coated with mucus,
their action is unequal. Drugs may also be inserted
in the form of ointment by a similar uterine applicator,
provided with a piston. The ointment of iodide of
mercury, or one containing iodoform with vaseline
(ʒj. ad ℥j.) may be thus used. Dr. Barnes recommends
such an applicator also for the introduction of strong
nitric acid, a few drops of the acid being placed upon
a sponge, and inserted in the tube of the instrument.
Iodoform in powder may also be introduced by means
of an applicator having both lateral and terminal
openings.

The last mode of intra-uterine medication to be
mentioned is that of the injection of fluids, and this
is the most dangerous of all. The danger lies chiefly
in the risk of the fluid making its way along the
Fallopian tubes, either from the force of the injection,
or what is more probable, from spasmodic contraction
of the uterus. This has been demonstrated by autopsy,
in cases where sudden death has followed the injection
of perchloride of iron, even though the Fallopian tubes
were not obviously more patent than normal. The
risk is not entirely obviated by securing full dilatation of the cervix—a precaution which should always
be taken—for the cervix generally contracts under
the stimulus of the astringent; nor by the use of a
double-action catheter, for the return canal may become
blocked by a clot. Intra-uterine injection cannot, however, be entirely dispensed with, and it is chiefly
called for in cases of alarming metrorrhagia, when a
sufficient bulk of fluid to arrest hæmorrhage cannot

be applied by means of a swab, or when, from enlargement and irregularity of the uterine cavity the swab cannot come into contact with the whole of it. The safest plan is to use rather a small tube, not larger than No. 12 catheter, so that the cervix, after full dilatation, may not so readily grasp it, and to inject only by hydrostatic pressure, applied by means of an elastic tube and funnel, elevated only very slightly. If milder fluids are used, intra-uterine injection is less dangerous, and the plan of treating endometritis by injections of a two per cent. solution of carbolic acid, after dilatation of the cervix, is highly praised by Schultze.

In fungoid endometritis the villous prominences may be destroyed either by the pressure of a sponge-tent introduced up to the fundus, by scraping the surface of the uterus by the blunt-wire curette (see p. 212), or by the application of a strong caustic, such as nitric acid. Of these, nitric acid has the greatest efficacy in modifying the nutrition of the mucous membrane. If the os is not fully patulous, the use of a tent may be chosen as the first measure, and it also forms a temporary plug. To obtain its influence upon the mucous membrane it should be long enough to reach the fundus, and should be rubbed down with sand paper till it has a uniform, slightly conical shape (see Fig. 11, p. 27), instead of bulging in the centre, like the tents commonly sold. The use of the blunt wire curette has the advantage of causing least disturbance, and may be tried first, if the cervix is open enough to admit it, and the symptoms not very urgent.

MEMBRANOUS DYSMENORRHŒA.

Causation and Pathological Anatomy.—In connection with endometritis may be considered the disorder called membranous dysmenorrhœa, which by some has been termed exfoliative endometritis, although it is still a matter of dispute whether its essential nature is inflammatory or not. It consists of the expulsion during the menstrual period, generally on the second or third day, of membrane either in shreds or forming a more or less complete cast of the uterus, which membrane, when examined microscopically, shows the structure of the uterine mucous membrane. Many other apparent membranes may be passed at a menstrual period, such as fibrinous clots, exfoliations from the vagina or cervix, or mucus coagulated by astringents; but such cases have no connection with the disease under consideration. It is also to be distinguished from cases in which repeated abortion occurs at intervals of little more than a month, a condition which may be due to an excess of menstrual nisus, or, perhaps, in some cases, to an imperfect fertility on the part of the husband. Some have supposed that all cases of so-called membranous dysmenorrhœa are to be thus explained; but it has been clearly shown that the complaint may occur in virgins.

Membranous dysmenorrhœa consists essentially in the menstrual decidua being thrown off in pieces of greater or less size, instead of being, as it should be, disintegrated and coming away in minute fragments. This result may depend upon excessive growth, too deep exfoliation, or an unduly fibrous character in the decidua, and the true explanation is not yet fully ascertained. The disorder may exist in any degree. If the fragments of supposed clot passed in cases of dysmenorrhœa are examined microscopically, it is not uncommon to find small shreds having the cellular structure of the uterine mucous membrane, and showing a few tubular channels, which are the gland apertures, generally divested of their epithelial lining. From this every grade may occur up to that in which a triangular cast of the whole uterus, showing orifices corresponding to the Fallopian tubes, is passed, although the slighter degrees generally escape the attention of the patient herself. In the complete cast the whole structure of the mucous membrane is to be seen, including enlarged orifices of glands, and an undue amount of fibrillar tissue is to be found among the cells. In some cases a cast is expelled every month, in others only occasionally, while smaller shreds of membrane come away at the intervening periods. Sometimes there is a history of similar shreds having been passed since the first commencement of menstruation, which renders it probable that the affection may depend upon some peculiarity in the structure of the mucous membrane in certain individuals.

Membranous dysmenorrhœa is usually associated with active hyperæmia, and other signs of chronic endometritis and metritis. There is often also true hypertrophy of the muscular walls owing to difficulty in the expulsion of the membranes. Dr. John Williams has maintained* that the pathology of the complaint is the presence of an undue amount of fibrous tissue in the uterine walls, and that the inflammation which usually accompanies it is secondary to the irritation produced by the shreds of membrane. Excess of fibrous tissue, however, is very common as a sequel of subinvolution with chronic metritis, without leading to any such result as membranous dysmenorrhœa. It seems more probable that in membranous dysmenorrhœa there is generally a congenital excess in the fibrillation of the uterine mucous membrane, and that this excess may be further increased from the effect of chronic endometritis.

Results and Symptoms.—The symptoms of membranous dysmenorrhœa, apart from those due to the hyperæmia or inflammation generally associated with it, consist simply of the pain and tenesmus evoked by the expulsion of the membrane. When the affection exists in any marked degree it generally gives rise to sterility. It is one of extreme obstinacy, and frequently persists for many years.

Diagnosis.—The more perfect casts are easily recognized by the naked eye, and the orifices of the uterine

* " Obst. Trans." vol. xix.

glands may be seen in them. Generally, however, the diagnosis must be confirmed by microscopic examination. If the membrane is thin, it may simply be spread out upon the slide. If it is too thick to show its structure in this way, sections of it, parallel to the surface, should be made by the freezing microtome. If the cast has the structure of uterine mucous membrane, it only remains to distinguish the case from one of repeated abortion. The latter is generally characterized by irregularity in the intervals of apparent menstruation, but the most crucial test is to try the effect of temporary separation between husband and wife.

Treatment.—The first indication is to secure a freely open and straight cervical canal of much greater dimensions than are needful in the normal uterus. By this means distress is much alleviated, and the tendency of the membrane to keep up inflammation by mechanical irritation is diminished. At the same time the hyperæmia or metritis should be treated by bromide of potassium, ergot, absorbents, purgatives, local depletion, or other suitable means. To arrest the tendency to formation of membranes, only such measures as tend to effect a profound alteration in the nutrition of the uterine mucous membrane are at all hopeful. Applications of iodine, carbolic acid, nitrate of silver, or nitric acid may be tried for this purpose, but even nitric acid has failed to cure. On the hypothesis that the disease is due to imperfect evolution of the uterus, the galvanic current, applied by introducing the rheophore into the uterine cavity, and the galvanic stem pessary have been tried, but without very encouraging results. The administration of arsenic has done good in some cases ; and this drug appears to have some selective and alterative action on the uterine mucous membrane. Pregnancy is likely to modify the mucous membrane more than any other influence. In some slight forms of the affection I have found it to be apparently cured after pregnancy and delivery, to re-appear after an interval of several years.

CHAPTER VII.

NEW GROWTHS OF THE UTERUS.

MUCOUS AND GLANDULAR POLYPI OF THE UTERUS.

Causation.—Just as, in cervical endometritis, a single mucous gland, when its orifice has become closed, readily elevates the loose mucous membrane, and becomes a minute projection or Nabothian gland, so the same process may be exaggerated. A fold of mucous membrane containing numerous glands may take part in it, while hyperplasia of the stroma of the mucous membrane takes place, and in this way a *mucous polypus* is formed. The same process may also occur in the glands of the body of the uterus, or those near the edge of the os, or on the outside of the cervix.

Pathological Anatomy.—Mucous polypi generally vary from the size of a pin's head up to that of a hazel-nut, rarely exceeding the latter dimension. They are made up of one or several mucous follicles, with a stroma of soft and delicate connective tissue containing many nuclei, the stroma predominating over the glandular portion. They generally grow from the cervical canal, and the pedicle then tends to become elongated, until the polypus appears outside the cervix. Sometimes they grow within the cavity of the uterus, and then generally do not reach the cervix. Polypi of this variety are often found unexpectedly at autopsies. Mucous polypi are covered by a thin and very vascular mucous membrane, and are generally bright or deep red in colour. Whether they grow from the cervix or body of the uterus they are often multiple, and, after the removal of one, others are apt to grow in the same individual. When the proliferation of the gland-follicles predominates over that of the cellular tissue, " *glandular polypi* " are formed. These are of two kinds :—(1) The " *cystic polypus*," formed by dilatation of a single follicle. These generally grow from the cervical canal, and are sessile, or nearly so, not larger than a cherry, and very fragile, being filled with mucoid fluid. (2) The " *channelled polypus*." These are generally attached at the lower part of the cervical canal, or outer part of the cervix. They contain large irregular cavities, communicating with each other, and opening on the surface, often by rather large mouths. These are lined by cylindrical epithelium, and contain mucoid fluid. The surface may be covered by cylindrical epithelium, or by squamous epithelium, if the growth has sprung from the outer part of the cervix, or again partly by squamous and partly by cylindrical epithelium. These polypi grow to a larger size than mucous polypi, and may attain a diameter of two inches or more. They correspond to the "follicular hypertrophy of the cervix" of Schroeder. If they reach or pass through the vulva they are apt to become ulcerated on the surface. In some cases a proliferating papillomatous growth springs up in the cavities, the tumour thus showing an approximation toward the malignant type.

Results and Symptoms.—Small mucous polypi may exist without any obvious symptoms, but more generally they produce leucorrhœa and sometimes menorrhagia or metrorrhagia, with occasionally dysmenorrhœa and other symptoms dependent upon hyperæmia. The hæmorrhage is sometimes altogether out of proportion to the size of the polypus. It is due not so much to bleeding from the surface of the polypus as to hyperæmia set up by the irritation of its presence. In the same way the polypus tends to keep up and increase that hyperplasia of the cervix with which it is often associated at its commencement. The symptoms are generally greater while the polypus is within the cervix than after it is extruded outside the os. Comparing ordinary mucous with glandular polypi, hæmorrhage is the more prominent symptom of the former, leucorrhœa of the latter.

Diagnosis.—The smaller mucous polypi, after they have passed outside the os, may sometimes escape detection by the finger from their extreme softness, but are easily recognized by the speculum. Polypi high up in the cervix, or in the cavity of the uterus, are generally only discovered when the cervical canal has been dilated for exploration as to the cause of hæmorrhage.

Treatment.—Small and soft mucous polypi may be twisted off with forceps, and the base touched with liquor ferri perchloridi fortior, solid nitrate of silver, or nitric acid. Those of larger size are best removed by the écraseur (Fig. 68, p. 248)—écraseur and wire both being of dimensions suitable to those of the polypus. This is preferable to cutting them off with scissors, since the hæmorrhage on cutting a mucous or glandular polypus is greater in proportion than that from a fibroid polypus. If the polypus is small, the loop is most easily adjusted by the aid of a speculum. After the removal the patient should be kept in bed for a day or two. If intra-uterine mucous polypi are detected as the cause of hæmorrhage, and are not large enough to be ensnared by a small écraseur, they may be destroyed by the pressure of a sponge tent, by the blunt wire or steel curette (Fig. 61, p. 196), or, if necessary, by the application of nitric acid, as in the case of the villous prominences of fungoid endometritis.

FIBROID TUMOURS OF THE UTERUS, OR MYOMATA.

Causation.—Fibroid tumours are among the commonest of uterine diseases. In most cases they date their origin to the period of active sexual life. Nothing certain is known as to their causation, but it depends in a measure upon hereditary predisposition, and they are specially frequent in the negro race. It may be presumed that all causes of hyperæmia of the uterus favour their growth, and that they may take their starting-point from any localized inflammation, the result of parturition, abortion, or any other cause. Dr. Emmet, however, gives statistics which tend to show that, taking into account the relative number of unmarried, sterile, and fertile women, unmarried women between the ages of thirty and forty are twice as liable to fibroids as the sterile or fertile; and also that fertile women are considerably less liable than the sterile. He considers that the perverted nutrition arises from superfluous nerve-force not expended in the natural way in the sexual relations and in pregnancy and parturition, so that even marriage without conception acts in some degree as a safeguard.

Pathological Anatomy.—A fibroid tumour, more accurately called a myoma or fibro-myoma, is composed of the constituents of the normal uterine tissue, involuntary muscular fibres and connective tissue, in varying proportions, but, for the most part, in the main of muscular fibres. It is analogous to the tumours which frequently enlarge the prostate in the male sex. In most cases the tumour consists of one or more rounded masses, separated from the uterine walls around it by a capsule of connective tissue; but occasionally the tissue of the tumour is completely continuous with that of the uterus, and this is especially the case with the softer and more rapidly-growing varieties. The muscular fibres are generally larger than in the unimpregnated uterus, but smaller than in

the pregnant organ, and they are larger in the softer
tumours than in the encapsuled variety.   The encapsuled tumours are tough on section, their substance
but slightly vascular, the cut surface white and glisten-
ing.   The size of each mass usually varies from dimensions only discernible by the microscope up to about
the size of a fœtal head.   Vessels usually enter the
fibroid only at the one point at which it is continuous
with the uterine wall.   The arteries are generally
comparatively small, although occasionally they may
be of large size.   These encapsuled tumours are more
frequently multiple than single, and twenty or more
may be present in one uterus.   Sometimes several
simple fibroid masses are united together within a
single capsule, and thus form a conglomerate fibroid
tumour, with a lobular irregular surface.   The individual masses then generally lose their spherical
shape, and become distorted by pressure.   In the
other and less common variety of fibroid tumour,
which is not encapsuled, the tissue may be loose, and
become œdematous by infiltration with serum, so that
the whole is fluctuating and semi-fluid to the touch.
Large collections of fluid may be formed by separation
of the fibres, and in this way is constituted the
fibro-cystic tumour.   There is no cyst-wall, and the
spaces are generally traversed by trabeculæ of cellular
tissue.

Varieties.—The growth of a fibroid tumour commences in the substance of the uterine wall, but, by
the effect of muscular contraction, it generally tends to
be squeezed out, either towards the outside or the
inside, according to the position of its starting-point.
Hence there are four varieties of fibroids, which are
called *subperitoneal* or *subserous*, when projecting from
the exterior of the uterus, whether pedunculated or
not ; *interstitial*, *intra-mural*, or *intra-parietal*, when in
the substance of the uterine wall ; *submucous*, when
they project internally ; *fibroid polypi*, when they have
become completely extruded on the internal surface,

so as to be attached only by a pedicle. As a rule, the whole uterus becomes hypertrophied, and its cavity enlarged, but in subperitoneal fibroids this is sometimes not the case, and it may even become atrophied. The subperitoneal fibroid tumours, especially, are more frequently multiple, and one or more varieties of fibroids may co-exist. By submucous or interstitial fibroids the uterine cavity is frequently much distorted. Fibroids occasionally grow in the cervix, but much less frequently than in the body of the uterus. Small fibroids growing in the anterior wall of the fundus tend to produce anteflexion, those in the posterior wall retroflexion of the organ. The softer varieties of fibroid tumour, especially when uterine hæmorrhage is a marked symptom, often undergo a manifest gradual increase of size before the onset of menstruation. This reaches its maximum shortly before the flow, or within the first day or two of its continuance, and the tumour then rather rapidly decreases again.

Besides the above-mentioned classification of fibroid tumours according to their position, another important division may be made into three classes, according to the nature of the growth. These are—(1) the encapsuled fibroid, which is almost always hard, and generally multiple; (2) the non-encapsuled fibroid, which is frequently soft and elastic, with fluid in its interstices, and which is generally also single, or else enlarges the whole uterus uniformly, though it may sometimes have irregular outgrowths when increasing rapidly; (3) the fibro-cystic tumour, which is a further development of the last variety.

Not unfrequently are found attached to the cervix small polypoid growths, whose tissue resembles that of the cervix itself, and is continuous with it, not enclosed in any capsule. Sometimes also the whole of one lip of the cervix becomes elongated in a polypoid manner. This condition usually results from the laceration of the cervix produced by parturition. Hyperplasia follows, and affects especially the portion of tissue intervening between two clefts, which afterwards may become constricted at its base, and so take a polypoid form. Polypi of this description, which are often associated with prolapse, are the "hypertrophic polypi" of Dr. Barnes.

Results and Symptoms.—The prominent symptom of those fibroids which enlarge the uterine cavity or project into it—that is to say, of the submucous and of many of the interstitial variety—is menorrhagia. This depends partly upon the increased surface, and partly upon the active hyperæmia of the mucous membrane due to the stimulus of the growth, and the passive hyperæmia which may result from pressure. In some cases the hæmorrhage takes the form of metrorrhagia. The bleeding takes place, for the most part, from the general mucous membrane lining the enlarged uterine cavity, but that which covers the fibroid tumour takes part in it also. By the same hyperæmia is produced leucorrhœa, and often congestive dysmenorrhœa. If the exit from the uterine cavity is obstructed, as it often is, by submucous fibroids or fibroid polypi, obstructive dysmenorrhœa is likely also to result, and frequently endometritis from the irritation produced by retained clots or secretions. Dragging pain is generally produced by the increased weight of the uterus, and the frequently associated hyperæmia or endometritis. Sometimes severe spasmodic pain arises from the contractions of the uterus, excited by the presence of the tumour. Fibroid tumours which lead to flexion may give rise to dysmenorrhœa, endometritis, hyperæmia, and the other possible results of that condition.

The remaining symptoms of fibroids are those due to mechanical pressure, and these are frequently the sole symptoms of subperitoneal growths. Vesical and rectal tenesmus are common, and sometimes retention of urine or extreme constipation is produced by direct pressure. In some cases this happens only when the tumour swells at or just before menstrual periods.

These symptoms are specially urgent when a fibroid growing from the posterior uterine wall is incarcerated in the pelvis. Sterility is a usual result, especially from submucous fibroids, and if pregnancy occur, there is great liability to miscarriage, and to hæmorrhage after delivery. A fibroid tumour in the pelvis may render delivery extremely difficult. Subperitoneal fibroids, however, of moderate size, if they do not obstruct the pelvis, are not inconsistent with natural pregnancy and delivery. When the tumour grows to enormous size, as is more likely to occur in the softer and fibro-cystic varieties, the circulation, respiration, and other vital functions may be interfered with, as in large ovarian tumours. Large fibroid tumours generally remain for a long period free from adhesions, but may eventually excite some degree of peritonitis, and become adherent to surrounding organs.

Natural Terminations.—In most cases the growth of a fibroid tumour is slow, and becomes limited after a certain time, so that even a very large tumour may be borne for many years without very serious result. After the menopause, growth is usually lessened or arrested, and not unfrequently the tumour tends to diminish. There are numerous exceptions, however, to this rule, especially in the case of the large and soft non-encapsuled variety of fibroid, and in the fibro-cystic tumour, which may continue to grow unchecked after the menopause. Extreme diminution, or absolute spontaneous disappearance of fibroids, has occasionally been recorded, especially as the result of involution after delivery. A fibroid polypus may eventually be spontaneously detached and expelled, or may slough away. A pedunculated subperitoneal fibroid may be separated altogether from the uterus by traction, if it has become adherent to other organs, or may even get entirely loose in the peritoneal cavity, where it remains without undergoing decomposition. Gangrene may also affect a submucous fibroid, especially after surgical interference. If the patient survive the risk of septicæmia, the tumour may in this way be got rid of, the disintegrated tissue being discharged in fragments by the vagina. This result depends mainly upon failure of the vascular supply, although inflammation may sometimes be instrumental in starting the process. Fibroids of the encapsuled variety sometimes undergo fatty degeneration or calcification, and the resulting calcified mass has occasionally been separated and expelled. In most cases fibroid tumours do not prove destructive to life, though death may result from sloughing and septicæmia, or from hæmorrhage or exhaustion, or more rarely may be brought about by the magnitude of the tumour. Those fibroid tumours, however, which cause excessive hæmorrhage, cannot be regarded as free from grave danger to life. Not only may death occur from exhaustion without any immediate hæmorrhage, but it may be brought about by secondary accidents consequent upon extreme anæmia, such as thrombosis and embolism. Again, even a tumour of moderate size, if impacted in the pelvis, may cause pressure on the ureters, wasting or inflammation of the kidneys, and death from uræmia. In some cases fatal peritonitis has been set up by the presence of a fibroid tumour. If a very large submucous fibroid, or fibroid polypus, be expelled through the os uteri, and become incarcerated in the pelvis, death may result from the effects of pressure, or from sloughing and septic absorption. In the substance of the softer or fibro-cystic tumours, effusions of blood may take place, or abscesses be formed and lead to a fatal issue. Fibroid tumours may be invaded by the extension of cancer, but primary carcinomatous degeneration in them is extremely rare, although such an occurrence has been recorded. One instance, at least, has also been recorded in which metastatic deposits containing involuntary muscular fibres were found in other organs, but this is also of excessive rarity. The softer variety of fibroid tumour may spread by continuity between the layers of the broad ligament to such an extent as to lead to a fatal result, but without any change in the histological character of its tissue.

Diagnosis.—Subperitoneal fibroids of small or moderate size will reveal their outlines to bimanual examination, and will usually be recognized as attached to the uterus, and movable with it. If a fibroid is reached by vaginal touch, and is about equal in size to the normal uterus, the distinction between the fibroid and the uterus must be made by the sound. If a fibroid exist in the anterior or posterior wall, producing flexion, it will still be felt as a prominence after the uterus has been restored, or its curvature reversed, by means of the sound. If a fibroid is fixed by adhesion in the pelvis, the diagnosis is more difficult, since it may be impossible to make out its attachment to the uterus. From a small ovarian tumour it is usually distinguished by its hardness; from a swelling due to hæmatocele, peritonitis, or cellulitis by its rounded and defined outline, connected with the uterus, and not merging gradually into surrounding parts.

If the enlargement of the uterus from the presence of a fibroid be externally uniform, the diagnosis from early pregnancy may usually be made by its greater hardness, generally less globular form, less variation of consistency, as well as by absence of softening in the cervix, and the association, not of amenorrhœa, but usually of menorrhagia. In molar pregnancy, or retention of a dead ovum, these distinctions may fail, and even the history be delusive; but the sound will generally reveal the presence of something in the uterus, and dilatation of the cervix will clear up any doubt.

From subinvolution and hyperplasia the diagnosis of fibroid tumour can sometimes be made only after dilatation of the cervix, the index finger being passed into the cavity of the uterus, when the localized character of the enlargement may be detected. When, however, the uterine cavity is lengthened to more than four inches, apart from pregnancy, the existence of a fibroid is probable, and the diagnosis is confirmed if the sound shows the cavity to be distorted and displaced to one side of the centre of the uterine mass. Large interstitial fibroid tumours are generally easily distinguished from solid ovarian tumours by the fact of the tumour forming one mass with the cervix, and moving with it, and by the great elongation of the cavity of the uterus. The distortion of the cavity, however, may render it impossible to pass the sound, while, in the case of ovarian tumours, the cavity of a uterus, closely connected to the tumour, may be lengthened to as much even as five inches. Large sub-peritoneal fibroids may often be distinguished by their multiple character, and hard, irregular, nodular outline. Even if the tumour be single, the presumption is in favour of its being uterine, if it is solid and hard. The attachment of the tumour to the front or back of the uterus may often be made out, either *per vaginam* or *per rectum*, especially if the cervix be drawn downward, in conjunction with the bimanual examination (*see* pp. 14, 15). Uterine tumours also remain longer free from adhesion than do solid ovarian tumours.

Fibro-cystic, or soft fibroid, tumours are often very difficult to distinguish from ovarian, but are in general of much slower growth. If their growth enlarges the whole uterus, the sound will usually decide the point. If they are subperitoneal, the chief point of distinction is that the distinct fluctuation is generally limited to special regions, while the rest of the tumour is hard or only semi-fluctuating. If a puncture be made in doubtful cases, the character of the fluid will generally decide. In the case of a vascular tumour, however, the puncture might prove more dangerous than an exploratory incision. In fibro-cystic tumours the fluid is usually clear, limpid, and yellowish, deposits spontaneously a coagulum of fibrin, contains albumen but not paralbumen, and under the microscope shows leucocytes or spindle-cells, but not the granular cells of ovarian fluid. In some cases the fluid may be blood-stained or purulent. For the characters of ovarian fluid, *see* pp. 297, 306. In some cases it may be impossible to distinguish with certainty between a uterine and ovarian tumour, except by exploratory incision. If an incision is made, the dark red colour of a uterine tumour distinguishes it from an ovarian, which is paler, or has a bluish tint.

Diagnosis of Fibroid Polypi.—Care should always be taken that an inverted uterus is not mistaken for a polypus. The criteria are given under the head of inversion of the uterus (p. 139). Before removal of a polypus the diagnosis should always be completed by making sure that the sound will pass by the side of the pedicle up to or beyond the normal length. Difficulty may arise from the polypus having become adherent to a part or even the whole of the margin of the os, but some point will almost always be found at which the sound can be forced through by moderate pressure. From a polypoid cancer a fibroid polypus is distinguished by its smooth pedicle, which can usually be traced up into the cervix ; by its generally smooth surface, although it may be sloughing or ulcerated in parts ; and by its less readiness to bleed on touching. Portions of retained ovum or clots may resemble a polypus within the uterine cavity, or presenting through the os, and when a portion of placenta has retained a partial connection with the uterus it has been termed by some a *placental polypus*. These are generally distinguished by the history, and by their easy removal by the finger or blunt curette.

Treatment.—In the large majority of cases palliative treatment will successfully carry on a patient up to the menopause, at which time symptoms are commonly, to a great extent, relieved, although the date of its occurrence is often in this disease deferred for several years beyond the usual period. All sources of hyperæmia, active or passive, should be avoided as far as possible ; and if a patient is single, and the tumour is of any considerable size, or causes any notable symptoms, marriage should be discouraged. Married women should be warned of the probably injurious effect of coitus when there is a great tendency to hæmorrhage, and in all cases special care should be enjoined at menstrual periods. Diet should be rather abstemious, and alcohol used very sparingly.

The objects to be aimed at by internal remedies are to alleviate the symptoms of hæmorrhage and leucorrhœa, and, if possible, to check the growth of the tumour or cause its diminution. The drugs which are most efficacious for the former purpose tend also, by restricting hyperæmia, in some degree to promote the latter, although it is only in exceptional cases that any notable diminution of the tumour can be hoped for. Those which have been found most useful are, in the first place, ergot, and, next to this, bromide and iodide of potassium. Of the two latter the bromide is more readily borne for a long period. Half-drachm doses of the extractum ergotæ liquidum, or Richardson's liquor secalis ammoniatus, may often usefully be given in combination with bromide of potassium. The most marked effects, however, of ergot are obtained when it is given subcutaneously, as practised by Hildebrandt, and in some cases a diminution in the size of the tumour may thus be obtained, while the hæmorrhage is generally more or less checked. The best forms of ergot for this purpose are sclerotic acid, in doses of half a grain or more; the extractum ergotæ liquidum, in doses of ten to thirty minims diluted with an equal quantity of water, as recommended by Dr. Atthill; ergotinine in doses of $\frac{1}{100}$ grain; and Bonjean's ergotin, in doses of three to five minims, dissolved in four or five parts of water. The discs of ergotin for hypodermic injection made by Savory and Moore may also be used, two or three discs being employed at a time, and this preparation appears to be one of the least irritating. The great drawback to the treatment is the risk of inflammation or abscess being produced at the point of puncture, and all the preparations are liable to cause at least some local induration and redness. All solutions should be freshly

prepared. The drug appears to be somewhat more efficacious if injected in the neighbourhood of the uterus,
and if injected deeply into the substance of the gluteal
muscles it is less likely to cause abscess than it would
be in the subcutaneous cellular tissue. Hildebrandt's
formula was three grains of Wernich's aqueous extract
of ergot with seven and a half minims of glycerine
and the same quantity of water, but the presence of
the glycerine appears to increase the local irritation.
The injections may be made every other day. Ergot,
given in any way, but more especially by injection,
often increases greatly the pain resulting from uterine
tenesmus, and the patient should be warned to expect
this. Subperitoneal fibroids are less affected by ergot
than those covered by a fair thickness of the uterine
wall. The softer variety of fibroid is that which is
most likely to show a marked diminution in size under
the influence of the drug.

Diminution of the tumour has also occasionally been
obtained after the use of baths or external compresses
containing bromine or iodine, in combination with the
internal administration of the water, but more frequently the advantage to be thus gained is limited to
mitigation of the symptoms. The waters of Kreuznach or Woodhall Spa are the most to be recommended.

In case of alarming hæmorrhage, the most effectual
plug is a sponge tent, which also produces a lasting
good effect by dilatation of the cervix. Enlargement
of the cervical canal is the first indication in case of
persistent hæmorrhage, since it is a necessary preliminary
to other means, and often by itself exercises an important influence. The full explanation of its mechanism
is not quite understood, but among its uses are that it
prevents any retention of blood or clots, and relieves
tension, sometimes allowing the uterine action to carry
on the extrusion of the tumour. It appears, also, that
there is such a nerve-relationship between the cervix
and body of the uterus, that dilatation of the cervix
tends to produce contraction of the body, just as the

expulsive action of the body is associated with physiological relaxation of the cervix. Dilatation may be effected either by laminaria tents, of which as many as possible should be introduced side by side, or by incision. Incision is preferable if one lip of the cervix be expanded over the surface of a tumour growing from the other side of the cervix. It may also be employed if dilatation produces only temporary benefit. After dilatation of the cervix, styptic applications may be used in the form of swabs for the arrest of hæmorrhage. Swabs sometimes fail on account of the tortuous and dilated character of the uterine cavity, and it may be necessary to have recourse to injections. For this purpose tincture of iodine, pure or diluted with one or two parts of water, may be used, or, if this fails, a solution of perchloride or subsulphate of iron. For the precautions necessary *see* pp. 219, 220. In cases of recurrent hæmorrhage the application of strong nitric acid may be tried if milder means fail.

If further surgical treatment be demanded after full dilatation of the cervix, the next step, in the case of an interstitial tumour, or one which has a very wide attachment to the uterine wall, should be to make a longitudinal incision across the face of the tumour deep enough to divide its capsule. The incision may be made by Simpson's metrotome (Fig. 17, p. 54). This allows the mucous membrane to retract, and so diminishes the hæmorrhage from its surface, while, if ergot be afterwards persistently administered, the uterus may extrude the tumour through the opening.

The operation of *enucleation* is one of the most dangerous in surgery when the tumour is interstitial or its attachment to the uterine wall very extensive. The most favourable cases for its application are those in which the tumour shows some tendency to become pedunculated, so that its surface of attachment is less than its greatest diameter; but it always involves a considerable risk of septicæmia and peritonitis, and should not be undertaken unless symptoms are urgent

and other means have had fair trial.
An indispensable condition is that the
tumour should be covered by a sufficient thickness of uterine wall to allow
it to be separated without risk of
opening the peritoneal cavity.

Unless the vagina be already capacious, it should be expanded, by
repeated plugging, or the use of
dilating bags, so as to allow it to admit
the whole hand, if required. The
cervix should also be fully dilated by
tents in the first instance, and, if necessary, by a hydrostatic dilator (*see*
p. 32) afterwards. The uterus is
then pressed down by an assistant, and
an incision made round the base of the
tumour, where this can be reached, so
as to separate the mucous membrane
covering the tumour from that of the
uterine wall, and divide its capsule, if
one exists. The best instrument for
this purpose is generally Thomas's
serrated spoon (Fig. 65), but long
scissors may also be used, if there
is space to adjust them accurately.
Powerful vulsellum forceps should then
be fixed into the tumour, and traction
made upon it, while its base is separated
from the uterine wall by the tips of the
fingers as far as possible. If bands of
tissue are met with too strong to be
separated by the fingers, the serrated
spoon should again be used to divide
them. The operation should not be
commenced unless there is a reasonable
prospect of removing the whole tumour.
If it is only partially separated from
the uterine wall, sloughing is likely to

Fig. 65.—THOMAS's Serrated Spoon for enucleating fibroids.

occur. The patient then runs a greater risk from
septicæmia than that incurred by oophorectomy (*see*
p. 243), or even by hysterectomy, in cases suitable for
that operation.

As a general rule, from the uncertainty which
exists as to the amount of uterine wall which may
be covering a tumour, it is preferable to defer operative
treatment until there is evidence that the uterus is
attempting to expel the tumour, in the fact that some
portion of it presents at the os, and that some spontaneous dilatation of the cervix is commencing. For
cases of this kind there is another method, specially
practised and recommended by Dr. Emmet, which is
generally preferable when it can be carried out, and
which is especially suitable for the case of rather
large tumours. In this the chief agent is strong
traction applied to the growth, the effect of which is
to excite contraction of the uterus. By the aid of
this the attachment is gradually stretched out, and
narrowed into the form of a pedicle, which may even
be found of small size by the time the operation is
finished. Ergot should be first administered, and the
os fully dilated. A powerful tenaculum is then fixed
into the presenting part of the tumour, and strong
traction made and continued for some time. The
tumour is then cut away as high as it can be reached
with strong curved scissors, and successive portions,
as they come within reach, are afterwards treated in
the same manner. If the line of attachment of the
tumour can be reached, it may be separated by the
serrated spoon, instead of by scissors, especially if
there is much tendency to bleed. In many cases there
is but little hæmorrhage, if the uterus is contracting
strongly. But if necessary a whip-cord ligature may
be passed temporarily round the upper part of the
tumour by the aid of Gooch's canulæ. These are two
straight metal tubes, through each of which one end
of the cord is passed, and drawn through tightly. The
ends of the tubes, united by the middle of the cord,

are passed up side by side, and are then carried round the tumour in opposite directions till they meet at the opposite side, and so place a loop around its neck. Finally the tubes can be fixed on to a stem having an adjustment for tightening the cords by means of a rack. After removal of the tumour by this method of traction, the uterus is to be syringed out with hot water, and afterwards some strong tincture of iodine injected, if necessary, to arrest hæmorrhage. Care must be taken not to produce inversion of the uterus, and then penetrate the inverted wall.

In the case of an irremovable fibroid there is yet another method of inducing a process of sloughing, which may be followed by a spontaneous enucleation through the artificial aperture. This is to burn a cavity in the tumour by the actual, benzoline, or galvanic cautery, a plan which Dr. Greenhalgh has found efficacious. It is most applicable to those fibroids which grow in the wall of the cervix.

If the tumour so far approximate toward the pedunculated form that a wire loop can be securely applied around its base, the best mode of removal is by means of the écraseur, fitted with a single steel wire of fair thickness. The écraseur itself must be so strong that there is no risk of its stem bending. The wire should not be too much annealed, and if the tissue to be cut through is very thick, steel piano wire, quite unannealed, is the best to use. This is extremely strong, though rather rigid to work with. A considerable stiffness in the wire loop, however, assists the operator in passing it up over the equator of the tumour, as it regains its shape and position within the uterus. An extra supply of wire in hank must be kept ready, in case the first loop should break. A loop at one end of the wire is fixed to the moving button of the écraseur (Fig. 68, p. 248); the other end is kept long and unattached, so that the size of the loop can readily be varied as required, till it has been got above the equator of the tumour, the stem of the

R

écraseur being pushed up as high as it will go between the tumour and the uterine wall in front, or wherever the attachment of the tumour reaches lowest. The slack of the wire is then drawn in, and the free end twisted round the crossbar of the écraseur by means of strong pliers. If the tumour is not completely cut through when the button is screwed up to the full, the wire must be removed from the button, and re-attached to it after the button has been screwed down again to the bottom. If the slice removed does not comprise the whole tumour, the écraseur is to be again applied, if possible, to what remains, in a similar way, until the whole is removed flush with the uterine wall. The compound wire of twisted strands should never be used to cut through thick tough tissue, since it is much more likely to break.

The galvanic écraseur may be used instead of the ordinary écraseur, but the pliant loop of platinum wire is much more difficult to pass into position beyond the reach of the fingers, than the stiff one of steel wire. Moreover, the batteries used for medical purposes have rarely power enough to heat sufficiently a loop of the requisite size to cut through a very thick tumour when in contact with moist tissues, and, if not heated enough, the wire will break.

It sometimes happens that a fibroid of very large size becomes partially extruded by uterine action into the vagina, or even appears at the vulva. The base of attachment may still remain large, and it will generally be impossible, after the os has become retracted, to determine its dimensions by any method of sounding. Sometimes it is not possible even to reach any point of the margin of the dilated os. This condition is distinguished from inversion of the uterus by the great size of the mass in the vagina, and by recognition of the fundus in its normal direction, but carried upwards by pressure. The functions of the bladder and rectum then become impeded, and if the pelvis is completely filled, nervous symptoms may arise similar to those resulting from impaction of the fœtal head in parturition. In any case, the patient is exposed to the risk of septicæmia from sloughing of the tumour. In this case also the tumour may be removed by the écraseur, fitted with a strong unannealed steel wire, in the manner already described. The loop is passed over the portion of tumour in the vagina, and, if possible, over its equator. It may then be possible to slip it up to its pedicle or base, the os uteri being generally so flattened out against the pelvic wall as not to offer an impediment. If not, as much of the tumour as possible must be taken off at first, and then the remainder, if necessary, in successive slices, the écraseur being passed up within the uterus.

When a fibroid tumour, which cannot be removed through the vagina without too great a risk, causes such severe hæmorrhage as to threaten life, or render life intolerable, one resource remains short of removing the whole uterus by gastrotomy, namely, the operation of *oophorectomy*. The object of this operation is to bring on artificially the menopause, and in the majority of cases this has been attained. In some instances indeed more or less regular hæmorrhage has persisted for a time, but it has been found more amenable than before to the influence of cold and other styptic remedies, which, after removal of the ovaries, may be freely used without fear of the ill-results of checking menstruation. The operation should be performed, like ovariotomy, according to Lister's antiseptic method in its fullest extent. A small incision is made in the median line, the mesovaria are transfixed by carbolized silk ligatures, which are cut short and dropped, and both ovaries are cut away. If menstruation is arrested, the fibroids generally diminish in size. In some cases, however, menstruation has continued after the operation, and the growth of the fibroid has not been affected. It appears probable that if any ovarian tissue be left, even on the distal side of the ligatures, its vitality may be maintained, and ovulation continue.

Those varieties of fibroid which are liable to continue growing after the menopause—that is to say, soft, non-encapsuled fibroids, and fibro-cystic tumours—are likely also to be unaffected by oophorectomy. Mr. Lawson Tait maintains that it is chiefly of importance to remove the Fallopian tubes rather than the ovaries. He holds that menstruation is regulated by some undescribed nerve-centres in connection with the Fallopian tubes, and declares that removal of the tubes, without the ovaries, arrests menstruation. Others believe that the greater effect produced by removing tubes as well as ovaries is due to the ligatures being placed more deeply in the broad ligaments, and so cutting off some of the arterial supply of the uterus.

Of cases of oophorectomy, or removal of uterine appendages, for myoma, tabulated by Dr. Battey, the mortality was 13 in 46, or 28·2 per cent. It may be hoped, however, that in future, in the hands of practised specialists, the results will be much better, since Mr. Thornton reports a series of fifteen cases without a death.

Subperitoneal fibroids are amenable to no surgical treatment except *removal by abdominal section*, but since they are, in general, comparatively innocuous in their results, this operation should, as a rule, only be undertaken when, by their increase in size, they directly threaten life, or absolutely incapacitate from its necessary avocations. Such a dangerous increase is more likely in the case of the softer or fibro-cystic tumours. In some cases also the operation may be indicated if a fibroid tumour gives rise to ascitic effusion. When a fibroid tumour has a thin pedicle, and is free from important adhesions, it may be removed without much greater risk than that of ovariotomy, but it is often impossible to ascertain beforehand the extent of its attachment to the uterus. The operation should be performed like ovariotomy, and the pedicle or pedicles, if of moderate size, transfixed and tied with carbolized silk. For cases, however, in which the attachment is broad, the method of ligature has not proved so successful as in the case of the pedicle of ovarian tumours. The chief risk is that the tissue is apt to shrink after a time, the ligatures to become loose, and hæmorrhage to occur. There is also the disadvantage that there is a broad surface left, to which intestines may become adherent, and thus a risk of intes-. tinal obstruction arise. If the pedicle, therefore, is broad, soft, and vascular, it appears to be safer to clamp it in the lower angle of the wound by Koeberle's serre-nœud (Fig. 66), if the traction so produced is not too excessive. Thick, soft iron wire should be chosen, so that it may not readily cut through the tissue. The wire is placed pretty closely round the neck of the tumour, and both ends fixed to the button of the instrument. The screw is then turned until there is sufficient constriction to prevent

Fig. 66.
KOEBERLE'S Serre-nœud.

hæmorrhage. After the tumour is cut away it may be tightened a little further, and if secondary hæmorrhage occur after an interval, from shrinking of the tissue, a

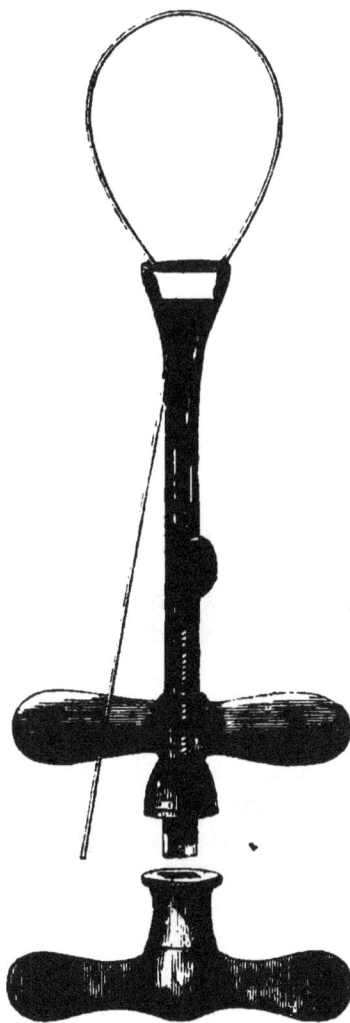

turn or two of the screw will always arrest it. Two
long steel pins, having guards to fix on their points
(Fig. 67), should be passed transversely through the
stump, immediately above the wire loop, to fix it in
the wound, some pieces of gauze being tucked under-
neath the ends of the needles. The stem of the serre-
nœud is then to be included in the antiseptic dressings.
If the intra-peritoneal treatment of the pedicle is
adopted, it would seem to be a good plan to cut away
the tumour by a V-shaped incision, so that the edges
of peritoneum may be brought together by sutures of
fine carbolized silk, above the ligatures placed round
the neck. In some cases of hard encapsuled fibroids
it has been found possible to enucleate the tumour

Fig. 67.
Guarded Pin for fixing pedicle in abdominal wound.

from the external surface of the uterus without any
hæmorrhage sufficient to require the application of
ligatures.

Hysterectomy, or the removal by abdominal section of
the whole uterus when enlarged by fibroid or fibro-cystic
disease, is a much graver matter, and as yet the majority of cases have been fatal, although Péan reported, in
1877, twenty-one cures out of thirty-one cases of operations performed by himself. It is possible, however, that
through the increasing success of abdominal surgery in
the hands of specialists in that department, the position of
the operation may in future be much changed. Spencer
Wells reports fifteen deaths in twenty-four operations
without carbolic spray; three deaths in ten operations
with the spray. Bantock, discarding the carbolic spray,
reports twenty-two hysterectomies with only two deaths,
with extra-peritoneal treatment of the pedicle; four
operations, all fatal, with intra-peritoneal treatment of
the pedicle. Thornton reports twelve complete hysterec-

tomies with five deaths; but none of those were fatal in
which extra-peritoneal treatment of the pedicle in one
clamp or serre-nœud was found possible. The great danger
lies in the bulk and vascularity of the pedicle formed
by the cervix and broad ligaments, and consequent risk
of primary or secondary hæmorrhage. As the operation has hitherto been performed, the essential condition
for its possibility is that a sufficient portion of the
cervix should be free from the growth to be converted
into a pedicle. If the intra-peritoneal treatment of
the pedicle is adopted, there is a risk that septic
infection may be conveyed to the peritoneum from the
vagina through the cervical canal, since the ligatures
inevitably become slack after some days, and are apt to
produce ulceration of the cervical mucous membrane.
The antiseptic method is thus necessarily imperfect.
This risk is avoided if the stump can be fixed in the
wound. With any form of clamp some vessels in the
broad ligaments close to the sides of the uterus are apt
to be insufficiently compressed. Hence the method of
circular constriction is preferable. A thick wire should
be placed round the cervix and broad ligaments, and
the stump secured with Koeberle's serre-nœud, as
described for the case of the stump of an external
fibroid tumour (*see* pp. 245, 246). Unless the menopause
has passed, the loop must be placed low enough to
allow the ovaries to be removed with the uterus. The
compression of the neck of the uterus often causes great
pain and collapse, so that opiates in full doses are
generally required, and stimulants may be called for.
If the pedicle is to be dropped, the broad ligaments
may be transfixed and tied in two or more sections
below the ovaries with carbolized silk, and the cervix
itself afterwards transfixed and tied in the same way,
care being taken that all the loops interlace. Schroeder
places a temporary elastic ligature round the lower
part of the uterus, cuts away the tumour by a V-shaped
incision, or enucleates it if extending into the pelvic
cellular tissue, and unites the edges of peritoneum

with deep and superficial sutures, which also arrest
hæmorrhage.   He reports nine deaths out of forty
operations of this kind.   This
method may render it possible to
remove a fibroid even when developed in the lower part of the
uterus.

Treatment of Fibroid Polypi.—
Removal by the écraseur is preferable to cutting away the polypus
with scissors, since the latter method
is not absolutely free from the risk
of serious subsequent hæmorrhage.
The method of ligature, which
involved the danger of septic
absorption, is now obsolete.   The
écraseur itself, and the thickness of
the wire, should be in accordance
with the size of the polypus.   It is
convenient to have an instrument
with a very strong stem, into
which terminals of various sizes
can be screwed (Fig. 68).   The
écraseur may be used either with
the wire-rope made up of several
strands of wire twisted, or with a
single wire.   The former, being
moderately pliant, is generally
easily adjusted over a polypus, and
may be used in preference whenever there is a pedicle only to cut
through, except in the case in
which the loop has to be adjusted
high up in the uterus somewhat
above the reach of the fingers.
In that case the stiffness of the
single steel wire is an advantage.
The single steel wire is also much the strongest.
When the polypus has passed through the os, the

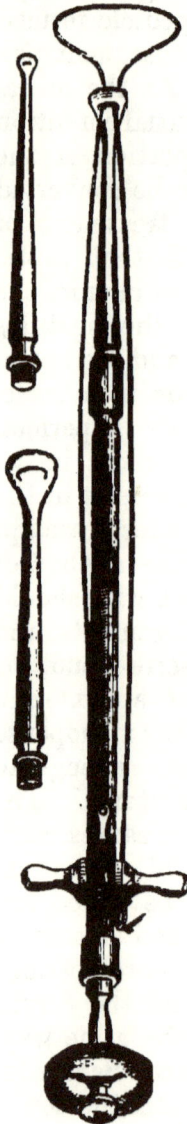

Fig. 68.—HICKS' WIRE ÉCRASEUR.

application of the noose is generally easy, and may be managed by the fingers without any speculum. The tip of the stem should eventually be passed up within the cervix, and the slack part of the wire drawn in before it is finally attached to the transverse bar of the écraseur, the other end having been previously secured to the travelling button. If a small portion of the pedicle is left, it generally shrinks up after removal of the main growth. An anæsthetic should not be given for this operation, if it is possible to avoid it, for dividing the pedicle of the polypus gives little or no pain, while pain is severe if the uterine wall be included in the loop, and thus an error may be revealed at the last moment. If a polypus is very large, difficulty may arise in its extraction, after division of the pedicle. In the absence of forceps specially constructed for the purpose, delivery may sometimes be effected by midwifery forceps, or, preferably, by passing the loop of the écraseur again over the polypus, and so dividing it into pieces. If this cannot be accomplished readily, the tumour may often be extracted, without laceration of the vulva, by the following plan. The portion presenting at the vulva is seized by strong tenaculum forceps, and an incision is made with scissors in the tumour in a spiral form, commencing near the point where it is seized, while traction upon the tenaculum is meanwhile maintained. In this way the tumour is gradually elongated and drawn through the vulva.

If the polypus is within the uterus, the cervix must be fully dilated before removal is attempted. In order to pass the loop of an écraseur over an intra-uterine polypus, it is desirable to depress the uterus as much as possible, by drawing down the anterior lip of the os with tenaculum forceps, while an assistant presses it down from above. It is often useful to fix a tenaculum in the polypus itself, make some traction by its means, and then pass the loop over the handle. Care must be taken, however, not to produce in this way partial inversion of the uterus, and the traction should therefore be relaxed before the écraseur is tightened.

Polypoid elongations growing from the os are easily removed by the écraseur, or, if their base is broad, they may be cut off with scissors, and the bleeding either stopped by cautery or by application of a styptic and plugging of the vagina.

CANCER OF THE CERVIX UTERI.

Causation.—Cancer of the neck of the uterus is a very common disease. It is more frequent even than cancer of the breast, and is the chief cause of the greater prevalence of cancer in the female than in the male sex. It most commonly occurs between the ages of forty and fifty, but a considerable proportion of cases are also met with between thirty and forty, while a few appear before the age of thirty, and others occur even up to advanced old age. Cancer of the cervix is extremely rare in virgins, and commoner among parous than among nulliparous women, while among the subjects of it a considerable number of women are found who have had large families. From this it may be inferred that inflammation of the cervix, induced by parturition or other mechanical causes, plays an important part in the causation of cancer, and that the so-called erosion or granular inflammation near the os, or within the cervical canal, may eventually, *in predisposed subjects*, go on to malignant degeneration, although in any given case of this common affection such a termination is an improbable one. This view is confirmed by the researches of Ruge and Veit (*see* p. 178), who found that in so-called erosion a gland-proliferation takes place, and that cancer may commence with a similar proliferation, with the addition of an exuberant growth of epithelium, partially or wholly filling up the acini. Lace-

ration of the cervix may thus predispose to cancer, by
giving rise to inflammatory irritation of the exposed
mucous membrane of the cervical canal. At a very
early stage of epithelioma it is not uncommon to find
evidence of a pre-existing laceration. That constitutional predisposition is also an important element in
the origin of cancer of the cervix is shown by the
comparative immunity of negroes, though it is not to
be inferred from this that the disease may not be
purely local at its commencement. Predisposition to
cancer in individual families has also some influence,
although it is only in a minority of cases that a
history of this can be traced.

Pathological Anatomy.—Using the word cancer in
its widest sense, to signify a growth having the clinical
characters of malignancy, namely, that it tends to spread
by contiguity into tissues of a different character from
that in which it originated, to return after removal,
and to infect the glands and distant organs, we must
include among the varieties of cancer affecting the
cervix—(1) true carcinoma, having a more or less
alveolar structure ; (2) epithelioma, or the ''cancroid''
of German writers ; and (3) many forms of sarcoma.

There are three possible modes of origin of true
carcinoma in the cervix—(1) from the proliferation of
the epithelium of glands, either primarily existing or
formed by ingrowth of cylindrical epithelium into
depressions under the influence of irritation ; (2) from
ingrowth of columnar processes from the deeper layers
of the squamous epithelium of the cervix, a process
which gives rise in the first instance to epithelioma ;
(3) from proliferation of the connective tissue cells and
their transformation into groups resembling epithelial
cells. It is at present, however, a matter of dispute
whether carcinoma ever originates in the third method.
It is an undoubted fact that it spreads at its circumference by the connective tissue cells becoming infected
by a kind of spermatic influence, and growing into
clusters of epithelium-like cells. The very commence-

ment, however, of growths in connective tissue is
scarcely open to observation, and it is possible that
some primary proliferation from an epithelial structure
always occurs, as is now held by Billroth, Waldeyer,
and others.

In carcinoma affecting the cervix the proportion of
cells to fibrous trabecular tissue may vary in any degree,
from the most rapidly growing medullary carcinoma,
consisting almost entirely of cells, up to the scirrhous
form. The occurrence of scirrhus at all approaching
in hardness to that which is found in the breast is,
however, very rare, although the more infiltrating form
of carcinoma often contains enough fibrous tissue to
give an impression of considerable hardness to the
finger. Medullary carcinoma is the commonest form
of all, and the varieties which are harder at first generally become softer in their later stages. The variety
of carcinoma is sometimes found in which the cells are
arranged in a columnar manner round the borders of
the alveoli, and this arrangement may be reproduced
in secondary growths. More frequently the alveoli are
very small and irregular, often elongated in shape, and
each containing only a few cells, so that, especially
near the growing margin, the growth may be difficult
to distinguish from a round-celled sarcoma. In the
more rapidly growing varieties, again, the alveoli are so
large that the cell-masses may be pressed out from the
cut surface in the form of soft plugs.

In cancer of the cervix uteri the distinction between
epithelioma and carcinoma is not so marked as in most
parts of the body. Study of cancer in this region is
especially calculated to show that epithelioma is apt to
merge into carcinoma, as soon as either the epithelial
masses no longer simply increase by their own growth,
but begin, by a kind of spermatic influence, to stimulate the nuclei of the adjacent stroma (which are
always abundantly proliferating in epithelioma) to
grow also into epithelioid cells, or else the epithelial
cells or their nuclei migrate along the lymphatic tracts

into the cellular tissue. Hence, there are many cases which it is not easy positively to classify either as epithelioma or carcinoma.

By examination of sections of thirty-four cases of cancer, of such a kind that it increased the bulk of the cervix, so that it was possible, at an early stage, to excise the cervix, more or less completely removing the obviously diseased tissue, I have arrived at the following results. The histological characters of those growths which are generally clinically regarded as epithelioma are very variable. It is only exceptionally that the bird's-nest bodies, or epithelial globes, whose presence proves the growth to have originated from squamous epithelium, are seen. Even when they are present, it is only just at the edge of the ulcerated surface that it is possible to trace any ingrowth of processes from the surface epithelium, and the cancer generally spreads for some distance beneath normal, or merely thickened, epithelium. Hence, a test which has been given for diagnosis of epithelioma of the cervix, namely, that the mucous membrane is bound down to the tissue beneath, is purely à priori and imaginary, for the ingrowing processes over a considerable surface, usual in epithelioma of the skin, here do not exist. The epithelial masses nearest the healthy surface generally consist of cells resembling those of the squamous epithelium, bounded by a regular margin of columnar-like cells, sharply demarcated from the surrounding stroma. The cells are also cemented together like those of the squamous epithelium, either by the delicate processes uniting cell to cell, and constituting the so-called " cog-wheel " cells, or, apparently, by adhesion of the whole cell-walls. The " cog-wheel " appearance, however, is never so manifest as in the normal epithelium, and frequently cannot be made out. In older portions or deeper parts of the same growth the cell-masses may be seen without any border of regular cells, and no longer sharply demarcated from the stroma. The cells also may be no longer clearly cemented, but more or less separate from each other like those of carcinoma. Not unfrequently may be seen in the stroma small detached groups of cells without intercellular substance, and having large nuclei similar to those of the cells of the larger masses, thus constituting an approximation to the alveolar arrangement of carcinoma.

In more numerous cases no epithelial globes are seen, but the large masses of cemented cells, often with regular borders, render it probable that these also commenced from squamous epithelium, and sometimes their continuity with it can be traced. It is not uncommon to see the cell-masses elongated into the form of more or less parallel columns, having borders of regular cells, and separated by narrow bands of stroma. In other parts of the growth may be seen an approximation towards the characters of carcinoma, like that already described. Sometimes the cells become elongated into a long spindle shape either at right angles to, or parallel with, the axis of the cell-columns. In the more rapidly growing forms of tumour the cells deviate in another manner from the characters of squamous epithelium, and not only cease to be cemented, but show proliferating nuclei, and become very various in shape and size. Sometimes the section shows an almost continuous mass of cells, cemented or not, traversed only at wide intervals by delicate bands of stroma, which carry the vessels.

In a smaller number of cases I have found evidence of the commencement of the growth by the degeneration of mucous glands. The epithelium of the glands proliferates, so as more or less completely to fill up the acini. In this way the alveolar arrangement of true carcinoma is at once reached, but at first the cells in the alveoli are cemented, or at any rate in close contact, though without any " cog-wheel " appearance, and generally have a border of regular columnar-like cells round the margin. Eventually the cellular tissue is infected by the growth in it of similar cells, or migration of cells from the primary alveoli. This is the only mode of origin accepted by Ruge and Veit,* but the material for their observations appears to have been limited, since they did not find epithelial globes in any instance, whereas I have found them in five cases out of thirty-four. More generally, at the very early stage of cancer I have found the glands simply proliferating in close contiguity to the epithelial masses without any evidence of malignant degeneration. But it is probable that the forms of cancer which infiltrate at an early stage, and do not enlarge the cervix, more commonly have their origin from glands. Certainly the alveolar arrangement of true carcinoma is generally more definitely seen in these forms.

In a small number of cases, less than one-tenth of the whole, the structure is that of sarcoma, or lymphosarcoma, originating in the cellular tissue. There is generally a large proportion of round cells in the growth, and its character at an early stage may resemble that of the stroma of the villous prominences seen in severe erosion of the cervix, or of the normal prominences within the cervical canal. Growths of epithelial origin are also sometimes surrounded by an extensive border of highly nucleated sarcoma-like tissue, so that such a growth might be mistaken for sarcoma on examination of only a small portion of it. There may also be considerable hypertrophy of the normal muscular tissue in the vicinity. In some instances even the epithelial cells themselves seem to have thrown out processes, and to have approximated toward the character of connective tissue cells ; so that it is a question whether some growths, whose histological character would be pronounced to be that of alveolar sarcoma,† may not have had their origin from epithelium.

As regards the character of the growth at a late

* " Zeitschrift für Geburtshülfe und Gynäkologie," Bd. ii. Hft. 2.

† i.e., made up of cells arranged in alveoli, but having intercellular substance between them, or connected by uniting processes.

stage, when it has proved fatal to the patient, the statistics of Mr. Arnott may be quoted. He reports fifty-seven cases in which autopsies were made at the Middlesex Hospital. Of these he was able, on microscopic examination, to speak positively as to the nature of the growth in twenty-two, of which twelve were carcinoma, eight epithelioma, and two spindle-celled sarcoma.

By the name "*cauliflower excrescence*" has been generally understood a sprouting, papillary growth from the cervix, readily bleeding, and so soft that after its removal or after death nothing but a broken-down pulpy mass, like a macerated placenta, may remain. Some have supposed this to be a special form of disease, and not necessarily malignant. Villous outgrowths, however, may be associated with different forms of growth in the subjacent tissue. Commonly the individual papillæ contain a central loop of vessels supported by delicate areolar tissue, and their substance consists mainly of round cells, though the superficial layers may be flattened. The cellular portion of the papillæ is generally continuous with the cell-masses of a carcinoma or epithelioma in the subjacent tissue, and the fibro-vascular tissue with the fibrous trabeculæ of the carcinoma, or the nucleated areolar tissue surrounding the cell-masses of the epithelioma, as the case may be. In one case I have found the papillæ to contain numerous epithelial globes. In other cases the individual papillæ themselves show the alveolar structure of medullary carcinoma. The papillæ may sometimes adhere by their outer surfaces, and thus form enclosed spaces. Cases have been recorded in which the structure was merely that of papilloma, and no evidence of malignancy could be detected. But the friable growths are probably always malignant, the epithelium tending to invade the subjacent tissues at the points intervening between papillæ. Thus, in a case recorded by E. Wagner,* a growth was removed,

* "Der Gebärmutter-krebs." Leipzig, 1858, p. 13.

which was considered, on microscopical examination, to be pure papilloma. Five months after the operation, however, the patient died from unmistakable cancer of the cervix. A strong tendency to outgrowth may, however, be associated with a very slight tendency to infiltrate; and, in such case, the growth may be eradicated completely by removal, and never recur. Friable growths of this kind are rare, but the term "cauliflower excrescence" has also sometimes been applied to the commoner and less pulpy outgrowth of the cervix, the surface of which is only slightly papillary, and which has a nearer actual resemblance to a cauliflower than the growth to which the name was first given. These growths are always malignant, and the variable character of their histological structure has been already described.

In the great majority of cases cancer of the cervix commences near the external os, and extends thence throughout the cervix, and upward along the cervical canal, generally affecting one lip of the os more than the other. Exceptionally it commences high up in the cervical canal, near the internal os, and the external part of the cervix may then show no sign of it for a considerable period.

The growth rapidly extends up the cervical canal to the internal os, and to the vaginal walls and cellular tissue around the cervix, thereby fixing the uterus. In its further advance it involves chiefly the body of the uterus and the anterior and posterior vaginal walls, usually descending furthest along the former. Sometimes the posterior vaginal wall in contact with the cervix becomes infected by contiguity, in the case of vegetating growths. Ulceration generally occurs early, with extensive sloughing of the morbid deposit, and the greater part of the uterus and vagina is often eaten away; so that a deep ragged cavity is formed. Perforation frequently occurs into the bladder, and more rarely into the rectum, so that an enormous cloaca may be formed of the three cavities. Metastatic deposits

S

may occur in the pelvic and lumbar glands, ovaries, peritoneum, and also in distant organs, as the lungs and liver, even in the case of spindle-celled sarcoma, the least actively malignant of the various forms of growth. Mr. Arnott found metastatic deposits in the glands in 50 per cent., in other organs in 41 per cent. of his cases. The growth often interferes with the ureters, leading to wasting or inflammation of the kidneys, and may even entirely occlude them.

Rodent ulcer of the cervix has been described as a disease distinct from cancer, and marked by its slow progress, and the absence of any perceptible amount of deposit, but recent investigations of its histology are wanting. Probably at least many cases so described have been instances of superficial epithelioma, in which ulceration preponderated over infiltration.

Results and Symptoms.—In the early stages of cancer the symptoms are frequently so slight that the disease commonly does not come under observation until it has reached a stage at which it is ineradicable. When an early symptom occurs it is usually that of hæmorrhage, often not profuse, but irregular, and frequently recurring. Hæmorrhage on coitus is not unfrequently the first symptom. Menstruation is also increased, and leucorrhœal discharge is generally present, sometimes slightly tinged with blood. A recurrence of uterine hæmorrhage, after the menopause has for some time passed, should always lead to the suspicion of cancer, and be regarded as an imperative indication for a vaginal examination. Early symptoms are more commonly present in the vegetating than in the infiltrating forms of cancer; in the latter of which there may be no hæmorrhage up to quite a late stage. Pain is usually absent or slight while the disease is confined to the cervix, and in some cases of soft cancer very little is felt up to quite an advanced period of the disease. In most cases, however, as soon as the growth has infiltrated the tissues round the uterus, severe lancinating pain is a marked feature, and renders cancer of the cervix one of the most terrible of diseases. It may generally be distinguished from the pain of chronic inflammation or engorgement from the fact of its being felt severely at night, and disturbing sleep, while the other is chiefly evoked by standing or locomotion, and is relieved by rest. Pain may be also produced in cancer by the soreness of the ulcerated surface exposed to friction, and when of this nature it may be much relieved after removal or destruction of the diseased surface.

As soon as ulceration has commenced, the discharge has generally a watery character, often tinged with blood, and soon acquires the most intense fœtor, which forms not the least among the sufferings of the unfortunate patient. Frequently shreds of gangrenous tissue come away with it. At the outset the patient may be apparently in the most florid health, but as soon as the ulcerative stage is reached cancer of the cervix very quickly induces loss of flesh and the well-known cancerous cachexia. As displayed in this form of cancer it depends, in great measure, upon the effect of repeated hæmorrhages, and upon a constant slight absorption from the foul discharges. Thus a great improvement may be effected in the general appearance by partial removal of the growth, leading to a temporary cessation of hæmorrhage and fœtid discharge. The cachexia shows itself mainly in a sallow, yellowish tint of skin, accompanied by emaciation, but this does not present anything absolutely characteristic; and a very similar appearance may be seen in other cases, especially in those of fibroid tumour or polypus, accompanied by hæmorrhage and sloughing. Digestive functions are impeded, and nausea and vomiting are frequent, being partly the effect of the disgusting smell of the discharge. Obstinate constipation may result from mechanical interference with the rectum, while occasional attacks of diarrhœa from reflex irritation are not unfrequent. Disturbance of the bladder occurs pretty early, and may be the first symptom which attracts attention.   At first there may be reflex tenesmus, then difficulty of micturition as the base of the bladder and the urethra become involved, and finally incontinence, from the existence of a fistulous opening. The duration of the disease commonly varies from one to two years after the recognition of its character ; more rarely the patient survives for three or four, or even a greater number of years.   Very rare instances have been recorded in which an apparent spontaneous cure has resulted after sloughing of the growth.   If the growth can be removed the course of the disease is much prolonged, and it is doubtless possible to eradicate it entirely in some cases, if the operation can be performed early enough.   Death occurs sometimes directly from hæmorrhage, but more frequently from gradual exhaustion and emaciation, aided not unfrequently by the effects of uræmia or kidney inflammation, set up by the interference of the growth with the bladder or ureters.   Intercurrent peritonitis or pneumonia may close the scene, or death may occur rapidly from occlusion of the ureters, producing complete suppression of urine, or suddenly from pulmonary embolism, usually the sequel of thrombosis of the pelvic veins in or near the growth.

Diagnosis.—When the disease has reached the ineradicable stage, the diagnosis should be easy, although mistakes have not unfrequently been made. The cervix is fixed, and dense inelastic induration may extend to the vaginal walls.   In the hard mass is felt an ulcerated cavity with hard nodular edges. Its surface gives to the finger a peculiar sensation of superficial friable softness, with extreme inelasticity of the tissue beneath.   Hæmorrhage is generally produced by touching the surface of the ulcer.   The fœtor of the discharge is a ready diagnostic sign in the later stages. In endometritis or vaginitis the discharge may be offensive enough to annoy the patient, but the intense and nauseating smell, hardly to be removed from the fingers even by disinfectants, belongs only to cancer, and to the decomposition of the products of conception, or the sloughing of a benign tumour, such as a fibroid or polypus. The two latter conditions can usually be easily distinguished by the history and physical signs. The absence of fœtor is, however, no disproof of cancer.

In the proliferating forms of cancer, the growth often attains considerable size before the cervix becomes fixed. There is then an unequal enlargement of one or both lips of the cervix, not nearly so hard to the touch as chronic hyperplasia. The surface is more or less villous or papillary, and readily bleeds on manipulation. The whole cervix is broadened, and tends to grow into a mushroom-like shape, with eversion of its edges— a valuable diagnostic sign. By speculum a bulging, irregular, mottled, deep-red surface of more or less extent is seen, which is destitute of the normal squamous epithelium. The speculum, however, should be used with great caution in cases of cancer of the cervix, on account of the severe hæmorrhage which it may induce. When it is required for diagnosis, or the application of remedies, it is usually best to employ either a Sims' (Fig. 7, p. 22), or Neugebauer's speculum (Fig. 10, p. 25), the first blade of which can be guided past the cervix by the finger. Nothing but the microscope can decide whether the growth is carcinoma or epithelioma, since both may have a papillary or villous surface. The exceedingly friable, villous, readily-bleeding surface of the true cauliflower excrescence can scarcely be mistaken for any other condition.

The diagnosis of the earliest stage of cancer may be one of very great difficulty. The most valuable assistance is to be found in the fact, that it usually commences *on the surface* near the os, or just within the cervical canal, and is associated with some degree of papillary growth, which leads to ready hæmorrhage on manipulation. If bleeding is produced by a gentle touch of the os by the finger, and not merely by rough handling, or the use of the speculum or sound, the suspicion of commencing cancer should be excited, especially if any papillary surface or inelastic nodules are felt around the os. Villous erosions, however, may also sometimes readily bleed, and in a very doubtful case, the only certain method of distinguishing is to remove a portion of tissue by scissors or sharp spoon, and examine it microscopically. For any certain conclusion, the fragment should be large enough to allow it to be hardened, and sections cut from it. As seen through a speculum, cancer, at a very early stage, may

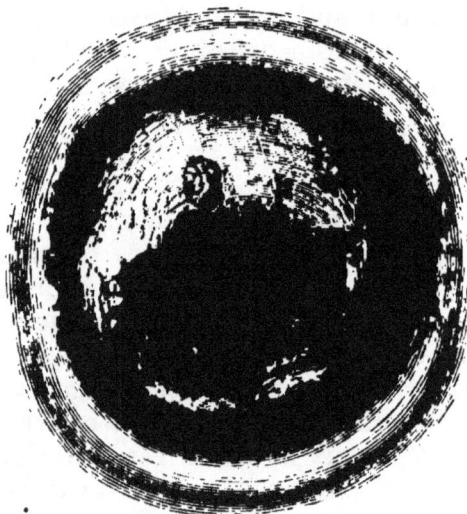

Fig. 69.—Cancer of the Cervix Uteri at a very early stage (after RUGE and VEIT).

sometimes be distinguished from any non-malignant ulceration or erosion by the presence of limited irregular prominences, which may be separate from the cervical canal, having a deep-red and papillary or villous surface. There may also be an excavated ulceration, with sharply cut edge, at the margin of the os. Fig. 69, showing an early stage of cancer, may be compared with Fig. 53, p. 184, and Fig. 54, p. 185, showing granular inflammation with laceration and simple erosion, respectively.

Hyperplastic induration of the cervix has, formerly, often been mistaken for cancer. It may be distinguished by the fact that the cervix is movable (unless fixed by inflammation) ; its irregularity is due to fissures radiating from the os, and its tissue has some elasticity with its hardness, while pain is usually increased during the menstrual flow, instead of being relieved by hæmorrhage, as is commonly the case in cancer. Generally, also, there is no irregular hæmorrhage, and menstruation is scanty rather than profuse, nor is there usually hæmorrhage on gentle manipulation, unless a severe erosion exists. In hyperplasia there is commonly a history of symptoms referable to uterine disorder of many years' duration, while the symptoms of early cancer are not likely to have existed many months. It is to be remembered, however, that cancer may supervene upon disease of a non-malignant kind.

Treatment.—If the disease is recognized before the uterus is fixed, no time should be lost in attempting to eradicate it by removal. Even though it usually recurs, the patient is relieved for a considerable period from hæmorrhage and discharge, and the course of the disease is much protracted. When the cancer forms a prominent mass, the operation is very easily performed by the galvanic écraseur. It is usually most convenient to adjust the loop by the fingers without any speculum, the stem of the écraseur—which should not be too short—being passed up in front of the cervix, and pressed somewhat upwards as the wire is being tightened. Caution should be used as to drawing down the cervix with a tenaculum, since .otherwise a portion of the pouch of Douglas or of the bladder may be brought within range of the loop. It is better, therefore, to apply the loop with the cervix in its natural position ; but after the wire has begun to cut for itself a groove, some very slight traction may be made with much caution upon the cervix, in order to bring as much as possible of the cervical canal below the plane of section, so that the stump remaining is represented by a cone, with its apex towards the fundus. If there is a difficulty in keeping the loop in position round the base of the cervix, a groove may be made by cutting through the mucous membrane with scissors at the junction of the cervix with the vagina. If the wire is sufficiently heated, and if it is wound up extremely slowly, so as to cauterize thoroughly the whole surface, there is generally no hæmorrhage whatever from the cut surface, but, to secure this, the battery power must be adequate. In the absence of the galvanic cautery, the amputation may be performed with the wire écraseur. It is well to apply afterwards the actual or benzoline cautery, even if there is no bleeding, in order to destroy, to a greater depth, the tissue to which cancerous infiltration may have extended. After the operation, antiseptic vaginal injections should be used until the superficial slough has separated. The cautery protects, to a considerable extent, against the absorption of septic material; but a certain amount of local inflammation, producing thickening around the uterus, is not uncommon after amputation of the cervix in cancer.

The rule of abstaining from operative interference when it is impossible to remove the disease wholly does not apply so much to cancer of the cervix as to that of other parts of the body, since a large part of the cachexia, whether due to hæmorrhage or septic absorption, depends upon the presence of the diseased surface in a situation exposed to friction and subjected to the influence of warmth and moisture. Whenever hæmorrhage is an urgent symptom, and a vegetating surface exists, removal of a portion of the growth will often arrest hæmorrhage and do away with the fœtor of the discharge for a considerable period. This may be carried out whenever the disease is not so advanced that a risk would be run of opening the peritoneal or other cavity. Any prominent mass round which a wire can be placed may be sliced off by the galvanic écraseur. To remove the deeper tissue the best instruments are the sharp spoons introduced by the late Professor Simon (Fig. 70), which are made of various shapes and sizes. In scraping away the tissue by their means a selective action is exercised, since the cell-masses of the cancer are readily removed, while the normal tissue is more resistant. The selective action will not extend, however, to the infiltration of scattered cells among normal tissues outside the borders of the growth, while, for the harder forms of cancer, containing a large proportion of fibrous tissue, these spoons are less effective. Caustics may also be used, either in the first place, or, preferably, after the more friable and manifestly cancerous tissue has first been scraped away. We may then employ either the actual, benzoline, or galvanic cautery, or chemical agents, as the potassa fusa, potassa cum calce, chloride of zinc, or alcoholic solution of bromine. Potassa fusa or potassa cum calce must be used with the precautions previously described. The solution of bromine, first recommended by Routh, and very highly lauded by Schroeder, is supposed to exert a special influence upon cancer cells, and is certainly an efficient caustic. One part of bromine dissolved in five parts of rectified spirit* may be applied on a tampon of cotton wool. This should be covered with a piece of gutta-percha skin, and a large tampon soaked in carbonate of soda placed in the lower part

Fig. 70.—SIMON'S Sharp Spoon for scraping cancer.

* The mixture should be made cautiously, the bromine being slowly added to the spirit, on account of the heat developed, and care should be taken not to inhale too much of the fumes, which are irritating to the lungs, and may even damage the sense of smell.

of the vagina to protect the intact mucous membrane by neutralizing the bromine which escapes. If the vagina be sufficiently protected, the caustic may be left in place from six to twelve hours.

If the attempt be made to eradicate the cancer by applying a caustic, after the removal of all the obviously diseased tissue, a powerful indiscriminate, rather than a selective, caustic should be chosen, in order to destroy all the neighbouring tissue into which a few cancer cells, or cancer germs, may have penetrated. One of the most effective is chloride of zinc, which may be used in a solution of 300 grains to the ounce, as recommended by Dr. Marion Sims. A tampon of cotton wool is soaked in the solution, squeezed rather dry, to prevent the action of the caustic extending too far, and packed into the cavity. A tampon soaked in oil, or a piece of gutta-percha skin wrapped in oiled lint, may be placed next, and then a larger tampon, soaked in carbonate of soda, in order to protect the vagina as far as possible. After about six hours, the lower tampons should be removed, and the vagina syringed with a solution of carbonate of soda. Used in this way, the caustic brings away, after the lapse of a week or so, a white cup-shaped slough, free from smell, and a quarter of an inch or more in thickness, from all the tissue to which its action has reached, sometimes including even the intact vaginal wall. It must not, therefore, be employed if there is only a very thin sëptum intervening between the cavity and the bladder or peritoneum. A good deal of pain is produced, and opiates are generally required after the application of the caustic.

After removal by the écraseur of a cancer apparently confined to the vaginal cervix, sections should always be made of the piece removed, and if a milky juice can be scraped from the surface up to the plane of section by the wire, or if the microscope shows the cancer to have extended so far, the stump should be scraped out with the sharp spoons, and bromine, chloride of zinc, or other caustic afterwards applied. In several cases thus treated, after the first section had failed to remove the whole of the disease, I have found that from one to three years afterwards there was no sign of local recurrence, although in two cases the cervix had closed by gradual contraction, and an operation became necessary to evacuate retained menstrual fluid. If the recurrence of any ulceration in the vagina can be averted, the patient is saved from much distress, even though the disease reappear in the pelvis.

Dr. Marion Sims has lately proposed to excise the cervix more completely than has hitherto been attempted. The operation is performed with the aid of Sims' speculum, and the semi-prone position. If there is any prominent growth from the cervix, it is to be first removed by écraseur or scissors. Then the diseased tissue is to be cut away piece-meal, by scissors or Sims' knife (Fig. 19, p. 56), so as to remove a conical portion from the uterus, going up to or even above the internal os, if necessary. This is to be continued until the remaining tissue appears to the sense of touch to be healthy. Then a tampon soaked with the liquor ferri subsulphatis diluted with two parts of water, is to be placed in the cavity, and the vagina firmly plugged below it. This operation would be useless if the disease had reached the outer margin of the cervix, but since the growth almost invariably reaches a higher level along the cervical canal than at the outer part of the cervix, it may succeed in removing the whole of the cancerous tissue in cases in which the ordinary removal by écraseur would fail. But it also involves a much greater risk to life. Arterial hæmorrhage is apt to be formidable ; the pressure of the plug may cause a sloughing through into the peritoneal cavity or bladder, and there is some risk of septicæmia from retention of discharge by the plug. I have modified this operation by excising the cervix with the curved knife of Paquelin's benzoline cautery constructed for this purpose, or cutting it out with scissors and arresting hæmorrhage by the blunt-ended cautery of the same instrument, which is more effectual for this purpose than the use of the cautery-knife alone. In this way the necessity for plugging may often be avoided, and the risk of septic absorption thus diminished. Another method is to cut a groove upwards from the vaginal junction all round the cervix with the knife, scissors, or galvanic cautery, and afterwards apply the wire loop at the bottom of the groove. Dr. Marion Sims recommends the application of chloride of zinc, in the mode already described, after removal of the plugs. If the cautery has been used, the slough should be allowed first to come away.

Total extirpation of the uterus, for cancer of the cervix, was performed through the vagina by Dr. Blundell and others. The operation has been revived by Freund of Breslau, who has introduced a carefully-devised method for removing the whole uterus by abdominal section under carbolic spray, and closing the peritoneal wound by sutures, bringing the ligatures down into the vagina. Out of his first ten cases, Freund reported five recoveries from the effects of the operation, but later operations have been more fatal, and the growth has recurred in a large proportion of those who survived. It does not appear, therefore, at present that this operation can be recommended.

The operation of extirpation through the vagina has also been revived, and appears somewhat less fatal than that by the abdomen, but the proportion of recurrences afterwards is still large. The uterus is separated by cutting upwards with scissors, the vessels being tied as divided, and the peritoneal wound is left open. Mr. Spencer Wells recommends that instead of ligatures, large long pressure forceps should be applied to each broad ligament and left in place several days, till all fear of bleeding is past.

Palliative Treatment.—In a considerable proportion of cases of cancer of the cervix there is no hope of benefit from even a partial removal of the growth. In the slower and more infiltrating forms of the disease, with little tendency to vegetation, hæmorrhage may often be kept in check by occasional application to the ulcerated surface of somewhat milder caustics than hitherto mentioned, such as strong nitric or carbolic acid, or a saturated solution of chromic acid. When the disease is too extensive, and ulceration too far advanced to allow any stronger caustic to be used, the condition of the surface may often be improved, and hæmorrhage and fœtor diminished, by the occasional application of the dried sulphate of zinc in powder. This may be kept in place by a tampon of cotton-wool, and left from twelve to twenty-four hours. If hæmorrhage is severe, the liquor ferri perchloridi fortior, which acts as a caustic of moderate strength as well as a styptic, or a paste made with the solid perchloride of iron and glycerine, may be applied from time to time. Besides the stronger caustic applications, great benefit may be derived from the constant use of astringent and antiseptic solutions, by which the surface of the growth is hardened. As an astringent and antiseptic combined, from one to two drachms each of tincture of iodine and of solid perchloride of iron, dissolved in a pint of water, form perhaps the most useful lotion. As simple astringents, alum, iron alum, or acetate of lead may be used. As an antiseptic, permanganate of potash is of little avail. Carbolic acid is perhaps the most powerful, but weak solutions of iodine or bromine, or a lotion containing two drachms to the pint of liquor sodæ chloratæ, or liquor calcis chloratæ, are also effective.

Chian turpentine, recently introduced by Mr. John Clay, as a cure for cancer of the cervix, has been found to fail, like other reputed cures for cancer. I have tested it in a considerable number of cases, and have not found any arrest of the growth in any one. In some cases at an early stage I have found that, over an interval of several months, the hæmorrhage and sometimes also the pain, have appeared to be diminished, and the surface of the growth has seemed to bleed less easily on touching. In other instances, the drug appears to be useless. It is probable that the benefit derived from it may be explained by its acting as a styptic, like other forms of turpentine, and diminishing the blood-supply. The Chian turpentine is likely to be best absorbed if given in emulsion rather than in pills. It may be dissolved in ether or hot rectified spirit, and the solution added to a mucilaginous mixture, flavoured with syrup of lemons or syrup of ginger. Mr. Clay recommends eight grains for each dose.

As regards the general treatment, total sexual abstinence should be strictly enjoined, diet should be light, and stimulants should be avoided, or used sparingly. Internal remedies do not exercise much control over hæmorrhage, but ergot and gallic acid may be of service, in conjunction with local measures. In most cases, as the disease advances, the most urgent indication is to alleviate pain. In the earlier stages, hyoscyamus, with camphor, cannabis indica, especially in the form of chlorodyne, belladonna, or conium may be tried, but generally their effect is not to be compared with that of opium and its alkaloids, and their chief use is for those cases in which the latter are not well tolerated. If opium and morphia are not well borne when taken by the mouth, they will often answer in the form of suppositories or subcutaneous injections. Battley's liquor opii sedativus or nepenthe is generally the most suitable form of opium for protracted use. The dose should not be increased too quickly at first, but, in a fatal disease, there should not be too much reluctance to establish an opium habit, and very large doses may be required before the close. Care should be taken, at the same time, to regulate the bowels, and secure that the fæces are soft. The general cachexia may be combated in some degree by tonics, especially quinine and iron, and gastric remedies are often required to alleviate indigestion.

CANCER OF THE BODY OF THE UTERUS.

Causation.—Cancer of the body of the uterus, while very much less frequent than that of the cervix, is yet not extremely rare. It does not show the same preference as cancer of the cervix for married women and those who have had many children, but is, on the contrary, more common in the nulliparous. True carcinoma of the body of the uterus occurs later in life than that of the cervix. It is rare under 40, and commoner between 50 and 60 than between 40 and 50. Sarcoma, however, occurs with a relatively greater frequency during the period of sexual activity.

Pathological Anatomy.—There are two chief forms of cancer of the body of the uterus, namely, true carcinoma and sarcoma, of which the latter is generally regarded as being much the most frequent. The round-celled sarcomata have clinically all the characters of malignancy, although their course is generally not so rapid as that of medullary carcinoma. Even the spindle-celled sarcomata, though much slower in growth, and deviating less from benign tumours, may lead to metastatic deposits in distant organs. Of sarcoma of the body of the uterus there are two varieties. The first, which is more frequently of the spindle-celled kind, arises in the muscular walls of the organ, often from degeneration of a fibroid, but is never encapsuled. It may grow into a polypoid form, and be only distinguishable from an ordinary fibroid polypus by its microscopic structure, and by the fact of its recurrence. Such tumours are described by the older writers under the name of "recurrent fibroid." More or less of muscular tissue may be contained in them, their structure being that of myo-sarcoma, or fibro-myo-sarcoma. The second variety of sarcoma grows from the internal surface, and is usually of the round-celled kind, probably having its origin in the round or elongated connective tissue cells of the mucous membrane. It rapidly assumes a fungating character, and readily breaks down, leading to hæmorrhage and fœtid discharge, and so assuming an obviously malignant character. True carcinoma may commence from the uterine glands, which at first may simply extend deeply into the tissue, while the glandular acini retain their character. Then proliferation of the epithelium takes place, filling up more or less the lumen of the acini ; the cells lose their character of cylindrical epithelium, becoming heaped up irregularly. Clusters of similar cells begin to form in the parenchyma around, or migrate thither from the acini, and thus the growth degenerates into carcinoma, a more or less alveolar arrangement being eventually produced. In some cases the growth may retain, in great measure, the character of gland-tissue, resembling the growths to which the title of cylinder-epithelioma or adenoid cancer has been applied, and which are found in the alimentary canal, and sometimes even appear in distant organs by metastasis. This cylinder-epithelioma appears to be the commonest form in which cancer commences from the internal surface of the body of the uterus. In several instances I have found it almost uniformly distributed over the whole of that internal surface, as if the mucous membrane had undergone enormous thickening. There is always, however, more or less deviation from the characters of normal glandular tissue. It is probable that true carcinoma may also have its origin in the parenchyma of the uterus ; but this question must be regarded as yet open to doubt. As in other organs, the carcinoma may either take the form of more or less isolated nodules, or may infiltrate the whole organ diffusely.

Remarkable cases sometimes occur in which, generally after the menopause, the uterus becomes distended into a globular cavity containing pus, without any obstruction to the outflow through the cervix. To this condition, as well as to that in which the uterus is filled with pus with occlusion of the cervix, the name of *pyometra* is applied. To the naked eye, at any rate, the appearance is that simply of inflammation of the uterine walls, but the cases are apt to run an apparently malignant course, and to end by perforation of the uterine wall, and the formation of a sloughy cavity among the intestines. In two cases I have found that, although microscopic sections showed for the most part merely infiltration with inflammatory cells, yet here and there evidence of malignant growth, sarcoma or carcinoma, in the uterine wall, could be detected. It would seem probable that in many, at any rate, of these cases the inflammatory condition is set up by the presence of cancer, although the symptom of hæmorrhage is absent.

Results and Symptoms.—In all forms of cancer of the body of the uterus the main symptom is usually hæmorrhage. Fœtid discharge occurs in carcinoma and round-celled sarcoma when disintegration of the surface has taken place, but is more usually absent in spindle-celled sarcoma, except at quite a late stage. Severe spasmodic or lancinating pain is an early symptom in many cases, but in others it is absent throughout or up to a late period, though usually present when surrounding organs are becoming infiltrated. In the latter stages cancerous cachexia becomes marked. The cancer may extend to the cervix, to all neighbouring organs, and by metastasis to different parts. It is liable also to break down into cavities which may penetrate the uterine wall, so that cyst-like spaces are formed, with gangrenous or semi-purulent contents. These may at first be limited by pseudo-membranes, but are apt to lead to perforation into the peritoneal cavity and fatal peritonitis. Otherwise death may be brought on gradually by hæmorrhage and exhaustion. The fatal result may also be due to peritonitis without perforation, or to septicæmia.

Diagnosis.—The disease in its early stage is very apt to be mistaken for a fibroid tumour, especially for a fibroid tumour complicated by fixation due to peri-

T

uterine inflammation. The differentiation may sometimes be made by the fact that in cancer there is
generally a frequent recurrence of hæmorrhage during
the intervals of menstruation, or a persistent blood-
tinged discharge. Profuse hæmorrhage is also likely
to be produced by the use of the sound. The fact of
the tumour commencing or growing rapidly after the
menopause would also be in favour of its being malignant; so, too, is the presence of ascitic fluid. If a
soft fungoid mass, not being the product of conception,
is felt within the uterus during the period of active
sexual life, the probable diagnosis is that of round-
celled sarcoma. The only certain mode of distinction,
however, in the earlier stages of cancer is to remove a
portion of tissue for microscopic examination. The
simplest mode is to bring away a small fragment by
the blunt wire curette (*see* p. 212) or in the eye of a
silver catheter which has been introduced and turned
round once or twice. If this fails, the cervix may be
dilated, and a fragment removed by finger, or by the
blunt or sharp curette. To enable any positive conclusion to be arrived at, the fragment should be large
enough to allow sections to be made of it, after harden-
ing, and the observer should have had experience in
examining sections of uterine mucous membrane.
healthy and diseased. In a late stage induration and
fixation of surrounding parts take place, and nodular
masses like glands may be detected.

Treatment.—If Freund's operation for extirpation
of the whole uterus, or the vaginal method of extirpation, should hereafter show a less formidable mortality,
cases of cancer, especially cylinder-epithelioma, of the
mucous membrane of the body of the uterus would be,
perhaps, more suitable for it than any others, while
the uterus remains perfectly movable, since it is generally some time before the disease reaches the outer
wall of the uterus, or leads to secondary deposits. In
general the treatment can only be palliative, especially
for relief of pain. Severe hæmorrhage may be checked

by occasional applications of nitric acid, or by scraping away the proliferating surface by Simon's spoons, or the sharp curette, the cervix being first dilated if necessary. Any polypoid masses of sarcoma or carcinoma should be removed by the écraseur.

TUBERCULOSIS OF THE UTERUS.

Tuberculosis of the uterus is rare, and is almos always associated with the same disease in other organs. Tubercle is deposited in the mucous membrane, and is transformed into a cheesy material, which breaks down and leads to ulceration. The whole interior of the body of the uterus may thus be converted into a ragged cavity, the disease generally not extending to the cervix. The Fallopian tubes are commonly affected in the same way, and tuberculosis of the peritoneum and ovaries is also frequently associated.

The affection often escapes notice in the more important disease of other organs. The local *symptoms* are purulent discharge, with occasionally hæmorrhage, but, as a rule, amenorrhœa rather than menorrhagia. The uterus is found uniformly enlarged, and may also be fixed. The *diagnosis* is assisted by evidence of tuberculosis elsewhere, especially tubercular peritonitis. The *treatment* can only be palliative.

CHAPTER VIII

DISEASES OF THE OVARIES.

MALFORMATIONS OF THE OVARIES.

THE ovaries may be congenitally absent, but this defect is almost always associated with absence of the uterus, and generally with want of development of the vagina, vulva, and breasts. More frequently, while the uterus is absent, the ovaries are developed. In the absence of the ovaries, a childish condition is generally perpetuated in the whole body, and the stature remains small, but in some cases there may be an approximation towards the male type. Sexual feeling is always absent.

Imperfect development of the ovaries is of much greater frequency. It may be associated with a rudimentary condition of the uterus, or, more frequently, with a small anteflexed uterus, and small vagina, while occasionally the uterus is well-formed. Menstruation and the general changes associated with puberty are either deferred, or entirely fail to appear. Menstruation, when it does commence, is scanty and irregular, and is liable from slight causes to be arrested for a long period or permanently, while the menopause generally occurs early. General development frequently either does not proceed much beyond the childish stage, or the body is muscular, with a tendency to production of hair on the chin and legs, and frequently a harsh voice. The pelvis is often uniformly small, or of a childish or masculine type. There is also usually a deficiency of sexual feeling, which may lead to unhappiness in married life. Even if the development of the ovaries is only so far imperfect or retarded as to lead to the postponement of menstruation more than three or four years beyond the usual time, a serious permanent result may follow. If the growth of the pelvis does not receive that stimulus, which it usually derives at puberty from the development of the ovaries and uterus, until the age has passed at which the growth of the bones in general ceases, it is apt to retain permanently the childish type, and hence parturition may be obstructed if pregnancy ever subsequently occurs.

It is often very difficult to make an absolute diagnosis of imperfect development of the ovaries, for although defective development of the breasts, and of the feminine characteristics in the body generally, frequently coexists, yet this is not always the case. Even in the absence of such defects a probable diagnosis may be made in cases of prolonged amenorrhœa for which no other cause can be discovered, especially if there is a total absence of the periodical feelings of uneasiness in the pelvis, breasts, and system generally, which are termed the menstrual molimen, and if there is evidence of sexual indifference.

Complete absence of the ovaries is difficult to distinguish from imperfect development, but may be diagnosed with probability, if, in a not very stout woman, no trace of the ovaries can be felt on bimanual examination with the aid of an anæsthetic, if sexual feeling is entirely absent, and if, after a fair trial, all treatment fails to induce menstruation. All further treatment should then be abandoned.

Treatment.—Since imperfect development of the ovaries is generally only a probable diagnosis to account for prolonged or permanent amenorrhœa, its treatment will be considered under the head of amenorrhœa.

ATROPHY OF THE OVARIES.

The physiological atrophy of the ovaries, which usually happens about the menopause, may occur prematurely, and lead to cessation of menstruation. This may be the result of acute ovaritis, or of pelvic peritonitis or cellulitis, more especially of peritonitis, from the effect of which the ovaries may be bound down in an abnormal position, and the natural liberation of the ovules prevented. Sometimes, also, it occurs without any local affection as the sequence of a serious illness, or from the effect produced upon the nervous system by a deep sorrow or other strong emotion. This is more likely to occur if the ovaries are from the first somewhat imperfect in development.

Treatment.—The treatment of atrophy, like that of imperfect development, of the ovaries will be described under the head of amenorrhœa.

PROLAPSE OF THE OVARIES.

In a normal condition, the ovary rests as far forward as its attachments will allow it, although posterior to the plane of the broad ligament, being kept in position by the pressure of the intestines, which fill the fossa behind it. Its most notable displacement is one in which it drops below its normal level, and too far backward, descending into Douglas's pouch, and at the same time, owing to its attachment to the angle of the uterus by the ovarian ligament, is necessarily brought nearer to the middle line. The causes of this displacement are (1) increased weight of the ovary due to hyperæmia, hyperplasia, or commencing degeneration of a cystic or any other kind; (2) laxity of the mesovarium, or of the broad ligament generally, and (3) retroversion, retroflexion, or prolapse of the uterus. By these displacements of the uterus the ovaries are necessarily carried backward and downward, and the coils of intestine which normally keep them in position are displaced.

Results and Symptoms.—Ovaries in this position are almost invariably affected by chronic hyperæmia or ovaritis, and not unfrequently they become fixed by inflammation of their peritoneal covering. The displacement, by rendering the ovary more exposed to pressure or traumatic influences, tends to promote or maintain the inflammation. Thus the symptoms of ovarian hyperæmia or inflammation are present, often in an acute degree, those specially intensified by the displacement being pain in defecation and pain on coitus.

Diagnosis.—When thus prolapsed the ovaries can be reached more or less easily by the finger in vagina or rectum. When the prolapse is slight the examination must be made in the lateral position, and the finger, with its flexor surface towards the sacrum, must be carried as high as possible posterior to, and a little to one or the other side of, the cervix. The ovaries are recognized by their size and their shape, somewhat globular, but having often nodular irregularities. They are generally also more or less movable, and have a peculiar tenderness, analogous to that of the testicle. Often sickening pain, and not unfrequently hysterical manifestations, are produced when they are pressed. In carrying the fundus uteri forward with the sound the displaced ovary is elevated to some extent. There may sometimes be a difficulty in distinguishing a prolapsed ovary from the retroflexed body of the uterus. The distinction may be made by restoring the uterus, if retroflexed, completely with the sound. An ovary, if present, will then still be felt posteriorly to the cervix, although generally elevated to some extent by the replacement of the uterus. If both ovaries are prolapsed, the double tumour makes the diagnosis more obvious. It is, of course, necessary to be careful not to mistake a mass of scybala in the rectum for a prolapsed ovary. In any case of doubt, digital exploration of the rectum will decide the point; and, if an ovary is only just within reach, it may often be more readily touched by this mode of examination than *per vaginam.*

Treatment.—When any degree of retroversion or retroflexion of the uterus is present it is important to remedy that displacement, if possible, but the presence of the tender ovary frequently renders it difficult to adapt any pessary which can be tolerated, even after hyperæmia has been treated by rest and local depletion. In many cases, however, a Hodge's pessary may be found by trial, which will restore the uterus and elevate the ovary in some measure. A thick instrument should be chosen, or one having a broad cylindrical expansion at its posterior part, like that of Dr. Thomas (Fig. 30, p. 83). If this cannot be tolerated, an elastic ring pessary (*see* p. 123) may prove useful. The general treatment should be that of ovarian hyperæmia and inflammation (*see* p. 289), special care being taken to render the fæces soft.

HERNIA OF THE OVARY.

The ovary may descend into a hernial sac, generally of the inguinal kind. A congenital, but very rare, form has been described in which one, or usually both, ovaries descend by a fault of development into the labium majus, or into a pouch of peritoneum which remains open in the inguinal canal, just as the testis descends into the scrotum in the male sex. The descended body is then generally irreducible, and other malformations of the genital organs often exist also. Acquired hernia of the ovary is generally associated with hernia of intestine or omentum, and is more likely to occur soon after delivery, when the attachments of the ovary are loose. The ovary is apt to become inflamed and degenerated in its abnormal position. In the case of an apparent hermaphrodite, a body of doubtful nature is more likely to be a testicle than an ovary. This holds true even if the external genital organs and general type are entirely feminine. Such cases appear to be usually instances of the rare condition of "transverse hermaphroditism," in which the external organs are female and the internal male. Thus, doubtful bodies congenitally placed in the inguinal canals or labia, may be assumed to be probably testicles, unless either they are found to undergo regular enlargement at monthly intervals, or an unequivocally developed uterus is discovered.

Treatment.—An acquired hernia may be reduced, if possible, and its return prevented by a truss. In congenital or irreducible hernia, the ovary should be protected by a concave shield. If the ovary becomes inflamed, and causes very severe distress, it may be excised, an operation which has been found necessary in several recorded cases. The operation should be performed with antiseptic precautions.

ACUTE OVARITIS.

Acute ovaritis (or oophoritis, as it has been called with greater philological propriety) is a rare affection. In its most severe form, leading to the formation of abscess, it is generally the result of septic absorption after delivery or abortion, and forms part of an acute inflammation of the broad ligaments and adjacent peritoneum. It may also, apart from parturition, be associated with pelvic peritonitis or cellulitis, especially but not exclusively, with that of septic origin. An abscess in the ovary may also follow operations on the uterus, intra-uterine application of caustics, or the use of intra-uterine stems. An abscess originating in ovaritis may in very rare cases run a chronic course and present signs similar to those of a small cystic tumour. It is much more common, however, for suppuration to occur secondarily to cystic disease. An abscess of the ovary may burst into the peritoneal cavity and lead to fatal general peritonitis.

A somewhat less acute ovaritis, not usually ending in abscess, may result from the extension of acute endometritis, especially that of gonorrhœal origin. The infection appears to extend directly to the ovary, when it is embraced by the Fallopian tube at a menstrual period. The ovaritis may also arise through the medium of pelvic peritonitis, which is itself a common result of gonorrhœal inflammation. Cases of acute ovaritis have been traced to exposure to gonorrhœal infection, even when there has never been any manifestation of acute vaginal inflammation. Acute ovaritis, not usually leading to suppuration, may also occur in the course of specific fevers, such as small-pox. As the result of acute ovaritis the tissue may be so disorganized that ovulation ceases, and permanent amenorrhœa is the result. The sub-acute forms are apt to leave a chronic ovaritis behind, and sterility is a common consequence of the peritoneal adhesions which remain around the ovary.

Diagnosis.—The symptoms of acute ovaritis are often merged in those of the septicæmia, peritonitis, or cellulitis, with which it is associated, and this is almost always the case in the most severe forms of the disease. In cases, however, in which the ovaritis is the prominent feature, a diagnosis may be made from the localization of pain and tenderness, and from the recognition, on bimanual examination, of a rounded swelling, not usually movable, in the position of the ovary.

Treatment.—In the septicæmic form no special treatment can be directed to the ovary. In simple inflammation, when acute ovaritis forms the chief part of the affection, perfect rest should be enjoined, and leeches may be applied to the groin, round the anus, or to the cervix uteri. Poultices or fomentations should also be applied, and opiates may be given with iodide of potassium. If an abscess is recognized special care should be taken to avoid any movement which might lead to rupture into the peritoneal cavity. If fluctuation can be felt from the vagina the pus should be evacuated by the aspirator.

HYPERÆMIA OF THE OVARY AND CHRONIC OVARITIS.

Causation.—Since the ovaries, like the uterus, are naturally subject to a periodical active hyperæmia, this hyperæmia is easily rendered excessive by various causes, and may pass into actual inflammation. It is still more difficult than in the case of the uterus to draw any positive line between hyperæmia and inflammation, the ovaries being less accessible to observation. The tendency to ovarian hyperæmia is often a constitutional peculiarity of the individual, and is probably associated with excessive development of the organ, or of the sexual emotion on its mental side. It is generally found in women of an emotional and hyperæsthetic temperament, with frequently a predisposition to hysteria. Such women begin to menstruate early in life, and their menstruation is habitually profuse, until after marriage and parturition, by which it is often rendered more normal. Unless brought on prematurely by some cause of uterine or ovarian degeneration, the menopause generally occurs late.

The most important causes of reflex ovarian hyperæmia are morbid conditions of the uterus, which is more liable than the ovary to displacement, and to disturbances of menstruation dependent upon malformation, and more exposed to traumatic influences. Chronic ovaritis may probably also be a sequel of chronic endometritis by direct extension of inflammation from the uterus along the Fallopian tube, without the occurrence of any acute ovaritis. When symptoms of obstructive dysmenorrhœa have existed from puberty, it is by no means uncommon for symptoms of congestive dysmenorrhœa, apparently ovarian in

character, or those of chronic ovaritis, to be added after
some years. In such instances it is impossible to
determine whether reflex nervous influence only comes
into play, or whether there is direct extension of
inflammation from the chronic endometritis which
generally exists. Ovarian hyperæmia may also be
produced by sexual excitement or excess. Masturbation is undoubtedly one of the causes of hyperæmia,
both of uterus and ovaries, but is much less common
than in the other sex. A similar effect may result
from imperfect coitus. This may be dependent either
upon premature emission on the part of the husband,
the result of former habits of masturbation or other
causes, or upon relative sexual frigidity or want of
general vigour on the part of the wife. It may also
occur if, from any cause, coitus takes place in such a
way that the clitoris does not receive sufficient excita-
tion. By the failure of the natural orgasm (analogous
to the orgasm of emission in the male) on the part of
the woman, which failure is by no means uncommonly
habitual, the normal sedative to sexual excitement and
congestion is removed. From such a cause, not only
does local congestion arise, but, more especially,
hysteria is apt to be produced. Celibacy must be
reckoned among the causes of ovarian hyperæmia,
since after marriage menorrhagia, congestive dysmenorrhœa, and other signs of ovarian irritability are
often relieved, even if pregnancy does not occur. But
on the other hand chronic endometritis or metritis
with sterility is often associated with ovarian hyperæmia or ovaritis, due in part to the want of that
physiological rest to the ovary which is afforded by
pregnancy. Swelling and tenderness of the ovaries are
not unfrequently also found in women who have had
children, sometimes apparently as the result of child-
birth. They may then be the sequel of the laceration
or inflammation of cervix, retroflexion, partial prolapse,
or other lesion of the uterus which may be the consequence of parturition. In these cases the affection

does not usually show the extreme obstinacy which it often manifests in nulliparous women.

Passive Hyperæmia of the ovary is produced by general causes similar to those which lead to the same condition in the uterus, especially by constipation. When the ovary has once become prolapsed its venous circulation is further interfered with, and it becomes more exposed to direct causes of irritation. Passive hyperæmia renders the organ more vulnerable to causes of inflammation, and tends to produce hyperplasia and enlargement. From induration of the superficial tissue the normal rupture of follicles may be interfered with, and inflammation thus secondarily set up, or the foundation of cystic degeneration laid. The importance of passive hyperæmia as a predisposing cause of chronic ovaritis is shown by the preponderance of that affection on the left side. This, like the usual occurrence of varicocele on the left side in the male sex, must depend upon the presence of the rectum and sigmoid flexure on the left side, whereby pressure upon the veins is liable to be produced, and upon the more indirect course of the left ovarian vein, opening into the renal vein instead of directly into the vena cava.

Inflammation of the ovary may result simply from an intensification of the reflex irritation which leads to active hyperæmia, or it may be produced indirectly by the hyperæmia leading to an excess in the normal slight effusion of blood on the rupture of a follicle. The results of this may be irritation and inflammation either directly in the ovary, or primarily in the adjacent peritoneum, and secondarily in the ovary. Chronic ovaritis is also frequently the sequel of acute or subacute ovaritis, especially that of gonorrhœal origin. It is a common result again of pelvic peritonitis, either by direct extension of inflammation, or from the obstacle to normal ovulation which thus arises, and the interference with the venous circulation. Inflammation may also be set up by irritation due to the presence of follicles, either simply distended with

limpid fluid or in a state of commencing cystic
degeneration.   Dr. Matthews Duncan declares his
belief that the most frequent cause of chronic ovaritis
is the use of alcoholic liquors, even when not taken to
excess ; and says that this view of the causation of the
disease is frequently corroborated, if not proved, by
the cure which follows upon the adoption of teetotal
living.

Pathological Anatomy.—That organic change in
functionally active ovaries, passing beyond the stage of
mere hyperæmia, is very common, is shown by the
frequency with which, after death, signs are found of
a very limited local peritonitis, apparently having had
its origin in those organs.   The ovaries themselves
also are often enlarged and nodular from irregular
hyperplasia, and in such cases frequently contain small
cysts containing limpid fluid, and formed by enlargement of the Graafian follicles, probably often the
consequence of a previous fibroid degeneration of the
stroma.   When the ovaries have been removed by
oophorectomy, on account of extreme nervous symptoms, dependent upon ovarian irritation, they have
frequently been found degenerated, and enveloped in
adhesions, even though no pelvic peritonitis had ever
been diagnosed.   Still more frequently they have been
found full of the small cysts containing clear fluid,
already mentioned, to be distinguished from the
ovarian cystoma, containing glairy fluid, which more
generally goes on to the formation of large ovarian
tumours.   It would seem that cysts of the former kind
not very unfrequently increase to such a size as to enlarge the ovary into a tangible globe, from two to three
inches in diameter, and that this condition may undergo
spontaneous cure by rupture of the cyst, without the
production of any serious symptoms.   Distinctions
have been made between follicular and interstitial
ovaritis, but they cannot practically be clinically
separated, although inflammation of the follicles is
doubtless generally the primary change, except in the

acute septic forms of the disease. In the advanced stage
of fibroid degeneration the ovary is small and contracted, and Graafian follicles, in any advanced stage
of development, are scarce or absent.

Results and Symptoms.—Pain in one groin does
not necessarily indicate ovaritis, but is a common
result of uterine disease. But in ovarian hyperæmia,
or inflammation pain in the groin, and extending
down the thigh, is a marked symptom, while there is
also tenderness in the ovarian region, and the muscles
on the affected side are rigidly contracted to protect the
tender spot. Menorrhagia is a usual symptom, except
in the later stage, when the ovary is atrophied, or when
it has been severely damaged by acute inflammation.
In other cases, however, the uterus may be imperfectly
developed, or be in the cirrhotic stage of chronic
metritis, and then menstruation may be scanty, while
the insufficiency of the flow is in part the cause of the
ovarian hyperæmia. A more extreme hyperæmia has
sometimes been observed in cases of entire absence of
the uterus. The pain in the ovarian region is usually
aggravated in connection with menstruation, and the
aggravation generally commences a few days or a week
before the period. It is often relieved by the flow,
provided that no cause of obstructive dysmenorrhœa
coexists. If there are prolapse and enlargement of the
ovary, pain on coitus is often a marked symptom, and
in this case defecation also is apt to be specially painful. In accordance with the physiological function
of the glands of the cervix uteri, increased secretion
of these glands, without any altered quality of the
secretion, may be produced by ovarian irritability or
undue sexual emotion. Such a condition, therefore,
may be a cause of leucorrhœa, without any morbid
change in the uterus itself.

The reflex nervous symptoms enumerated under the
head of corporeal endometritis and metritis (*see* pp. 207,
208) are still more marked, in susceptible subjects, in the
case of ovarian hyperæmia or chronic ovaritis. The chief

effects produced are nausea, vomiting, flatulence, or other gastric neuroses, pain under the left breast or at the top of the head, and, above all, hysteria. Hysteria, while largely dependent on constitutional proclivity, is commonly due, in the first instance, to some actual source of pain, which, in predisposed subjects, leads to such a state of irritability that after a time the slightest stimulus of a physical or mental kind is sufficient to evoke hysterical manifestations. The prime source of irritation is not necessarily in the sexual system, since hysteria may sometimes be induced before the age of puberty, or after the menopause, from the effect of a wound or injury. But in the hysteria of young adults, some source of irritation in uterus or ovaries, or some mental condition connected with the sexual emotions, such as a disappointment in love, or, in married women, an absence of perfect satisfaction in the marital relations, appears to be the commonest cause. In the extreme forms of hystero-epilepsy with hallucinations recorded by Professor Charcot, the connection with the ovary is reported as being a constant one. It is probable also that, when uterine disturbance is the prime factor, reflex ovarian irritation is often an intermediate step in causation. Thus, in the not uncommon case of retroflexion and engorgement of the uterus, with prolapse of the ovaries, in a hysterical subject, nervous manifestations are usually more easily produced by pressure upon an ovary than by that upon the uterus. Hysteria is not necessarily the result of sexual excitement or sexual abstinence, for it may occur for the first time in a married woman as a sequel to uterine displacement, resulting from parturition, or to peri-uterine inflammation, as well as from causes altogether independent of the sexual system. But in the strong emotional susceptibility of hysterical subjects, the sexual emotion usually takes part; and this is often the case in a special degree when ovarian hyperæmia is the starting point of irritation.

Diagnosis.—Pain and even tenderness on external

pressure in the ovarian region is not sufficient ground
for positive diagnosis. If a vaginal exploration be
made very gently and carefully, it will be found that,
although a general hyperæsthesia often exists, a special
and extreme tenderness is manifested when pressure
is made upon the ovary. If the ovary is more or less
prolapsed, so as to be reached by the finger in the
vagina somewhat behind the cervix, this tenderness
is easily recognized, and the ovary is often felt to be
enlarged and nodular. If not, the ovary may often, on
bimanual examination, be caught between the two
hands in its normal position, and made out to be
enlarged and tender but generally movable. Sometimes the rigidity of muscles prevents this, and, unless
an anæsthetic be given nothing more than increased
resistance and excessive tenderness localized in the
ovarian region can be detected. Frequently rectal
exploration allows the finger to reach the ovary more
fully than is possible by vaginal touch.

Treatment.—All postural causes of passive hyperæmia, such as prolonged standing or sitting, should be
avoided. Long-continued practising on the piano, and,
still more, playing on the harmonium, or the use of the
treadle sewing-machine, are especially injurious. Any
sources of undue emotion should be removed as far as
possible, and rest practised in moderation. It is of
importance to relieve the portal system and render the
fæces soft by saline laxatives, and it is often of advantage to secure an action of the bowels in the evening,
that there may be no source of venous congestion
during the night. In the case of married women, strict
moderation in coitus should be enjoined, but in ovarian
hyperæmia temperate use of the sexual function is
generally more salutary than total abstinence, provided
that no sufficiently acute inflammation is present to
cause distress in intercourse. If local pain and tenderness are severe, they may be relieved by depletion of the
cervix uteri, and, in this case, leeching is more effective
than puncture. The effect, however, is not so direct as

U

in the case of hyperæmia of the uterus; and the depletion should not be too often repeated, lest it have deteriorating effect upon the general health. Counter-irritation is usually preferable to depletion, and may be carried out by the application of blistering fluid to the groin over the tender regions, repeated at intervals. In milder cases the linimentum iodi may be painted repeatedly over the same spot of skin, as long as it can be tolerated. Dr. Barnes recommends, as a still more efficacious counter-irritant, the application of caustic, such as the potassa fusa cum calce, to the cervix uteri (*see* p. 152). Counter-irritants often tend also to relieve reflex nervous symptoms, such as vomiting, the second impression having apparently an inhibitory effect upon the primary irritation. To relieve reflex vomiting it is often best to apply the counter-irritant over the epigastrium.

Of internal remedies, the most valuable for curative effect are the iodide and bromide of potassium. The former, when long-continued in sufficient doses, tends to cause atrophy and absorption of ovarian, as of other glandular tissues. Bromide of potassium relieves active hyperæmia of the pelvic organs in general, and acts also as a sexual sedative. It is also better tolerated for a long period than iodide of potassium. When gastric and intestinal neuroses are present these drugs may be combined with bitter stomachics and laxatives.* Small doses of perchloride of mercury, administered for a long period, may also be tried as an absorbent (*see* formula, p. 214).

In the more chronic stage, a general tonic treatment is desirable, especially for the cure of the nervous or hysterical symptoms, which often persist after the prime irritant cause has, in great measure, been removed. Cold baths are specially efficacious, and when aided by the change of scene afforded by a course of treatment

---

* The following is a useful formula :—℞. Magnesiæ Sulphat. gr. xxx. ; Potass. Bromid. gr. xx. ; Tinct. Gentianæ co. ℨj. ; Aq. ad ℥j. ter die.

in a hydropathic establishment, have often an additional effect. Sometimes sea-bathing, or the addition of salt to the baths, proves still more beneficial. Cinchona or quinine may be combined with bromide or iodide of potassium. Iron must be avoided if any considerable menorrhagia, or active hyperæmia, is present, but for the cure of nervous symptoms it is of great value, and it is specially useful if menstruation is scanty. It is often better borne if combined with bromide of potassium, and a laxative, should be added if necessary. Alcohol should be used sparingly, and all sedatives, and especially opiates, should be reserved as much as possible for paroxysmal attacks, when they may be used in the manner described under the head of congestive dysmenorrhœa. Great relief, at such times, is afforded by warm hip-baths, and still more efficacious is the whole bath, in which the patient should remain for a considerable period. Poultices or fomentations may also be employed when pain is acute. The use of alcohol for the relief of any nervous symptoms, or at any other time than with meals, is specially to be discouraged.

In a few cases of extreme nervous disturbance arising from ovarian irritation, especially those in which reason appeared to be threatened, the operation of oophorectomy—that is, the removal of the functionally active ovaries, either by vaginal or abdominal section— has been performed by Dr. Battey and others. Dr. Battey chooses vaginal section by preference, but the other method gives a greater probability that the whole ovary can be removed, even if adherent, and allows of more perfect antiseptic precautions. Experience has shown that the operation is likely to fail in curing unless both ovaries are removed.

The mode of operation recommended by Dr. Battey is to make a small incision, posterior to the cervix, by the aid of Sims' speculum, until the pouch of Douglas is opened. Each ovary in succession is drawn down into the vagina, has a temporary ligature placed round the mesovarium, and is then crushed off by an écraseur, no permanent ligature being left. Dr. Battey reports thirteen cases operated on by the vaginal method, with three deaths; three by the abdominal method, with no deaths. In all his cases in which both ovaries were completely removed, nine in number, menstruation ceased; in the others it continued as before. In 123 tabulated cases of oophorectomy, or removal of uterine appendages, not including operations performed for the relief of myomata, but including twenty-two operations in which both ovaries were not completely removed, there were fourteen deaths in 100 operations by the abdominal method, and four in twenty-three operations by the vaginal method, or 17·4 per cent. The mortality hitherto, therefore, is greater than might have been expected, but it is notably greater in the hands of surgeons who have operated only a few times; while in those of some specialists in abdominal surgery it is very much less than the general average. Thus, Dr. Savage, of Birmingham, records a series of twenty-six cases, including four performed in cases of myoma, without a single death. The abdominal method is chosen by Lawson Tait and Savage, and will probably be that generally followed. It would seem that the operation has been in many instances performed too readily in cases of dysmenorrhœa, or complaint of severe pain in hysterical women. But, in view of the greatly improved rate of mortality, it may be allowed that, if there is opportunity for the operation to be performed by a specialist practised in abdominal surgery, it is worthy of consideration in some very exceptional and extreme cases in which the usefulness and enjoyment of life are entirely destroyed by symptoms which are believed to be referable to morbid ovulation. It should be an indispensable condition that the symptoms are manifestly associated with menstruation, so that there would be a strong probability that the menopause occurring naturally would effect their cure, and also that the patient is made fully to understand the significance of the operation. If the abdominal method be adopted, a short incision is made in the median line through the abdominal wall, sufficient to admit two fingers; the ovaries are sought for, and drawn forward successively into the incision by the fingers, any adhesions being separated by aid of the sense of touch. The mesovaria are then transfixed and tied by carbolized silk. The fact of the ovaries being prolapsed would be in favour of the vaginal method, but this should not be adopted unless there is a strong presumption that the ovaries are free from adhesion.

CYSTOMATA, OR CYSTIC TUMOURS OF THE OVARY.

Causation.—Cystic tumours are the most frequent and important of new growths in the ovaries. The origin of cysts in this situation, as in other places, has been attributed by some to the formation of a space in the interstices of the stroma, or to the enlargement of a single cell. The special frequency, however, of cysts in an organ which normally contains physiological cysts, namely, the Graafian follicles, and the rarity of the commencement of ovarian cysts except during the years of active sexual life, are sufficient to indicate that, in the great majority of cases at any rate, the cysts originate from abnormal growth either of the actual Graafian follicles, or of the embryonic structures from which they are developed, and are therefore a form of adenoma. In multilocular tumours the ovum or its remnant has actually been detected by Dr. Ritchie and others in many of the smaller cysts, whose size does not exceed that of a cherry, and whose contents are a limpid fluid, and thus the mode of origin of the cysts is demonstrated as regards these instances. In the larger cysts, and those having colloid contents, the ovum can never be detected, and even the smallest cysts differ from the more developed Graafian follicles in the fact that their lining epithelium consists of a

single layer of cells. It is maintained by Waldeyer
and also by De Sinéty and Malassez that ovarian cystomata are not developed from Graafian follicles at all,
but from ingrowing epithelial processes or tubes derived from the surface epithelium of the ovary, from
which epithelium the ova themselves are developed.
It is supposed that many cystomata have their origin
in fœtal or very early infantile life by a deviation from
the normal process of formation of Graafian follicles
out of epithelial processes of this kind, and that there
may be also an abnormal development of processes from
the surface epithelium in later life. De Sinéty and
Malassez describe the ingrowths as having a glandular
form, resembling cylinder-epithelioma (see p. 272). I
have found a similar glandular growth commencing from Graafian follicles, every stage being visible,
from that of a single pouch or diverticulum in the
wall of the otherwise spherical follicle. It would
seem, therefore, that cystomata may originate either
from a very early stage in the development of Graafian
follicles, as is probably the case with cysts containing
mucoid or colloid fluid, or from somewhat more advanced follicles. Small colloid cysts may be found
in the ovary even at birth, but appear usually not
to undergo enlargement before the age of puberty.
Nothing certain is known as to the cause of commencement of cystic growth, but it is probable that in some
cases it is due to fibrous hyperplasia of the ovary—
usually the result of previous hyperæmia or ovaritis—
preventing the maturation or rupture of the follicles,
or to their becoming developed too far from the surface
to allow of their reaching it. It is possible that the
premature death of the ovum, by preventing maturation
of a follicle, may lead to cystic degeneration. In the
normal condition, however, many follicles become atrophied without ever having ripened, and this cannot
therefore be the sole condition present. It is possible,
again, that the failure of the follicle to rupture may be
the result of an insufficient menstrual hyperæmia in

the ovary, such as occurs in chlorosis and other forms
of anæmia.

Pathological Anatomy.—An important practical
distinction is to be made between tumours consisting
mainly of one, or of a very few, large cysts, and those
made up of a great number of small ones. It is very
rare, however, for a true ovarian cyst to be actually
unilocular, and the two classes are rather to be termed;
paucilocular and multilocular cysts. At an early stage
of degeneration the cysts are almost invariably multiple,
and the large cyst generally arises by the breaking down
of the partitions between a number of smaller ones, or
by the growth of one cyst at a very much more rapid
rate than the rest. Thus a large number both of
paucilocular and multilocular tumours are merely aggre-
gations of simple cysts.

From these are to be distinguished the *proliferous
cysts*, in which there is a further departure from the
normal conditions of growth, and an approximation
towards malignancy. In these tumours secondary
cysts are formed all over the walls of the primary cysts,
instead of being merely developed out of the primary
ovigenous layer. The growth is at first of a glandular
character. The epithelial lining dips into the cyst wall
in the form of crypts, and these become closed cavities
which are afterwards distended by secretion. Some
describe the glandular formations as commencing in
the form of a closed cavity, beneath the superficial
epithelium. In another variety of proliferous cyst the
proliferation takes the form of a growth of papillary
processes from the cyst-wall. These are covered at
first with cylindrical epithelium, which becomes irregular
and multiform in its proliferation, and is often heaped
up in projecting masses, like bunches of grapes, which
are easily detached. On rupture of the cyst, either from
excessive papillary growth or any other cause, the
exuberant epithelium becoming detached may convey
cancerous infection to the peritoneum. This form of
growth has been called *cystoma proliferum papillare.*

At the same time the depressions between the papillæ generally tend to invade the tissue beneath in the form of branching glandular crypts. Thus is formed a tissue resembling cylinder - epithelioma, identical with that which appears to be the first stage in the formation of proliferous cysts. It may either be sharply limited, or may invade deeply the connective or sarcomatous tissue between the cysts. It is probable also that the acini may be totally filled by proliferation of the cells, and the adjoining cellular tissue infected, so as to constitute a true carcinoma. It has been supposed by some that the secondary cysts are generally formed by the union of adjacent papillæ, and this view was maintained by Dr. Wilson Fox. But the probability would be greatly against papillæ so uniting as to form a completely closed cavity, and moreover proliferous cysts do not always show a papillary surface.

The cyst walls of an ovarian cystoma are covered on their peritoneal surface with an epithelium like that of the peritoneum, and internally generally by cylindrical epithelium, which in the larger cysts is often converted into a single layer of flattened cells. The structure of their substance is of the connective tissue type, and varies from a fibrous or areolar tissue with few nuclei, such as is found in the walls of large and slowly growing cysts, to a more vascular and imperfectly formed tissue, which must be regarded as sarcoma, generally of the spindle-celled variety. By rapid growth of this tissue the thickness of the cyst-walls may become great in proportion to the dimensions of the cysts ; and, when the proportion of this solid matter is considerable, the tumour becomes a *cysto-sarcoma*. In proportion to the relative amount of solid material is the tendency towards malignancy in the tumour, and this is more manifest if the tissue has anywhere the character of round-celled sarcoma. Such form of growth frequently affects both ovaries together, and it tends to invade other tissues, when adhesions have occurred, and to recur in the pedicle, or by metastatic deposits, after removal. That, in comparison with other sarcomata, it does not earlier show a malignant character, and that it may be eradicated if removed early enough, probably depends upon the isolated position of an ovarian tumour, while free from adhesion.

The contents of the cysts vary from a gelatinous or colloid substance, which will not flow through a canula, to a clear and limpid fluid. In most cases the fluid is somewhat viscid, and its colour is often brownish or greenish. In the multilocular, and especially in the proliferous cysts, the fluid is usually more gelatinous. Frequently it varies greatly in the different cysts of the same tumour, and generally it is more viscid or gelatinous in the smaller cysts than in the larger. The more viscid fluids contain albumen and its derivatives, also albumen, paralbumen, metalbumen, and peptone. They also contain a considerable proportion of mucin. This is distinguished from the albuminous series of substances by its not being precipitated from its solutions by tannin or by neutral metal salts, and by its swelling up in water. The so-called *colloid* tumours are made up of a number of very small cysts containing colloid or mucoid fluid. In some cases cysts filled with gelatinous material rupture even while comparatively small, apparently from the tension produced by the abundant production of such material, which may then be found free in the peritoneal cavity. When this is the case secondary colloid degeneration in the omentum and other parts is apt to result. In some such instances gelatinous material is found not only in the cysts but among the fibres of the cellular tissue of the tumours, a condition which seems to be a further indication of a tendency toward malignant infection of adjacent parts. A clear limpid fluid, of specific gravity below 1010, containing only a trace of albumen, may be found in a true ovarian cyst, and even, in rare cases, in the several cysts of a multilocular tumour. If, however, a cyst containing such fluid is unilocular, it is more likely to be of parovarian origin (*see* p. 299).

In some instances the fluid of a cyst has escaped through the uterus, and such a discharge may happen on repeated occasions. In such cases the cyst is generally a *tubo-ovarian* cyst, which is described as originating in the following manner. A Graafian follicle ruptures while the point of rupture is enclosed within the pavilion of the Fallopian tube, adherent at its margins to the ovary. The communication between the follicle and the tube fails to close, and the follicle undergoes cystic dilatation. The pavilion, or a portion of the canal, of the tube then contributes to the formation of the resulting cyst, and the tube generally allows the passage of fluid only occasionally. A pseudo-cyst may also be produced by adhesion of the pavilion of the tube to the ovary, and distension of the cavity by serum or pus, or an ovarian cyst may rupture into the dilated extremity of the tube. A few cases have been recorded in which simple cysts containing limpid fluid have been found attached to the peritoneum without connection with the ovary, and these have been ascribed to cystic growth of an unimpregnated ovum, which had become attached to the peritoneum like the ovum in abdominal fœtation.

Ovarian tumours generally become pedunculated as they enlarge. The pedicle, which may be long and slender, or short and broad, contains a portion of the broad ligament stretched out, the ligament of the ovary, the ovarian vessels, and the Fallopian tube, which is generally much enlarged, and extended, more or less, over the surface of the tumour. In some cases, however, a cyst, having all the characters of an ovarian cyst, occupies the same position as the so-called " cysts of the broad ligament," having no pedicle, but descending deeply between the folds of the broad ligament. This condition may arise from a follicle having made its way, not to the surface of the ovary, but through the mesovarium into the broad ligament.

Parovarian Cysts, which constitute the most important variety of what have been called cysts of the broad ligament, are formed by distension of one of the tubules of the parovarium, or organ of Rosenmüller, a small body which is the relic of the ducts of the Wolffian body, and is situated in the thickness of the broad ligament, between the outer extremity of the ovary and the Fallopian tube. Their growth is slow, and they often do not increase beyond a small or moderate size, but sometimes they grow large enough to distend the whole abdomen. They are generally found in young women. The contained fluid is limpid, like water, of low specific gravity, generally below 1005, and contains only a trace of albumen, which is usually precipitated only by nitric acid, and not by heat alone. The cysts are almost always unilocular, but rarely may be made up of several, having thin septa, more than one tubule having become dilated. The cysts, in some cases, become pedunculated, but are more likely than true ovarian tumours to descend deeply between the layers of the broad ligament. The ovary is often found distinct, with its mesovarium intact. The Fallopian tube is usually more extended over the cyst than in the case of a true ovarian cyst, and may reach over three-fourths of its semi-circumference. The cyst-wall contains involuntary muscular fibres, which are not usually found in the wall of true ovarian cysts; it is generally separable into two layers, and is often lined by ciliated epithelium. Parovarian cysts are not unfrequently cured by a single tapping.

Results and Symptoms.—Ovarian cysts, as they grow large, are liable to become adherent to surrounding parts, especially to the omentum and abdominal walls, but sometimes also to the pelvis, intestines, and even the liver and stomach. The more solid tumours generally acquire adhesions more readily than those consisting mainly of a few large cysts. Nutrition of ovarian cystomata is apt to fail, especially in the case of multilocular tumours, and the cysts then undergo a partial necrosis, but without putrefaction, so long as air is excluded. The walls of the cysts may become softened, and the fluid within them may contain shreds of broken-down tissue. More complete death of the tumour results if the pedicle becomes twisted, so as to compress the vessels contained in it. Even after this accident, however, more or less vitality may be maintained through the medium of vascular adhesions. Twisting of the pedicle, sufficient to produce strangulation, is said to be more common with cysts of the right ovary, and Mr. Lawson Tait attributes its causation to the alternate filling and emptying of the rectum. Any tangential pressure produced in this way would obviously exercise greater leverage on a tumour attached on the right side.

Inflammation in the tumour may be set up by necrotic changes, or other causes, and then the cysts may suppurate, or lymph be effused within them. As the effect of inflammatory or necrotic changes, general peritonitis is apt to be set up, and the surface of the tumour may then become completely adherent to all the surrounding parts. Without the occurrence of any acute peritonitis, more or less ascitic fluid is often poured out by the peritoneum in consequence of the irritation caused by the presence of the tumour, and occasionally this fluid is copious though the growth is only of small size. In some cases, the walls of a thin cyst give way from distension, or from the effect of some strain or violence. If the contained fluid is bland, it is absorbed by the peritoneum, and in this way a spontaneous cure sometimes results, while, in other cases, the fluid again collects after a time. If the fluid has undergone inflammatory or necrotic change, or if it is from any cause irritating, as the thicker fluids are apt to be, severe peritonitis is set up, and often proves quickly fatal. In rare cases, after inflammation and suppuration of a cyst, it may discharge either into the intestine (generally the rectum) or externally. Adhesions form in the first place, and perforation afterwards occurs at some adherent spot. Still more rarely discharge takes place into the vagina or bladder. After admission of air, as by tapping, or in consequence of communication with the intestine, septic inflammation of a cyst may be set up, and it may then become distended by fœtid gas. Hæmorrhage may take place into ovarian cysts, either after strangulation of the vessels by twisting of the pedicle, or spontaneously from papillary growths. Death may then result from loss of blood or shock, or inflammation may be set up in the cyst, or in the peritoneum after rupture of the cyst. In rare cases death occurs from intestinal obstruction, ileus being produced either by the effect of adhesions, or from the intestines having simply become twisted in consequence of the pressure. Cure of an ovarian cyst, of any considerable size, by absorption is doubtful, though tumours diagnosed as ovarian cysts sometimes disappear. But, in those cases, rupture or perforation may have occurred, or the tumour may have been a pseudo-cyst (see p. 309).

In the earlier stages of an ovarian tumour, menstruation is often irregular and painful, and sometimes excessive. In the later stages it is often diminished, and amenorrhœa is common if both ovaries are affected. The general symptoms are often slightly marked, and frequently nothing is noticed except the increase of size. While the cyst is small and remains in the pelvis, trouble in defecation and micturition may be produced by pressure. These are relieved when it rises into the abdomen, but progressively increase if it happen to become fixed by adhesions while still small. In other cases more or less pain is felt in the tumour, and attacks of pain may also indicate the occurrence of local adhesive peritonitis set up by its presence, though such a local peritonitis often runs a very latent course.

As the tumour becomes very large, its pressure interferes seriously with vital organs, especially the heart and lungs. There is general wasting, and the face acquires a peculiar expression of combined emaciation and anxiety. The urine becomes scanty from pressure on the renal vessels, and sometimes albuminuria may be produced, although usually not till a very late stage. In some cases, especially when extensive pelvic adhesions exist, there is pressure upon the ureters, and consequent damage to the kidneys. Swelling of the legs is frequently produced by pressure, and the œdema may extend to the abdominal walls and back.

When inflammation or necrotic change has occurred in the tumour, hectic fever of an irritative kind is set up. The occurrence of such fever, in the absence of sufficient pain and tenderness to indicate acute general peritonitis, is an evidence of changes in the tumour, and an indication for early removal if practicable. When pregnancy occurs in conjunction with an ovarian tumour, considerable increase of danger arises, if the tumour is large, from the excessive distension, and also from the risk of strangulation by twisting of the pedicle. If the tumour is small it is apt to occupy the pelvis, and impede delivery.

All varieties of ovarian tumours may grow to an enormous size. In the majority of cases of considerable tumours, not subjected to curative surgical treatment, death occurs within three years, although small or moderate tumours may remain quiescent for a long period. Exceptionally even in the case of large tumours the course may be protracted for many years, and in some instances the operation of tapping has been repeated very many times.

Diagnosis.—When the presence of an abdominal tumour is suspected, the patient should be placed on her back on a hard couch, the head on a low pillow, the skin of the abdomen uncovered, and the knees drawn up so as to relax the abdominal muscles. The examination should be made first by abdominal palpation and percussion, afterwards by bimanual exploration.

Phantom Tumours due to flatulent distension of
intestines, deposit of fat in the abdominal walls,
omentum, and mesentery, or muscular contraction, are
generally easily distinguished by the resonance of the
abdomen, and by no tumour being felt between the
hands on bimanual examination. In most of these
cases, if relaxation of the muscles be obtained by
distracting the patient's attention by conversation, or
by the administration of an anæsthetic, the hand may
be pressed down sufficiently to feel the promontory of
the sacrum, and the absence of a tumour ascertained.
When there is a great deposit of fat, the percussion
note may be partially dull, but not absolutely so. In
such cases the layer of subcutaneous fat can be grasped
with two hands and lifted; and the umbilicus is
depressed. Whether or not any other tumour is discovered, special care should be taken to discover the
presence or absence of *pregnancy*, by looking out for
all the signs of that condition, particular regard being
paid to the consistency of the cervix, the condition of
the breasts, and the size of the uterus.

When an ovarian tumour is of great size, and of the
multilocular variety, but contains one or more large
cysts, diagnosis is generally easy. The outline of the
tumour is more or less irregular, both to the eye and
the touch, and the irregular prominences are often
found to move downward on deep inspiration. There
is dulness over the whole tumour, and resonance in the
flanks, while the margins of the tumour can often be
felt by pressing the hand flat upon the surface. If one
hand be laid upon the abdomen, and a gentle flip be
given by a finger of the other hand, a fluid thrill can be
felt over some part of the tumour, but not throughout
its whole extent, while over the area not reached by
the thrill, more resistant portions, or solid masses, can
often be felt. This vibratile thrill is conclusive of the
presence of fluid, while mere fluctuation may sometimes
be transmitted by an elastic semi-solid tumour.

When a considerable tumour consists mainly of one

large cyst, so that a uniform fluid thrill is felt over the whole of it, and no firmer portion can be detected, the essential point is to distinguish it from *ascites*, and the differentiation is sometimes a difficult one. In most cases the diagnosis may be made from the shape of the abdomen and the results of percussion. In ascites the abdomen spreads out more laterally, while in ovarian cysts it is more prominent in front, and tends to overhang the symphysis pubis. In ascites there is dulness in the flanks in the dorsal position, and resonance in front, while the areas of dulness and resonance alter according to position, so that, in the lateral or upright position, the portions of the abdomen uppermost at the time become resonant, and the dependent parts dull. In a large ovarian cyst there is dulness over the whole centre of the abdomen, and resonance only very far back in the flanks, coming further forward on the side opposite to that on which the tumour originated. When an ovarian cyst, after entry of air, contains gas as well as fluid, there will be resonance in front, but this condition is easily distinguished by the succussion splash produced on shaking the patient.

Exceptions to the usual rule occur in the case of ascites when the intestines are bound down by adhesions, or the mesentery shortened, so that they cannot reach the surface, and also when the abdominal distension is very extreme, in which case also the abdomen becomes more prominent in front, so that its shape resembles that usual in the case of ovarian tumour. More or less resonance may also be produced by the large intestines fixed in either flank, and then the signs correspond almost exactly to those usual in the case of an ovarian cyst. Again, in an ovarian cyst a coil of intestine may be adherent in front, and so give resonance in that position ; but as a rule this is only found when inflammation has occurred after tapping. In order to solve the difficulty an attempt should be made to detect the upper margin of a tumour, by laying the hand flat upon the surface at about the upper limit of dulness, and pressing it in during expiration. Trial should also be made, whether on deep inspiration any inequality can be seen or felt moving downward beneath the abdominal wall, or any friction sound heard. When the fluid is within an ovarian cyst, it is generally possible to reach from the vagina the lower segment of the cyst, most frequently in front of the cervix, and feel an impulse transmitted from above. In ovarian dropsy, the uterus is generally drawn some-what upward, and its mobility often somewhat impaired, while in ascites it is low down and movable.; In ascites the fluid thrill extends further round toward the back, over the area in the flanks where partial resonance may exist, while in ovarian dropsy it is limited at the same line as the dulness, unless ascitic fluid is present in addition to the cyst. In some cases the only certain test is a preliminary tapping. By this the character of the fluid is ascertained, and frequently after its removal the presence of secondary cysts or solid matter is revealed. If desired, a small quantity of fluid may be drawn off by the hypodermic syringe for examination, but the preliminary tapping is generally desirable in the case of a possibly unilocular cyst of slow growth, and occurring in a young woman, since such a tumour may be cured by tapping alone.

The difficulty of diagnosis between ascites and ovarian dropsy is especially likely to arise in the case of cancer of the peritoneum, in which the intestines are apt to be held back by shortening of the mesentery, and solid masses may be felt in the abdomen, somewhat resembling the firmer portions of an ovarian tumour, but usually more movable in the fluid, and felt only by dipping for them. In peritoneal cancer hard masses are also often felt behind the cervix, and may usually be distinguished from the lower portions of an ovarian tumour by their nodular character and fixation to the pelvic wall.

Distinctions between Ovarian and Ascitic and other Fluids.—The varying physical characters of ovarian fluid have already been mentioned (*see* p. 297). The dark ropy fluid of high specific gravity (1018 upwards) is found only in ovarian cysts. Ascitic fluid is limpid and yellowish, containing a considerable quantity of albumen, and having usually a specific gravity of from 1010 to 1015, characters which distinguish it from most ovarian fluid. It may also be generally recognized by its property of depositing fibrin spontaneously in a very delicate layer upon the surface of the glass, though fibrin is occasionally found also in ovarian fluid, after inflammation of the cyst. Ovarian fluid, which physically resembles ascitic fluid, may be distinguished by the chemical character that it contains paralbumen* as well as albumen, while in the more viscid kinds of ovarian fluid, mucin (*see* p. 297) may also be detected. Ovarian fluid may also generally be recognized by its microscopical characters. It usually contains epithelial cells of various forms, cholesterine, leucocytes, and often large granule masses. But of special importance is the granular ovarian cell (Fig. 71, *a*), described by Dr. Drysdale, of Philadelphia, as being pathognomonic of ovarian fluid. This contains a number of fine granules, but no nucleus. Its size varies from $\frac{1}{3000}$ to $\frac{1}{2000}$ inch, but the size commonly met with is about that of a leucocyte. From a leucocyte, however, it is distinguished, according to Dr. Drysdale, by the addition of acetic acid. This renders the leucocyte very transparent (Fig. 71, *c*), and nuclei, varying in number from one to four, become visible, while the ovarian cell is simply rendered rather more transparent, and its granules

---

* Add nitric acid to form a precipitate; shake up the fluid, and then boil it with about an equal quantity of acetic acid. Boil another portion similarly with a like quantity of water for the sake of comparison. If the precipitate by nitric acid is partially dissolved, or gelatinized, by the acetic acid, the presence of paralbumen is shown.

more distinct. From the larger granule masses it is distinguished by being unaltered on the addition of ether. Dr. Drysdale is said himself not to fail in diagnosis, but other observers have reported similar cells found in fluids not ovarian. Ascitic fluid under the microscope generally shows only cells of the peritoneal endothelium, leucocytes, and fibrin. For the characters of the fluid of a fibro-cystic tumour *see* p. 234; for those of parovarian fluid, *see* p. 299. If the fluid is found purulent, when no previous tapping has taken place, it is more likely to come from a dermoid cyst.

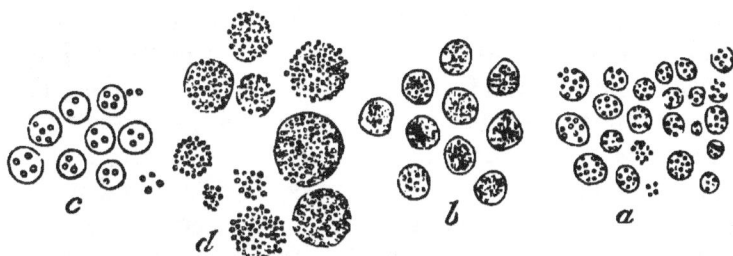

Fig. 71.—Microscopic Characters of Ovarian Fluids.
a, granular ovarian cell of Drysdale; *b*, leucocyte before addition of acetic acid; *c*, leucocyte after addition of acetic acid; *d*, compound granular cell or inflammatory corpuscle of Gluge.

If the existence of a large cyst is established, it is necessary to distinguish between an ovarian tumour and several other conditions which have sometimes been mistaken for it. *Pregnancy with hydrops amnii* is usually distinguished from ovarian dropsy alone by the condition of the cervix, but has not unfrequently been mistaken for an ovarian cyst associated with pregnancy, and sometimes even the uterus has been punctured owing to a mistake in diagnosis. It may be possible to make out the body of the uterus as distinct from the cyst on bimanual examination; but this may fail if distension is extreme. The varying consistency of the tumour of hydrops amnii, from alternate contraction and relaxation of the uterus, will be distinctive, if observed, but this also sometimes is absent, if the hydrops be at all excessive. The cervix uteri in hydrops amnii is generally more pervious to the finger than in the earlier months of normal pregnancy, and its condition may be an important aid in diagnosis. It should also be ascertained whether the lower segment of the uterus, felt through the anterior vaginal wall, has the characters usual in normal pregnancy. A *distended bladder* is distinguished by the use of the catheter, which should always be employed if an apparently central and unilocular cyst of moderate size be discovered. *Hydronephrosis* and *pyonephrosis*, as well as solid tumours of the kidney, generally malignant, have led to mistakes. The tumour formed by any of these commences from the region of the kidney, pushes the colon in front of it, and rarely completely fills the abdomen towards the opposite groin, or comes into close relation to the uterus. The fluid of hydronephrosis may sometimes, but not always, be recognized by the presence in it of urea or creatin. In renal tumours a history of disturbance in the urinary system or the presence of albumen, pus, or blood in the urine may sometimes guide to a diagnosis. *Hydatid tumours* would more often lead to error, but for the fact that they very rarely occur in the pelvis, and generally commence from the region of the liver, or the upper part of the abdomen. The best test is the microscopical examination of the fluid, which, from its physical character, may be mistaken for that of a parovarian cyst. I have met with one case in which the cyst had none of the tenseness usual in hydatid tumours, and no sign of hydatids was discovered on microscopical examination of the fluid. The cyst afterwards suppurated, and was eventually removed by Mr. Jacobson in Guy's Hospital. It was still thought to be parovarian, until, at the autopsy, its association with smaller hydatid cysts was discovered. In a case under my own care, the cyst, having become inflamed, closely simulated a large ovarian tumour, and a thick brownish fluid was evacuated by tapping. Decomposition then occurred, and hydatids escaped when the cyst was laid open. The patient recovered after insertion of a large drainage tube, and frequent injections with a weak solution of iodine. Multiple hydatid cysts of the omentum. may also closely resemble a multilocular ovarian tumour.

Encysted Serous or Purulent Fluid may closely resemble an ovarian cyst. Such a collection may attain to considerable size, especially when due to cancer or tuberculosis of the peritoneum. Sometimes large encysted collections of pus are formed in a late stage of septic peritonitis. I have met with one case in which a large pus-containing cyst behind the uterus, and reaching as high as the umbilicus, was secondary to cancer high up in the rectum. The resemblance will generally be to an ovarian cyst associated with peritonitis, and fixed by adhesion. If the pseudo-cyst is irregular in shape, sending prolongations among the intestines, it may often be distinguished by the fact that the fluid thrill extends beyond the area of dulness ; but, in general, the diagnosis must be made in great measure by the history and course of the affection. When the effusion is serous, and due to simple peritonitis (the serous perimetritis of Dr. Matthews Duncan), it may be distinguished by its sudden appearance in connection with acute inflammatory symptoms, and by its gradual diminution and disappearance. If the fluid be drawn off, it will generally be found to contain flakes of lymph, and will spontaneously deposit a coagulum of fibrin. *Retro-peritoneal cysts* have occasionally been found, and are to be distinguished by their place of origin, and want of connection with the uterus. *Advanced extra-uterine fœtation* will be recognized by a history of pregnancy not ending in delivery. The hard parts of the fœtus will be felt in the midst of the fluctuating cyst, or the fluid will have been absorbed, and the whole tumour be firm and irregular.

When a considerable tumour exists, but no manifest

fluid thrill can be felt over any large part of it, the point requiring most care is the diagnosis, whether the tumour is *uterine or ovarian*. For this a careful exploration by the sound of the position of the uterus and its mobility, with an exploration of the mode of attachment of the tumour to the uterus, is necessary. The method of diagnosis has already been described (*see* pp. 14, 234). *Cancer springing from the omentum* or elsewhere in the abdomen, or even medullary cancer of the kidney, may form a large tumour, sometimes reaching into the lower part of the abdomen, and forming deposits in the pelvis. Such a tumour is not fluid, though it may be soft and semi-fluctuating, and is often associated with ascites. Some portion of it is generally fixed by adhesion. If the ascitic fluid be examined, it may indicate the presence of blood, obviously or by spectroscope, or show the clusters of cells described at page 313. The age of the patient, and the amount of emaciation and cachexia present, will greatly aid the diagnosis. *Leukæmic enlargement of the spleen* may form a tumour, which becomes displaced into the lower part of the abdomen, but is easily recognized by its sharp hard border, generally broken by a depression and usually looking upward, or upward and toward the right, the convex surface of the spleen having rotated downward. The distinctive edge, however, may be obscured to a great extent, if the growth in the central part of the spleen has assumed a spherical form. *Fœcal accumulations* are distinguished by their position, unconnected with the uterus, by their doughy feel, and by the effect of purgatives and enemata.

Diagnosis in the Early Stage.—In the early stage of an ovarian tumour, while it is smaller than a fœtal head, the method of diagnosis is somewhat different. If the tumour is free from adhesion, the diagnosis is usually easy. A well-defined, globular, and elastic tumour is felt, on bimanual examination, behind, in front, or at one side of the uterus, movable to some extent, but tethered to that organ. A parovarian cyst, or hydro-salpinx, may be confounded with an ovarian cyst, and can hardly be distinguished except by examination of the contained fluid. *Pyosalpinx* is also difficult to distinguish, but is accompanied by signs of local peritonitis. A *subperitoneal fibroid of the uterus* is generally known by its hardness, and by its mode of attachment to the uterus, especially if the examination be made by finger in the rectum, the patient being placed under an anæsthetic, and the cervix drawn down by tenaculum (*see* pp. 14, 15). If a small ovarian tumour is surrounded by adhesions, especially when it lies behind the uterus, it may be very difficult to distinguish it from *hæmatocele*, or a swelling due to pelvic peritonitis. The best guide is the history and course of the affection. With *pelvic cellulitis* an ovarian tumour is less likely to be confounded. In *extra-uterine fœtation in the early months* there will usually be a history of amenorrhœa for a time, with general signs of pregnancy, although sometimes menstruation is never interrupted. In any of the forms of tubal pregnancy, the amenorrhœa, if any occurs, is apt to be followed by irregular hæmorrhage, and violent spasmodic pain, while pain sometimes exists from the very first. The uterus will be notably enlarged ; and strong arterial pulsation will be felt near the cyst, which usually occupies the retro-uterine fossa. Sometimes ballottement in the cyst, or signs of fœtal life may be detected, or the diagnosis may be decided by the expulsion of a decidua from the uterus. After the death of the fœtus, the tumour tends to diminish in size, and to become harder. *Small hydatid cysts* in the pelvis are very rare in Britain, and can scarcely be diagnosed, except by examination of the fluid.

Diagnosis of Adhesions.—The general mobility of a tumour of moderate size may be tested by grasping it with both hands through the abdominal wall, and moving it from side to side. In the absence of adhesions in front, if distension is not very great, the abdominal walls can be freely moved over the tumour, and also lifted up, to some extent, from its surface. Frequently parts of the tumour may be observed to glide downwards under the surface on deep inspiration. All these signs are more easily detected in multilocular tumours, or those containing solid matter, than in those which are nearly unilocular. Measurements should be made of the distance of the umbilicus from the anterior superior spines of the ilia, and from the edges of the ribs on each side. Any marked inequality in these distances not accounted for by the shape of the tumour renders the existence of adhesions in front probable. A friction sound on respiration indicates the absence of any firm adhesion over the area where it is heard, and is generally due to prominent vessels or other inequalities on the surface of the tumour.

If a portion of the tumour descends into the pelvis behind the cervix or elsewhere, and cannot be pushed up, it may either be adherent or simply fixed by pressure. The degree of fixation is estimated by observing whether the mass yields at all to pressure from the vagina or rectum. If the whole pelvic roof is indurated, and the uterus fixed, firm adhesions in the pelvis are indicated. If the uterus cannot be moved by means of the sound separately from the tumour, it may be inferred that the tumour, if ovarian, is closely connected with the uterus; and this presumption is increased if the uterus be much drawn upward, and its cavity elongated. In estimating the attachment of the tumour to the uterus, rectal exploration should be used, and it is often of advantage to put the attachments on the stretch by drawing down the cervix (see p. 14). The probability of intestinal and other adhesions posteriorly can only be inferred from a history of peritonitis. Omental adhesions cannot be diagnosed, but are of little consequence.

Diagnosis of Malignancy.—The greater the propor-

tion of solid material in a tumour, the greater is its
tendency towards malignancy likely to be. The age
of the patient is a useful guide, and valuable evidence is
afforded by a cachexia and emaciation out of proportion
to the size and duration of the tumour. A very large
amount of ascitic fluid, in combination with a comparatively small and firm tumour, renders probable
at least some approximation towards malignancy. If
signs of pelvic adhesion are found, with much solid
matter at the base of the tumour, and nodular masses
like glands behind or around the cervix, and if much
cachexia is also present, the evidence of malignancy
is very strong. If grape-like clusters of cells, of very
varying shape, many of which have multiple nuclei,
are found in fluid withdrawn from a cyst, they indicate
that proliferating papillary growths are present, and
that the tumour should be immediately extirpated,
if possible. If similar groups of cells are found in
ascitic fluid, they indicate that cancer has reached the
peritoneum from the ovary, or from some other source,
and that ovariotomy will probably be too late to prevent
recurrence. A similar indication is given if blood is
detected, either obviously or by spectroscope or other
evidence, in ascitic fluid.

Treatment.—The great success of ovariotomy has
reduced the treatment of ovarian tumours, in the large
majority of cases, to ovariotomy as a curative, and
tapping as a preliminary or palliative measure. Since
tapping is not entirely without risk, especially if not
performed antiseptically, it is better, if a tumour is
positively diagnosed as ovarian and multilocular, and
if it is not of such enormous size as to cause excessive
pressure, to perform the major operation, without preliminary paracentesis. If, however, the tumour has, by
its pressure, caused great oedema of legs and abdomen,
and much interference with the functions of kidneys,
heart, and lungs, and if it appears to contain one considerable cyst, it is generally preferable to tap as a

preliminary measure, in order to give the organs time
to recover themselves, and to prevent the shock of the
operation being increased by too extreme a change
in abdominal pressure. If a cyst is apparently unilocular, and the patient is young, and the tumour of
slow growth, it is desirable to tap once before proceeding to ovariotomy, since such a cyst may prove to
be parovarian, and may then perhaps be cured by a
single tapping. Paracentesis may also be required in
cases where the diagnosis is doubtful, and as a
palliative measure in those in which ovariotomy is
decided to be impracticable.

The Operation of Paracentesis.—If ovariotomy is to
be performed with antiseptic precautions, any preliminary paracentesis must be so also. The instrument used should be about ¼-in. in diameter, and may
be either an ordinary trocar and canula, or Sir Spencer
Wells' syphon canula, which has a bevelled extremity,
and a cutting edge for half its circumference, so that
it cuts a valvular opening. The latter instrument
diminishes the risk of any fluid escaping into the
peritoneal cavity, during the momentary interval
between the puncture and the withdrawal of the
trocar.

For the operation the patient should be placed on
her side, with her head rather low, to avoid the
occurrence of faintness from sudden diminution of
abdominal pressure. One or more pails must be provided to receive the fluid. A small incision should be
made with a scalpel just through the skin, before the
trocar is inserted. If the fluid is conducted into the
pail by means of a long tube, the antiseptic method
is not carried out, unless a second spray-producer is
used to protect the extremity of the tube, as well as
that for the surface of the abdomen. I have generally,
therefore, used no flexible tube, but merely conducted the fluid into a pail by means of a mackintosh.
If, however, the carbolic spray is not used, a long
india-rubber tube should be previously attached, either

to the syphon canula, or to a short tube which is fitted into the ordinary canula at the moment when the trocar is withdrawn. In this way the entry of air is rendered less probable, and the syphon action assists the evacuation of the fluid. If a large amount of ascitic fluid is present, in addition to a tumour, it is better to make a small incision through the abdominal wall with a scalpel, until the peritoneum is just divided, and then to pass in through the opening a large gum-elastic catheter, which should be a new instrument and thoroughly disinfected with carbolic acid. In this way there is no risk of wounding tumour or intestines ; the ascitic fluid is drawn off first, the outline of the tumour can then be exactly explored, and it also can be afterwards tapped, if necessary.

Paracentesis is generally an innocuous operation, but occasionally it is followed by inflammation of the cyst, or of the peritoneum. This usually arises from access of air in absence of antiseptic precautions, and is especially likely to occur if the operation is performed in a place exposed to any septic influence.

Indications for Ovariotomy.—It has generally been considered preferable not to operate until the tumour has attained some considerable size, and has begun seriously to inconvenience or disable the patient, or to tell upon her general health. A person in active health usually bears a severe operation worse than one accustomed to an invalid life. Moreover, the healthy peritoneum is more prone to inflammation, especially from a septic cause, while after great and prolonged distension, and protracted slight irritation, it acquires a certain tolerance. This rule, however, may now be modified to a considerable extent in consequence of the very successful results attained in uncomplicated cases of ovariotomy. In any case the operation should be performed before there is sufficient distension to embarrass greatly the lungs, heart, or kidneys, or produce œdema of legs and abdomen. In the case of single women, to whom the increase of size is an annoyance, it may justly be performed earlier. If symptoms of inflammation or partial necrosis of the cyst supervene its removal should not be delayed. Sometimes the occurrence of hæmorrhage into a cyst, or the commencement of inflammation or necrosis in it, due to twisting of the pedicle, is indicated by the onset of severe pain in the tumour, or pain followed by vomiting. Ovariotomy should then be performed immediately, if practicable. If acute peritonitis occurs it may be desirable to wait awhile, until the symptoms have subsided; but ovariotomy, if practicable at all, should generally be performed within a few weeks, before the adhesions have become so firm as to render their separation very difficult. If there is an indication by any of the signs previously enumerated (p. 312) that the tumour is of a kind tending towards malignancy, it should be removed, if possible, before any infection of other tissues has occurred.

The chief *contra-indications* are signs of malignancy in the tumour, accompanied by extension of the growth to surrounding parts, or evidence of very unyielding and extensive pelvic adhesions, especially when the lower portion of the tumour is solid, or when the uterus is involved in the adhesions and completely fixed. Coils of intestine are sometimes recognized as adherent in front of the cyst, especially when a previous tapping has been followed by peritonitis; and this condition would indicate the probability of extensive visceral adhesion in other parts also. Elongation of the uterus, and its close connection with the tumour, render it likely that the operation will be difficult, but do not necessarily contra-indicate it. Adhesions to the abdominal wall are of comparatively little moment. In doubtful cases the age and general condition of the patient, and the state of the kidneys and other viscera, are important elements in the decision.

The complication of an ovarian tumour with pregnancy often proves a serious one; since the growth of the tumour is apt to be stimulated, distension may become so great as to necessitate some interference, and, moreover, the pressure of the growing uterus occasionally produces twisting and strangulation of the pedicle. Ovariotomy during pregnancy has proved remarkably successful in the hands of Sir Spencer Wells, and abortion is by no means a necessary sequel of the operation. If the case is otherwise suitable for ovariotomy, that operation may be performed, at any rate in the earlier months. I have once successfully removed an ovarian tumour about the middle of the sixth month without miscarriage following, but in other recorded cases which had passed beyond the fifth month, miscarriage has followed sooner or later, and this adds to the risk, if it happens shortly after the operation. In the case of adhesions near the uterus bleeding is likely to be formidable in the later months, and if the case has advanced much beyond the fifth month, it is perhaps preferable, supposing interference to be necessary, either to induce labour, or to tap, if that operation is likely to tide over the difficulty.

The Operation of Ovariotomy.—The room where the operation is performed, as well as the operator and his assistants, should be perfectly free from septic contamination, and no one should be present in it who has within the last few days attended any post-mortem or dissecting room, or seen any case of infectious disease, especially erysipelas, septicæmia, or pyæmia. It can hardly be considered as yet proved that it is wise, even with the antiseptic method, to omit this precaution. Dr. Keith considers the use of carbolic acid as an absolute safeguard against ill-effects from any previous contamination, but others have adduced some evidence tending to show that there may be contagious material so virulent as even to resist the effect of Lister's method as ordinarily employed. The room should be warm, but need not be overheated. The patient is placed in the dorsal position, close to a good light, the shoulders being slightly raised. In a private house two dressing tables may be conveniently used, one of which is placed crosswise, to support the head and shoulders. The abdomen should be washed with soap and water, and afterwards with carbolic lotion. The hair on the pubes should be shaved, to facilitate the close application of antiseptic dressings, if the antiseptic method is to be used, and the bladder evacuated. A belt or bandage should be passed over the knees and secured under the table, in order to prevent movement, if the patient should partially recover from the anæsthetic. For cleanliness it is convenient to cover the abdomen with a piece of mackintosh, in which an oval aperture of sufficient size has been cut, and its edges spread with adhesive plaster, to attach it at all points to the skin. In this way all fluids are conducted into the vessel ready to receive them.

The question whether the use of the carbolic spray in ovariotomy is generally desirable cannot be regarded as yet finally settled. Sir Spencer Wells, after his adoption of Lister's antiseptic method for this operation, reported an improvement of two or three per cent. in the mortality, even as compared with his most recent and most successful series of cases in which it was not employed; while his results in the removal of fibroid tumours by abdominal section, showed an even more marked improvement. On the other hand, Keith and other distinguished specialists have shown that extremely good results may be obtained either with or without the carbolic spray. In such hands the mortality of late has not been greater than from four to ten per cent. In large general hospitals, where previously the mortality was little less than fifty per cent., and where septic influence is more difficult to exclude, the more or less complete adoption of the antiseptic method appears to have effected a distinct improvement in the results. The experience gained by operating frequently is, however, an important element in success, and it cannot be doubted that a patient who undergoes the operation at the hands of a surgeon not a specialist has her chance of death notably increased. So far the conclusion appears to be that, at any rate, two advantages are gained by the use of the carbolic spray : first, that it enables the operation to be performed under somewhat less disadvantageous conditions in large hospitals, or other places where there is a liability to septic influences; and secondly, that by its use favourable results are obtained without drainage in cases complicated by extensive adhesion, such as without it can only be attained by drainage of the peritoneal cavity, and the more troublesome after-treatment thereby made necessary.

For ovariotomy, if the Listerian method is to be adopted, one or two steam spray-producers are required, capable of throwing a very fine and wide-spread cloud of spray, and of working for at least two hours, if necessary, without the boiler becoming exhausted. They are to be supplied with a solution of pure carbolic acid,* of the strength of 1 in 20, or one sufficient to produce a spray containing about 1 in 40, after dilution by the steam. A spray of thymol (1 in 1,000) has been used as less irritating than carbolic acid to the peritoneum ; but it appears that thymol is not so absolutely reliable an antiseptic as carbolic acid. The silk or gut for sutures should be wound upon reels, and with the instruments should be kept immersed in trays filled with carbolic solution (1 in 20). Sponges, after disinfection in carbolic solution (1 in 20), are to be kept immersed in one of the strength of 1 in 40. As an anæsthetic Sir Spencer Wells prefers the bichloride of methylene, administered by Junker's inhaler, as being safer than chloroform, and least likely to cause vomiting. Mr. Keith chooses ether, with the object of

---

* The so-called " absolute phenol," manufactured by Messrs. Bowder and Bickerdike, of Church, Lancashire, is the best form to use.

avoiding vomiting, but it is apt to irritate the lungs if any bronchial complication exists. Others prefer the mixture of alcohol one part, chloroform two parts, and ether three parts. Chloroform is apt to depress the pulse, if the operation prove long and difficult. Besides the assistant who administers the anæsthetic, one other at least is required, who stands on the left side of the patient opposite the operator, and is ready to protect the intestines with soft sponges, and prevent their protruding. There must also be nurses to wash sponges in warm carbolic solution (1 in 40), and hand them back to the operator. Sponges and hæmostatic forceps should be counted before and after the operation, to ensure that none are left in the abdomen.

The incision is to be made in the linea alba. It should be at first of moderate length, not exceeding four or five inches, nor passing above the umbilicus. It may afterwards be extended upwards, if required. Sir Spencer Wells has found the mortality to be notably less when the tumour could be extracted through an incision not exceeding these dimensions. It is preferable, however, to extend the incision rather than to separate out of sight adhesions to omentum or intestines, which are likely to cause subsequent hæmorrhage. Bleeding from vessels in the abdominal incision should be checked before the peritoneum is incised. For this purpose, Sir Spencer Wells' hæmostatic pressure-forceps, having a catch at the handle, are very convenient, and the operator should be provided with a considerable number of them. These are left attached while the operation proceeds. Ligatures of fine carbolized gut may also be used, if necessary. The fibrous structure of the linea alba and the peritoneum are then successively pinched up and divided upon a director. If the incision is extended downwards near the pubes, it is preferable to use two fingers as a director, to avoid any risk of injuring the bladder. The cyst is generally recognized by its bluish appearance, but when it is firmly adherent to the abdominal wall, difficulty may be found in ascertaining when the peritoneum has been divided, and the operator may peel off the parietal peritoneum, mistaking it for the cyst-wall. In case of great doubt two methods may be followed—(1) to ascertain whether the supposed cyst-wall can be peeled off at the situation of the umbilicus; (2) to extend the incision till the point is reached where the cyst leaves the abdominal wall.

Before tapping the tumour, the hand may be introduced between it and the abdominal walls, and swept round on all sides, to separate any adhesions existing

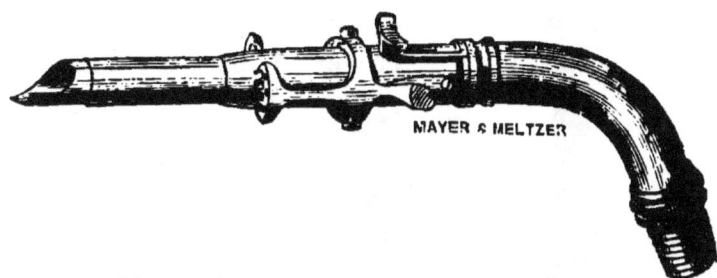

Fig. 72.—SPENCER WELLS' Ovariotomy Trocar.

in front. If, however, the adhesions be found very firm, and especially if the cyst-wall be thin or friable, it is better first to empty the tumour. A large trocar (Fig. 72), having claws attached, to fix the cyst-wall to it, is now plunged into the tumour. In case of a multilocular tumour the spot should be chosen where the largest cyst appears to be situated. After puncture of the cyst the inner tube is pushed forward to guard the point of the trocar. Care must be taken to prevent the fluid entering the peritoneal cavity; but, with the antiseptic method, it is better not to attach any flexible tube to the canula, but simply to let the fluid flow down over the mackintosh into a vessel ready to receive it. If the nature of the tumour still remains doubtful, after its exposure, a small trocar should be used in the first place. As the tumour empties, it is

Y

drawn forward in the grasp of the claws of the trocar, so as to bring the opening outside the abdomen. If secondary cysts remain large enough to prevent the withdrawal of the tumour, it is best to remove the trocar, enlarge the opening, and fix at each side of it a pair of cyst-forceps (Fig. 73) to hold it forward. The hand is then introduced through the opening, and the secondary cysts broken up by the fingers, or the trocar guided to tap them.

For the ligature of bleeding vessels or bands of adhesion of no very large size within the peritoneum, it appears preferable to use ligatures of carbolized or chromicized gut, not excessively fine. Many operators, however, prefer fine silk, which should be carbolized, if the operation is performed antiseptically. Special care is required to guard against subsequent hæmorrhage in the case of omental adhesions; and, if there is a broad surface adherent, it is a good plan to divide it into sections, and tie each separately before dividing the omentum with scissors. Other strong bands of adhesion may be tied before division in a similar manner. Adhesions to intestine, stomach, or liver require special caution in their separation, which should be effected, if possible, by gentle traction, or by the finger-nail, or handle of the knife. If, however, a portion of cyst-wall be very

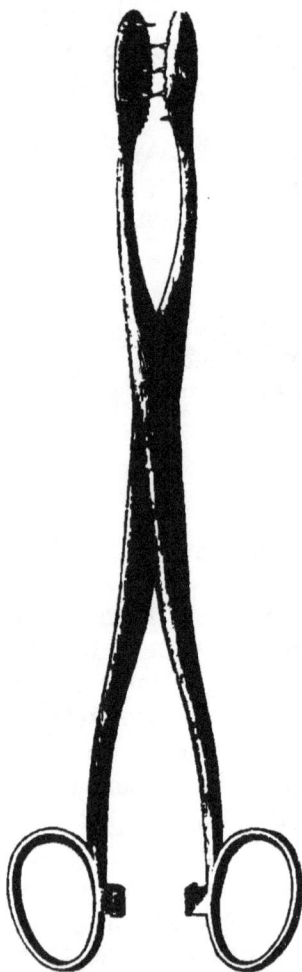

Fig. 73.—NELATON'S Cyst Forceps.

firmly adherent, it is preferable to cut the rest
of the tumour away from it, and leave it attached.
Firm adhesions within the pelvis are likely to lead to
the greatest difficulties, and, in some cases, render it
impossible to complete the operation. If they cannot
be separated by the fingers, the plan of enucleation
should be tried. A transverse incision is made through
the outer wall of the cyst, a little above the adherent
surface; the fingers are introduced through the opening, and the cyst-wall splits into two layers, of which
the outer is left attached, and treated as a pedicle, by
ligature or otherwise. After removal of the tumour
minute care must be taken to arrest all hæmorrhage.
Oozing from even a considerable vascular surface,
such as that produced by detachment of an adhesion
to the uterus, may be stopped by successively seizing
the bleeding points by tenaculum forceps and tying
them with carbolized or chromicized gut or fine silk.
In extreme cases the galvanic, benzoline, or actual
cautery have been used, but are very rarely necessary.

In the *treatment of the pedicle* the choice lies
between the extra-peritoneal and intra-peritoneal
methods. In the former the pedicle is fixed by a
clamp, or some equivalent means, in the lower angle
of the wound, and the tumour cut away. Before the
introduction of the antiseptic method in ovariotomy,
the treatment of the pedicle by clamp, except in cases
in which it was too short or broad to allow a clamp to
be applied, had given the best results in the hands of
most operators. With the antiseptic method, however, the intra-peritoneal treatment of the pedicle by
ligature with carbolized silk was introduced. This is
now established as being the best method, and is
adopted even by those who do not use the spray.

The plan adopted is to transfix the pedicle with a
ligature of strong, but not excessively thick, carbolized
silk,* and tie it in two or more sections. A long

* The silk is carbolized simply by soaking it for twenty-four
hours in carbolic solution (1 in 20).

piece of the silk may be threaded in an aneurism needle for transfixing the pedicle, and care should be taken to avoid puncturing any large veins which may be visible. The whole pedicle should afterwards be encircled by one of the ligatures, and the ends are then cut short. Care must be taken that the loops cross each other. Sufficient capillary circulation is afterwards restored to maintain the nutrition of the stump beyond the ligature.

The intra-peritoneal method has the advantage of shortening the cure, and avoids the leaving of any weak point in the abdominal wall, such as may afterwards give rise to ventral hernia. The division of the pedicle by cautery has been found by Dr. Keith to give excellent results. For this purpose a temporary clamp is used, having a shield of non-conducting material to protect the soft parts beneath. The pedicle is divided, and then thoroughly seared down to the level of the clamp.

If the cyst is found to descend between the layers of the broad ligament, and below the level of the fundus uteri, so that its lowest portion cannot be drawn out, the choice will generally lie between extracting it by enucleation, if this is practicable, or tying it in segments as low down as possible, and leaving a portion of its surface attached. If, however, a cyst is completely unilocular, as, for instance, a par-ovarian cyst, or if no secondary cysts can be detected in the portion of wall which cannot be extracted, the aperture made by the trocar, or, in the latter case, the walls of the cyst, after the upper portion of it has been cut away, may be stitched to the edges of the abdominal wound, and the cyst treated by drainage (*see* p. 326), after which it will shrink up.

Before the abdominal wound is closed, all parts of the peritoneum, especially its dependent portions, must be thoroughly cleansed by a succession of soft sponges, wrung out of warm carbolic solution (assuming the operation to be performed under the spray), from all

fluid, blood, or clots; and upon the effectual performance of this "toilette" the success of the
operation largely depends. While the sutures are
inserted into the abdominal wound a large, flat sponge,
previously selected for the purpose, should be placed
beneath it, to prevent any blood from the punctures
entering the abdomen, and to avoid any risk of the
intestines being wounded. The sutures may be of
carbolized silk, or, for greater security against their
producing any irritation, either of silver wire, or of silkworm gut, as used by Dr. Bantock. Sir Spencer
Wells' plan is to thread a silk suture with a straight
needle at each end, and, holding the needles in forceps,
to pass them from within outwards, including about
half an inch of peritoneum, and bringing them out
rather close to the margins of the wound. Silkworm
gut may very advantageously be used in the same
way, the only disadvantage being that the pieces
are necessarily somewhat too short for convenience.
The edges are thus accurately adapted without any
superficial sutures. In bringing the wound together,
care must be taken to express air as completely as
possible, and, if practicable, to draw down the
omentum so as to cover the intestines. The antiseptic
dressings are most easily kept in perfect apposition
to the skin, if the deepest layer of carbolized gauze
is moistened with carbolic oil (1 in 20). Some, however, prefer to use a layer of thymol gauze next the
skin, to avoid irritation. The dressings are secured by
broad strips of strapping reaching to the sides, but not
round the back, and over these a broad band of flannel,
previously passed under the back, is secured by safety
pins. The dressings can then be changed without
disturbing the patient.

The use of *abdominal drainage* has been, in great
measure, superseded by the antiseptic method, since
not only is a moderate effusion of sanguineous serum
in the abdomen thus rendered generally innocuous,
but, if the discharge is allowed to escape and soak

the dressings, the completeness of antiseptic precaution is apt to be destroyed. If the carbolic spray is not used at the operation it is desirable, whenever the adhesions separated are very extensive or very vascular, so that much oozing of blood or serous fluid may be expected, or if any inflammatory or septic fluid has escaped into the abdomen, to place in the lower angle of the wound one of Keith's glass drainage tubes (Fig. 74). It should be chosen of such a length that its lower extremity rests at the bottom of the pouch of Douglas. A suture through the abdominal walls, at the position of the tube, may be placed ready, so that, when the tube is removed, it can be tied, and the opening by this means closed. Dr. Keith's method of managing the drainage tube is as follows. A sheet of thin pure india-rubber, from one to two feet square,

Fig. 74.—KEITH's Drainage Tube for Ovariotomy.

has a small hole made in its centre, by which it is fitted tightly over the glass tube, which is closely surrounded with the antiseptic gauze, if this is employed. Several sponges wrung out of carbolic solution (1 in 20) are then placed over the end of the tube, and the india-rubber sheet is wrapped over them. The sponges can then be changed, at first every few hours, without interfering with the rest of the dressings. After a few days, when the discharge has nearly ceased, the tube may be removed. The use of drainage through the pouch of Douglas into the vagina, and that of antiseptic injections into the abdomen, have been found not to be advisable.

If drainage is not used the dressings may generally be left undisturbed till the sixth, seventh, or eighth day, when the sutures are removed. Towards the end of the first twenty-four hours a little milk or barley-water may be given; but, if vomiting is troublesome, recourse should be had early to nutrient enemata, and the stomach left at rest. Morphia or opium is to be given by the rectum or subcutaneously in sufficient amount to keep pain in check after the operation. In case of considerable elevation of temperature—a condition generally due to peritonitis of septic origin—advantage has often been found from the use of Thornton's ice-water cap, whereby a current of cold water is kept constantly circulating round the head. Leiter's temperature regulator, in which tubing of soft metal is arranged in the form of a cap, may be used in the same way.

Other Modes of Treatment, in the present day, are to be thought of only when ovariotomy is judged impracticable, or when, after exploratory incision, it is found that the tumour cannot be removed. Insertion of a large drainage tube through an incision in the linea alba, and washing out the cyst frequently with antiseptics,* is a plan which may be adopted if a cyst is found, on tapping, to contain pus; or if, after incision, it proves to be inseparably adherent to the abdominal wall. If the tumour consists solely or mainly of one cyst, this method may lead to cure. If the cyst fails to close up under this treatment, it may be desirable, if possible, to make a counter opening by the vagina. Injection of iodine may be tried in a tumour mainly unilocular, if ovariotomy is contra-indicated. Three ounces of tincture of iodine, with an equal quantity of water, may be injected into the cyst, and, after a few minutes' interval, withdrawn again as far as possible.

Electrolysis has in some cases produced diminution, and even reported cure, of ovarian tumours, and may

---

* A weak solution of iodine (Tr. Iodi ℥ij. ad Aq. Oj.) may be used. If, however, the carbolic spray be employed, it is best to use for washing out the cyst a solution of sulphurous acid of the strength of one part of the pharmacopœial acid in twenty or forty.

be tried when the radical operation is impossible. It is not, however, without risk of setting up inflammation, and the tumour is apt to enlarge again after apparent arrest or cure. A battery of eight or ten good-sized cells should be used, and one pole introduced into the tumour by means of an insulated needle, the other electrode being applied to the abdomen or vagina. It is not yet fully ascertained which pole it is best to insert, but probably the positive pole is more effectual. The applications should last from five to ten minutes, and may be repeated at intervals of one or several days, if no irritation is set up. The treatment should be continued for some months. If a cyst is very tense, preliminary tapping is desirable, otherwise oozing may take place through the punctures.

DERMOID CYSTS OF THE OVARY.

Pathological Anatomy.—The term of dermoid cyst, though not a strictly appropriate one, has been applied to cysts of the ovary whose inner surface is a structure resembling skin, generally containing sebaceous and sometimes sweat-glands, and often provided with numerous hairs growing in hair-follicles. In the cellular tissue beneath, true bones are often formed, frequently having teeth, and sometimes more or less resembling some definite bones of a foetus. Teeth may also be found separately in the tissue, or be cast off into the cyst, within which a large number of them may be accumulated. They are sometimes well-formed, but more frequently rudimentary. In rare cases striated muscular fibre, or grey nerve-substance has been found in the cyst wall. The cyst is generally single, or appears to be divided into compartments by the growth of septa from its walls. The contents of the cyst are generally a thick gruel-like material, made up of sebaceous secretion and cast off epithelial cells, with the addition frequently of cast-off hairs, sometimes to a considerable amount. Cholesterine may also be present in large quantity.

Causation.—The far more frequent situation of these cysts in the ovary than anywhere else shows that they cannot be the produce of an included twin ovum, for such an ovum would be attached to some more external part. The occurrence in them of structures other than epithelial, as well as their situation, distinguishes them from cysts formed by abnormal epidermic involution, like the dermoid cysts of the orbit. Their origin can only be ascribed to an abnormal formative energy of one of the ovules in the ovary, constituting an imperfect degree of parthenogenesis or development without impregnation of fœtal structures from an ovular cell, or from other cells, of the parent organism. In comparatively very rare cases similar tumours have been formed by erratic development of tissue in other parts of the body. They have been found especially in the testicle of the male, and this circumstance appears to be a confirmation of the view that cells specialized for the purpose of reproduction are most prone to such an abnormal development.

Dermoid cysts of the ovary have often been found in very young children, but they more frequently come under observation within a few years after the age of puberty. It is probable that, in at any rate a large proportion of cases, the tumour commences in fœtal life or soon after birth, while formative energy is specially active, and that it takes on a more active growth when the ovary becomes developed at the age of puberty. Dermoid cysts are occasionally found associated with ordinary ovarian cystomata, and it is possible that, in such cases, the presence of the dermoid tumour may have been the starting point of irritative stimulus.

Results and Symptoms.—Dermoid cysts are slow in progress, and do not generally pass beyond a small or moderate size. In exceptional cases they may attain a large size, from the accumulation of many years' secretion or from suppuration of the cyst. They are more prone than ovarian cystomata to undergo inflammation and suppuration, and to contract adhesions with surrounding parts. Fistulous openings may then be found, communicating with the rectum, bladder, surface of abdomen, or other parts, through which the contents of the cyst, with hair, teeth, and bones, may be discharged. Rupture into the peritoneal cavity is rare. It is seldom that the cyst is completely evacuated spontaneously, but suppuration and discharge may continue for years.

Diagnosis.—The existence of a dermoid cyst may be suspected if a tumour of slow growth be found, which first attracted notice soon after the age of puberty, or has existed indefinitely, especially if hard masses like bone can be felt in parts of it, while no very manifest or superficial fluid thrill can be detected. A similar opinion may be held if, in a tumour like that just described, signs of inflammation appear, while the tumour rapidly increases and fluctuation is developed. Positive diagnosis can generally only be made after spontaneous or artificial evacuation of some of the contents.

Treatment.—If the tumour, after becoming inflamed, points externally, it is desirable to make an incision, or, if a fistulous opening exists, to enlarge the opening and evacuate the contents as completely as possible. Tufts of hair may be extracted by a small blunt hook. A large drainage tube should afterwards be inserted, and the cavity washed out regularly with a weak solution of iodine or other antiseptic. Dr. Barnes recommends light cauterization of the interior of the cyst, to alter its character, and cause it to contract. If the tumour is enlarging rapidly, without having formed fistulous openings, or if the commencement of inflammation in it is suspected, it should be removed, if possible, by ovariotomy. This operation has even, in some cases, been successful after a fistulous opening had existed for a long period.

FIBROID TUMOURS OF THE OVARY.

A true myoma or fibro-myoma having its origin in the ovary is very rare, although it has occasionally been observed. In some instances, tumours of this structure, which appear to belong to the ovary, may be outgrowths from the uterus, or may originate in the muscular fibres spreading out from the uterus into the broad ligaments. The proportion of muscular fibres is less when the growth is of ovarian origin than in uterine tumours; and, in some instances, fibrous tissue largely preponderates. Some solid tumours of the ovary belonging to the sarcomatous group also consist. mainly of fibrous tissue, and do not show malignancy, but these are also rare.

Treatment.—Non-malignant solid tumours of the ovary are generally slow in progress, and but rarely require interference, but ovariotomy may in some cases be called for, if the tumour attain a large size, or be the source of great pain qr distress. In most cases it will be impossible, before the operation, to distinguish such a growth with absolute certainty from a fibroid outgrowth from the uterus.

CANCER OF THE OVARY.

The ovary occupies a rather exceptional position among organs of the body, as being a not unfrequent seat of secondary, as well as of primary cancer. Primary cancer of the ovary is rarer as an independent disease than in conjunction with, or as a complication of, cystoma. When it occurs it often affects both sides, and sometimes appears in early life. Primary growths in the ovary, which are clinically malignant, may have the structure either of sarcoma, originating in the stroma, or of carcinoma, which probably in most cases, if not always, has its origin from the

Graafian follicles. Sarcoma is the commonest, and it is even held by some that all malignant growths in the ovary are of this character. The sarcoma may be of any variety, from the spindle-celled to the round-celled encephaloid form, the round-celled being the most frequent, while myxomatous or myxo-sarcomatous tissue is occasionally seen. The degree of malignancy varies according to the structure, but, in most cases, the prognosis is similar to that of carcinoma. True carcinoma of the ovary is generally of the encephaloid form, but occasionally the scirrhous variety occurs, sometimes consisting almost entirely of fibrous tissue, only a few cell-masses being discoverable here and there. All solid growths in the ovary, whether fibroid, sarcomatous, or carcinomatous, commonly enlarge the whole organ equally.

Diagnosis.—The solid character of a growth in the ovary should always excite the suspicion of its being malignant; and the suspicion is increased if both ovaries are affected, if pain is severe, if the growth is rapid, if a large quantity of ascitic fluid is present, or if the cachexia and emaciation of the patient, or local or general œdema, are greater than can be accounted for by the size of the tumour. The patient's age is also an element in diagnosis. If the tumour becomes fixed to the uterus and surrounding parts, and nodular masses are felt in its neighbourhood, the diagnosis becomes pretty certain. Examination of the ascitic fluid may also give distinctive signs (*see* p. 313).

Treatment.—While the character of a solid ovarian tumour is only suspicious, and while it remains apparently free from fixation, it may be desirable to remove it by ovariotomy, and the prognosis will be more favourable if it turns out to be sarcoma and not carcinoma. In general, palliative treatment only is admissible.

TUBERCLE OF THE OVARY is very rare, and is almost always associated with tubercle elsewhere, especially in the uterus and Fallopian tubes.

CHAPTER IX.

DISEASES OF THE FALLOPIAN TUBES.

THE CONGENITAL ANOMALIES of the Fallopian tubes have been considered in connection with those of the uterus (see p. 42).

SALPINGITIS, OR INFLAMMATION OF THE FALLOPIAN TUBE, commonly arises by extension of inflammation from the lining membrane of the uterus. Acute inflammation, proceeding to the formation of pus, is generally the sequel either of acute septic inflammation in the uterus, puerperal or otherwise, or of extension of gonorrhœal contagion. A collection of pus in the Fallopian tube is liable to lead to sudden and rapidly-fatal peritonitis, either through extension of inflammation by continuity to the ostium abdominale of the tube, through the outflow of pus by the same orifice, or through escape of pus after ulceration or rupture of the tube-wall, if the fluid is at first retained through want of patency in the tube. Peritonitis may probably also arise by transmission of inflammation, or by transudation of the fluid under pressure, through the walls of the tube. The more subacute or chronic form of inflammation in the tube is also very likely to set up a local peritonitis and consequent adhesions.

PYO-SALPINX, or distension of the Fallopian tube with pus, may remain as a chronic result of acute or sub-acute salpingitis. It appears to originate most frequently by extension of endometritis, especially endometritis of gonorrhœal origin, from the uterus. The distension in such case is moderate in degree, not nearly so great as sometimes results in cases of hydrosalpinx ; and adhesive peritonitis is frequently set up around the tube. Mr. Lawson Tait has of late removed the uterine appendages in a considerable number of cases of this kind. The operation appears justifiable in the hands of a practised specialist, if a positive diagnosis can be made, although it is by no means impossible that spontaneous recovery from such a condition may take place. Positive diagnosis, however, is the chief difficulty in such cases, unless it is to be made by opening the abdomen on suspicion in numerous instances. The indications most probably pointing to pyo-salpinx would be the discovery of a limited local swelling on one or both sides of the uterus, accompanied by signs and symptoms of persistent localized peritonitis, especially if there is a history, or probability, of previous gonorrhœa.

The most important disease of the Fallopian tube is TUBAL FŒTATION, but since this is discussed in works on Obstetrics, it will not be considered here. Among other affections to which the tube is liable are small FIBROID TUMOURS, which are of little importance, and TUBERCLE of its lining membrane, often associated with tubercle of the uterus.

The **Diagnosis** of inflammation of the Fallopian tube can generally only be a probable one, based upon the existence of a possible cause and the occurrence of the consequent peritonitis.

OBSTRUCTION OR OBLITERATION OF THE FALLOPIAN TUBE is of frequent occurrence. It is most commonly due to inflammation of the neighbouring peritoneum, and may be produced either at the abdominal extremity of the tube by adhesion of lymph to the fimbriæ of its pavilion, or at any part of the tube in consequence of constriction or contortion resulting from bands of adhesion. Such a condition is the commonest cause of incurable sterility. Partial obstruction may also result from the presence of a small polypus, or other tumour, at the uterine extremity of the tube. A condition of partial obstruction is the usual antecedent of arrest of the impregnated ovum in the tube, and consequent tubal fœtation, generally ending in rupture, very frequently followed by death from hæmorrhage. A total or partial obstruction, or the mere fixation of the tube, may lead to one form of inflammatory dysmenorrhœa, in which the ovum, with a small quantity of blood effused on rupture of the Graafian follicle, falls into the peritoneal cavity, instead of being conducted into the tube.

DILATATION OF THE FALLOPIAN TUBE, in slight degree, may result from chronic inflammation of its lining membrane, which may produce also narrowing at other parts. In more considerable amount, it may be produced by any cause of obstruction to the outlet of menstrual blood from the uterus, especially when the obstruction is absolute from occlusion of the hymeneal orifice, or vagina ; and it is apt also to be associated with the stretching out of the tube over uterine or ovarian tumours, or its fixation by old adhesions. Such an abnormal patency of the tubes constitutes the great danger which attends the injection of medicated fluids into the uterus. By allowing reflux of the menstrual blood, it is also believed to be one of the causes of periuterine hæmatocele (*see* p. 358). A considerable number of cases has been recorded, in which it has been concluded that the uterine sound could be passed for two inches or more along the Fallopian tube. In some instances, this has doubtless occurred, but some supposed cases are probably open to the explanation that the sound was first passed through a soft uterine wall, and that the opening remained for some time patent, so as to allow the sound to pass repeatedly in the same direction.

HYDRO-SALPINX, OR DISTENSION OF THE FALLOPIAN TUBE WITH FLUID, is the result of stricture or obliteration of the tube at two points, one of which is generally the outer extremity. It appears that the obstruction is usually complete on the abdominal side of the distended portion, but is not necessarily so on the side towards the uterus, so that an occasional discharge of fluid through the uterus may take place. Accordingly, it is held by Schroeder, that the normal destiny of the small amount of mucus secreted by the tube is to be discharged through the ostium abdominale into the peritoneal cavity. The direction, however, of ciliary action, and of the movement of the ovum, would seem to indicate that the secretion normally is discharged into the uterus. Both Fallopian tubes are not uncommonly distended together. The distended tube is thrown into convolutions, the distension increasing toward its outer extremity. The size it attains is commonly only moderate, not exceeding that of a fœtal head, although a few cases have been recorded in which very large dimensions were reached. The fluid within it is usually clear, limpid, and yellowish, containing a considerable quantity of albumen. In some cases blood is present.

Diagnosis.—The distinction between a distended Fallopian tube and a small ovarian tumour will often be difficult. At an early stage the position of the tube will be more anterior, since a small ovarian tumour usually falls down behind the plane of the broad ligament. It may be possible to distinguish the convoluted outline of the tube, increasing in size outwards. When, however, the distension has passed beyond a certain point, the tumour will appear more globular. The occurrence of symmetrical swellings on both sides would be in favour of the existence of distension of the tubes. The nature of the contained fluid, if evacuated, may greatly assist the diagnosis.

Treatment.—If the swelling causes any grave symptoms, and a probable diagnosis is made, the tube, with the ovary, may (by a skilled specialist) be removed by abdominal section without great risk to life.

Hæmato-salpinx, or Distension of the Fallopian Tube with Blood, may result from hæmorrhage into the dilated tube, especially under the influence of the menstrual nisus.   In some cases of hæmatometra, arising from occlusion of the genital canal, the tubes have also been found distended by retained blood, although cut off from the uterus by atresia of their orifices.   Again, when a dilated Fallopian tube has been enclosed in clamp after ovariotomy, hæmorrhage has occasionally taken place from it at menstrual periods.   Hæmorrhage into the Fallopian tube does not, however, appear to be a normal event in menstruation, although, when any morbid condition of the tube is present, it may readily occur during the menstrual hyperæmia.

CHAPTER X.

DISEASES OF THE UTERINE LIGAMENTS AND OF THE ADJACENT PERITONEUM AND CELLULAR TISSUE.

INFLAMMATORY and other affections of the uterus and its appendages are apt to give rise to inflammation of the cellular tissue in the vicinity of these organs, especially in the broad ligaments of the uterus, where it exists most abundantly, and also to inflammation of the peritoneum covering the inflamed tissues. In the very acute and septic form of metro-peritonitis, inflammation extends to the whole peritoneum, and is often rapidly fatal; in the much more frequent cases, however, in which inflammation is less severe, it generally remains limited to the peritoneum of the pelvis and its vicinity. We have thus a pelvic or periuterine peritonitis and a pelvic or periuterine cellulitis. The terms of "*perimetritis*" to denote the former, and "*parametritis*" to denote the latter have been introduced by Virchow, and widely adopted. The former was suggested by the analogy of the word pericarditis signifying inflammation of the serous covering of the heart, in contra-distinction to which Virchow proposed the terms paracarditis, paratyphlitis, paranephritis, and parametritis, to denote inflammation of cellular tissue near the heart, cæcum, kidney, and uterus respectively. The words are thus divorced from their literal etymological sense, since, as they are now employed, perimetritis may be on one side of the uterus, and parametritis may extend all round that organ. They have, therefore, rather tended to introduce confusion. They have also the disadvantage of rendering ambiguous the adjective "perimetric," which would otherwise be preferable to the hybrid word "periuterine" in the simple sense of "around or near the uterus." On account of this ambiguity, I have retained the word "periuterine," although it would be etymologically more correct to use the word "circumuterine."

Neither pelvic peritonitis nor pelvic cellulitis often exists altogether independently of the other affection, for inflammation of the peritoneum extends more or less to the cellular tissue immediately beneath it. Again, the lymphatics bear an important part in all inflammation of cellular tissue, and these communicate freely with the peritoneal cavity. Not only, therefore, does cellulitis usually extend to the peritoneal covering of the part immediately affected, but, especially when of septic origin, it is apt to kindle a peritonitis which passes beyond that limit. The terms, are, therefore, to be applied, not in an exclusive sense, but according as the affection of one or other structure is predominant.

Very diverse opinions have existed as to the relation and relative frequency of pelvic peritonitis and cellulitis. It was formerly assumed by many authorities that, when a swelling was detected on vaginal examination, the existence of cellulitis was established. It was proved, however, by Bernutz, from the evidence of numerous autopsies, not only that a localized swelling, tangible *per vaginam*, may be due to the effusion of lymph or serum and gluing together of intestines produced by peritonitis, but that this swelling may be situated at one side, or in front, of the uterus. Evidence derived from autopsy affords, however, no information as to the relative frequency of the two affections, since peritonitis is much more likely to prove fatal than cellulitis. Bernutz also went to an extreme in almost denying the occurrence of periuterine cellulitis, except in the form of phlegmon of the broad ligament. It is now, however, agreed by most authorities that, apart from parturition or abortion, or operations upon the cervix uteri, pelvic peritonitis is much more common than cellulitis, and that, in a large proportion of the cases which do not end in suppuration, inflammation of the peritoneum is the preponderating element. In many instances, however, the characters of the two affections are largely combined.

PELVIC PERITONITIS, OR PERIMETRITIS.

Causation.—Pelvic peritonitis may originate by contiguity from inflammation of the uterus, ovaries, or Fallopian tubes, or may be secondary to cellulitis. It is frequently the sequel of suppression of menstruation, due to the effect of cold, in which case the primary condition is an acute or subacute endometritis and metritis, of which the arrest of menstruation is the consequence, while the inflammation extends through the whole substance of the uterus to the adjoining peritoneum. Peritonitis may also be the result of endometritis by extension of inflammation along the lining membrane of the Fallopian tubes. This is especially common as the consequence of gonorrhœa, whose frequency as a cause amongst all non-puerperal cases of pelvic peritonitis has been estimated as high as 50 per cent. Thus prostitutes almost invariably suffer, at some time, from this disease, which generally renders them permanently sterile. Amongst other causes of pelvic peritonitis are replacement of the uterus by the sound, in some cases even the simple introduction of the sound, the use of an intra-uterine stem pessary, applications to the interior of the uterus, and operations upon the body of the uterus. Peritonitis may also arise through septic absorption by the lymphatics, and thus may be the consequence of the use of tents, or operations upon the cervix or vagina;

but, in this case, the inflammation is more likely to become general.

From menstrual disturbances pelvic peritonitis may result, not only through the medium of endometritis, but by direct reflux of blood through the Fallopian tubes, when the outlet of the uterus is obstructed in consequence of stenosis or flexion, especially if menorrhagia co-exists. In the same way intra-uterine injections may be an exciting cause. Escape of blood or other fluid from any other cause, as from rupture of a vein, of an over-congested Graafian follicle, or of a cyst in the Fallopian tube or ovary, may equally set up peritonitis. External violence, cold, and sexual excess act mainly through the medium of the inflammation which they may produce in uterus or ovaries. In puerperal cases the starting point is usually an inflammation of the uterus or cellular tissue, due to a traumatic or septic cause, or a combination of the two. The peritonitis may, however, be kindled into activity by the effect of cold, of premature exertion, or of emotion. In other puerperal cases again, the peritonitis is part of a general peritonitis, due to some zymotic or other form of blood poison. Ovarian tumours frequently set up peritonitis ; fibroid tumours do so less frequently ; cancer or tubercle of the uterus or ovaries is sooner or later accompanied by such a result. A pelvic peritonitis may also be a part of a general peritonitis not originating near the uterus, and may then lead to the same results with respect to the pelvic viscera as the localized disease. Thus the signs of pelvic inflammation may attract attention in cases of tubercular or cancerous peritonitis.

Pathological Anatomy.—In the active stage of inflammation, plastic lymph is poured out on the surface of the peritoneum, and leads to adhesion between the pelvic viscera. In acute cases there is also an effusion of serous, of sero-purulent, or, in the septic forms of peritonitis, of purulent fluid. In the majority of cases, however, the peritonitis is mainly or solely of the adhesive form. The semi-fluid lymph tends to gravitate into the pouch of Douglas, where it forms no tangible swelling, so long as it remains fluid and free, but is generally converted into a firm mass, fixing the uterus, as the lymph consolidates. Within spaces formed by adhesion between coils of intestine, or between intestines and other viscera, serum may be poured out in considerable quantity, and a limited and rounded swelling may thus arise, which sometimes very closely simulates a true cyst (see p. 309). Suppuration may also take place in similarly limited spaces, though much less commonly than in the case of cellulitis. The pus thus collected may remain quiescent for a considerable time, and rarely escapes into the general peritoneal cavity. As a rule, the abscess perforates sooner or later, opening in most cases into the rectum or sigmoid flexure. Perforation into the vagina is less common, and that into the bladder still more rare. Perforation on the external surface is also comparatively uncommon. When it does occur the most frequent site is the flexure of the groin. Sometimes the abscess opens at more points than one. In some instances, especially after labour, but much more rarely than in the cases of pelvic cellulitis, the inflammation may subside near the uterus, but at some more or less distant point go on to the formation of serum or pus, or the production of an apparent tumour by agglutination of intestines. This condition Dr. Matthews Duncan describes by the name of remote perimetritis. In the later stage of the more common affection, or adhesive peritonitis, the lymph becomes organized into bands of adhesion. As times goes on, after subsidence of the inflammation, these are generally partially absorbed, and become lengthened and attenuated, especially in those situations where most motion naturally takes place. Some degree of distortion and fixation of the parts involved is, however, generally permanent in some situations, especially about the Fallopian tubes and ovaries ; hence sterility is a common sequel. The uterus is apt to be fixed, temporarily or permanently, in any abnormal position it may have had at the outset of the inflammation, and it is also liable to distortion gradually produced by traction, in consequence of the shrinking of plastic lymph.

Results and Symptoms.—In the more acute forms of pelvic peritonitis, the symptoms resemble those of general peritonitis, and differ only in the fact that the pain and tenderness are more or less localized in the lower part of the abdomen, and that the general symptoms are less intense in degree. Frequently there is a rigor at the commencement, and pain may be severe at first, accompanied by extreme tenderness in the hypogastrium. The pulse becomes rapid, and acquires more or less of the peritonitic quality. There is often considerable rise of temperature, but its elevation is generally less in proportion than that of the pulse, and sometimes even a normal temperature exists in severe septic forms of inflammation, so that the temperature alone is an unsafe guide as to severity. Frequent micturition, with severe vesical tenesmus, is a common symptom. The bowels are usually constipated (except in septic and general forms of peritonitis), and there is much pain on defecation. The abdomen is frequently tympanitic, and it is not uncommon to find a transient tenderness over its whole surface, which may shortly subside, and leave only localized symptoms. Rigidity of the abdominal muscles over the region of tenderness is almost invariably present. In the more severe forms of inflammation, nausea and vomiting are common, and the features become pinched and anxious.

Cases of adhesive peritonitis, however, are not unfrequent in which the inflammation is chronic and almost latent from the first, especially when the exciting cause is some continuous, but not very intense, source of irritation, such as endometritis or ovaritis, or when the attack is a recurrence of some old-

standing inflammation. This may happen even in the
gonorrhœal form of the disease, although, under such
circumstances, the attack is commonly more acute. In
the chronic cases the symptoms are extremely in-
sidious, and the patients may go about their occupations as usual, suffering only from a gradually increasing
pain in hypogastric or inguinal regions. In some instances, the only complaint made is that of bladder
irritation, although, on examination, the whole pelvis
is found to be filled with inflammatory induration.
'The pulse, in these cases, is usually found to be rapid,
often from 100 to 120, although elevation of temperature may be slight or absent.

The sequelæ of the disease are of an extremely
chronic kind, and those who have once suffered from
it are generally liable to relapse on slight provocation,
especially from the effect of cold, or imprudence at
menstrual periods. In many cases the uterus eventually
recovers its mobility, and almost all remnant of swelling in the pelvis disappears. When, however, the
whole roof of the pelvis has become hardened into a
board-like mass, and no sign of commencing absorption
appears within the first few months, this condition
may remain a permanent one. Years may elapse
before the utmost degree of relative cure which can
be hoped for is attained, and patients may remain
invalids for, at any rate, the remainder of the period
of active sexual life. After an attack of peritonitis,
the recurrence of menstruation is frequently deferred.
The ensuing period is apt to rekindle inflammation,
but, if this does not happen, it is often followed by
relief. Protracted, or even permanent amenorrhœa, or
scanty menstruation, is, however, a frequent sequel.
Dysmenorrhœa is also commonly produced from the
interference with the functions and vascular supply of
uterus and ovaries caused by the adhesions.

Diagnosis.—In the more acute forms of the affection,
the symptoms readily show the existence of periuterine
inflammation, and the chief point of difficulty is to

determine whether peritonitis or cellulitis is the main element. Assistance may be derived from the consideration that, when there has been no antecedent parturition, abortion, or operation on the cervix, and when the exciting cause lies in the body of the uterus, ovaries, or Fallopian tubes, rather than in the cervix, especially if that cause be gonorrhœa, the inflammation is more likely to be peritonitic. In peritonitis, also, tenderness is more acute, and vomiting and other symptoms pertaining to the digestive functions are more likely to be prominent than in cellulitis with little or no complication of peritonitis. In cellulitis, on the other hand, the initial rigor and elevation of temperature are more marked, in proportion to the other symptoms.

On vaginal examination in the earliest stage, while the exudation is still fluid, merely tenderness and slight increase of resistance will be discovered around the cervix. The uterus will be very tender on pressure, and still more so on displacement. After consolidation of the exudation, one of two conditions may be found :—

(1) In the first, the inflammation, while limited to the pelvis, is general throughout that region. This constitutes the most typically recognizable form of pelvic peritonitis. The cervix uteri is then central, or slightly pushed forward, low down in the pelvis, and firmly fixed, not displaced to either side. Induration extends all round it, and forms a complete roof to the pelvis of uniform hardness. At the posterior part, where lymph gravitates into the pouch of Douglas, it descends somewhat lower, and forms a more distinct mass. The induration can be reached from above the pelvic brim, but does not form an apparent tumour rising into the abdomen, or extending into the iliac fossa, nor does it descend so low upon the vaginal walls as that formed by cellulitis sometimes does. On rectal examination, a hard mass can be felt enveloping the cervix posteriorly, and extending at the sides of the

rectum to the pelvic wall, while its upper limit can generally be scarcely reached.

(2) The second condition is that in which there is a localized focus of inflammation, which may extend or not above the pelvic brim. Portions of intestine, matted together by adhesions, and often containing impacted fæces, may then form an apparent tumour, which may reach as high as the umbilicus. In this case the uterus may be pushed forwards, backwards, or to one side. If serum or pus be effused in a limited space, the tumour may be apparently cystic. The mass thus formed by adhesions may generally be distinguished from the swelling formed by cellulitis by the following characters. In cellulitis the swelling, if of any considerable dimensions, is always on one side, tending toward the iliac fossa, rarely rises more than two or three inches above Poupart's ligament, and has a strong tendency to suppurate. In peritonitis the swelling is later in its appearance, and may be more nearly central, and rise to a higher level, while, if it is situated near the groin, the abdominal walls are more movable over it than over a swelling formed by cellulitis. Such a swelling may generally be distinguished from an ovarian or fibroid tumour by its fixity and by the history of its first appearance after the onset of acute inflammatory symptoms (see p. 309). If situated at the side, or in front of the uterus, it may be pressed downward by the effusion of serum in its midst, so that it may become difficult or impossible to distinguish it from cellulitis. As a rule, however, it does not descend so low upon the vaginal wall. From an early extra-uterine fœtation it is distinguished by the absence of the characteristic signs of that affection (see p. 311).

A peritonitis affecting the pelvis which forms a part of cancerous or tubercular peritonitis, or originates in perityphlitis, is distinguished by recognition of the signs of the primary disease. An induration produced by diffuse cancer of the pelvis may closely resemble that of pelvic peritonitis, and may be only distinguishable by the amount of cachexia present, and by the course of the case, especially by the absence of inflammatory symptoms at the outset, and a downward course afterwards. The diagnosis may be assisted by the age of the patient, and sometimes by the detection of nodular masses, like enlarged and indurated glands. The discovery of the slight increase of resistance or diminution of mobility of the uterus, which may be the sole remnant of a bygone pelvic peritonitis, often requires a highly practised touch.

Treatment.—In the acute stage, when pain is severe, provided that the patient is not already anæmic, from six to twelve leeches may be applied to the groin or hypogastrium. Perfect rest is to be maintained, and hot linseed poultices or fomentations kept applied to the abdomen. Sufficient opium or morphia should be given to keep the pain in check, and may conveniently be administered by the rectum or subcutaneously. Opinions differ widely as to the use of mercury. It appears to be of little value as an antiphlogistic in the early stages, but in severe prolonged cases, not of a septic or purulent character, it may be tried if the disease appears not to yield to other remedies, care being taken to keep short of salivation, and not to act upon the bowels. The plan adopted by some is to give three grains of hydrargyrum cum cretâ, with five grains of Dover's powder, in pill or powder two or three times a day. It is generally preferable, however, to use the mercury locally in the form of ointment, equal parts of mercurial ointment and belladonna ointment being applied on lint over the seat of inflammation. It is recommended by Dr. Thomas that all other drugs should be avoided, and opium or morphia given in large and repeated doses, frequently as much as half a grain of sulphate of morphia being administered every two or three hours for a considerable time. Milk and beef-tea may be given as diet, ice being added to the former, if vomiting is urgent. Vomiting is often relieved in these cases by subcutaneous injection of morphia. The treatment of severe septic forms of inflammation has been considered under the head of metritis (*see* p. 172).

In less acute forms of pelvic peritonitis, opium or morphia in smaller doses may be given at first in combination with salines, as citrate and nitrate of potash, or acetate of ammonia. In the later stage iodide of potassium, in doses of from five to ten grains, is useful as an absorbent, and may often, with advantage, be combined with quinine or other tonic. After all febrile symptoms have passed absorption is not promoted so effectually by the administration of any drugs as by securing the best possible condition of general health, and by all ordinary means which tend to promote the activity of processes of nutrition. Small doses of perchloride of mercury may, however, in some cases be tried (*see* formula, p. 214), since this drug appears to be much more effective as an absorbent, in removing inflammatory products, than as an antiphlogistic during the acute stage of inflammation. As the inflammation is subsiding repeated counter-irritation is of great value. Vesication by blistering fluid, over a surface about four inches square, may be repeated at intervals of about ten days, or a blister may be kept open by savine ointment. A milder remedy is the linimentum iodi, painted daily over the same surface, as long as the skin will tolerate it. Absorption is also stimulated by hot vaginal injections, used in the manner previously described (*see* p. 188). The heat should be moderate at first, and should be increased gradually up to about 110° F. Hot hip-baths, or, still better, whole-baths, are also of value, and are efficacious to relieve pain. When the absorption of inflammatory material does not proceed satisfactorily the use of salt water often proves efficacious, from its greater stimulating power.

When a case has reached the chronic stage, or is

chronic from the commencement, complete confinement to bed is not advisable. Sufficient air and exercise to maintain the general health should be allowed, especially carriage exercise, if the motion can be borne without pain. A large amount of rest, however, should be taken in the horizontal position, and cold or exertion should be specially avoided at menstrual periods. Marital intercourse is to be forbidden in acute stages or while it produces pain, and should be greatly restricted for a long period. It is not, however, always desirable to prohibit it entirely, especially if there is any ovarian engorgement or inflammation, such as often results from the hindrance to the function of the ovary produced by adhesions. Recurrent pain, accompanied by tenderness of the uterus, may be relieved by a few leeches applied occasionally to the cervix. At this stage tonic treatment, especially the administration of iron and quinine, is beneficial, and change of air, or a sea-side residence, often proves useful. If relapses are found to occur from the effect of cold, spending the winter in a warm climate is to be recommended.

It is not generally desirable to evacuate collections of pus or serum, unless manifest pointing has occurred. If, however, an abscess remains long in a stationary condition, and the general state of the patient is unsatisfactory, it may be evacuated by the aspirator. The most favourable condition for opening an abscess or collection of serum *per vaginam*, is that in which an elastic swelling is felt behind the cervix. The puncture should then be made, if possible, in the median line, care being taken to avoid any artery which may be felt pulsating. In the absence of the aspirator, a grooved needle may be used to confirm the presence of fluid, and afterwards serve as a guide for a guarded straight bistoury or tenotomy knife. A small puncture having being made, it may be enlarged by introducing a pair of dressing forceps, and separating the blades, according to the method recommended by Mr. Hilton for the evacuation of deep abscesses. If pus again collects after the use of the aspirator, a free opening may be made in a similar way. If an abscess is pointing externally it should be opened under carbolic spray and a drainage tube inserted according to Lister's method, the end of the tube being cut off level with the skin, and secured by two loops of carbolized silk, passed through the end of the tube, and laid flat upon the skin beneath the gauze dressings. If an abscess has spontaneously opened externally, or has been opened without antiseptic precautions, and pus continues to be poured out from an extensive cavity, a large drainage tube should be inserted to the full depth of the cavity. The end of the drainage tube may be immersed in carbolic lotion, and the cavity may be washed out by means of a funnel with a weak solution of iodine (Tr. Iodi 3ij. ad Aq. Oj.) or sulphurous acid (Acidi Sulphurosi 3iv. ad Aq. Oj.). Carbolic dressings may be used, although it is difficult to render aseptic a large and irregular cavity. Carefully adjusted pressure by pads of cotton wool may assist in causing the cavity to close. If the abscess cavity is found to descend close to the vagina, it may in some cases be desirable to make a counter opening at its lowest point by cutting from the vagina upon the point of a probe passed into the abscess. This should not, however, be done until the plan of using a large drainage tube and antiseptic irrigation has had a full trial.

PELVIC CELLULITIS, PERIUTERINE CELLULITIS, OR PARAMETRITIS, WITH PELVIC LYMPHANGITIS.

Causation.—The chief causes of pelvic cellulitis are parturition, abortion, applications of caustic to, or operations on, the cervix uteri or vagina, inflammation of the uterus, especially the cervix uteri, and inflammation of the ovaries or Fallopian tubes. In a very large proportion of cases the cause is parturition, and the mode of origin may then be, in whole or in part, directly traumatic, from the pressure and bruising to which the cervix and cellular tissue are exposed, or from lacerations of the cervix. Thus puerperal cellulitis is much more common on the left side, on account of the usual direction of the occiput toward the left and the common deviation of the fundus uteri toward the right, both which causes tend to make the pressure greater on the left side of the pelvis. In the majority, however, both of puerperal and non-puerperal cases, the main element is septic absorption from some cut or lacerated surface, or from an abrasion, such as may be produced by the use of tents. Not only the loss of epithelium, but the injury to the tissue predisposes to septic absorption. For the damaged tissue, having its vitality lowered, does not, like healthy tissue, resist the entry and multiplication within it, even of organisms which may be generally or frequently present in the lochial or vaginal discharge. Thus an inflammation set up mechanically by very difficult instrumental delivery often acquires a more or less septic character, apart from any conveyed contagion. The determining cause of the acute outbreak of inflammation is often the effect of cold, mental emotion, or premature exertion after parturition, or after some operative interference. Cellulitis may result from menstrual disturbances, but much less frequently than peritonitis; and, when it does so, it is probably for the most part by extension of inflammation from the ovaries or Fallopian tubes into the adjoining tissue, while there is commonly a complication with a ' notable degree of peritonitis. Cellulitis may also be set up by sexual excess or external injuries, especially if any previous disease of the uterus or its appendages exists. A similar inflammation of cellular tissue again may take its origin from cancerous or syphilitic ulceration of rectum or vagina, or from disease of the bladder.

Pathological Anatomy.—Pelvic cellulitis is an inflammation or phlegmon of the areolar tissue in the pelvis, in the vicinity of the uterus or its appendages.

This areolar tissue is most abundant in the broad
ligaments. It also exists in plenty in front of the
lower half of the uterus between it and the bladder,
for a smaller space at the posterior part of the cervix
(*see* Fig. 24, p. 68, and Fig. 25, p. 69), as well as around
the vagina, bladder, and rectum, and in the sheaths of
the psoas and iliacus muscles and the muscles of the
abdominal wall. Between the uterus and its peritoneal
covering at front and back areolar tissue is so scanty
that cellulitis can scarcely occur there. The term
"periuterine cellulitis" has been used in a sense
limited to inflammation immediately adjoining the
uterus at its sides, front, or back. An abscess after
parturition, however, may appear at a distance from
the uterus, as in the groin or abdominal wall, while no
remnant of inflammation can be detected round the
uterus, and the wider term of "pelvic cellulitis"
therefore appears preferable. Those cases, however,
are to be distinguished in which inflammation merely
extends into the pelvis from outside, as in a psoas
abscess.

In the majority of cases, especially those of puerperal origin, the inflammation is chiefly situated in one
or the other broad ligament, whence it tends to spread
into the iliac fossa, and along the sheaths of the
muscles to the groin and the adjoining portions of
the abdominal wall. This form of cellulitis has been
distinguished by the name of "*phlegmon of the broad
ligament.*" In other cases, however, especially those
arising from lesion of the cervix, the tissue in front
of, or behind, the uterus may be chiefly or solely
affected, and the inflammation may also descend along
the walls of the vagina or rectum, or may occupy chiefly
the tissue at the base of the bladder.

Since septic absorption is so generally an element
in cellulitis, the lymphatic vessels play an important
part in it, as in all inflammations of cellular tissue.
In some cases enlarged lymphatic glands in the pelvis
may be detected as rounded masses in the midst of

inflammatory thickening, and these may form foci of inflammation or abscess formation.

In the earlier stage of cellulitis a swelling is produced by effusion, first, of serum, and secondly, of lymph in addition to serum, into the areolar tissue. This may end in resolution, or in the formation of an abscess, which is a much more frequent result than in peritonitis. It is commonest in puerperal cases, and has been estimated by some authorities as occurring in more than 50 per cent. of these ; but this is probably too high an average, if mild as well as severe cases are included. By far the most frequent spot for the abscess to open is the groin or iliac region, the pus generally making its way mechanically along the course of the psoas and iliacus muscles. It may also open externally above the pubes, beside the anus, or, very rarely, pass through the sciatic or obturator foramen. Internally, discharge into the rectum and that into the vagina are about equally common; that into the bladder is also frequent, but rather less so. Discharge into the peritoneal cavity is fortunately very rare. Internal evacuation is commoner in non-puerperal cases, in which the abscess is generally nearer to the uterus, or, in rare instances, may be situated between uterus and bladder. When an abscess resulting from parturition appears at a distance from the uterus, as in the inguinal canal, in the sheaths of the abdominal muscles, or near the sacro-iliac joint, and the vicinity of the uterus is found free, the explanation is probably for the most part that the inflammation has terminated in resolution near the uterus, and has proceeded to suppuration at a distant point only. In some cases, however, the distant abscess may be due to conveyance of septic material by lymphatics, without any perceptible intermediate inflammation. To this manifestation of inflammation at a distant point, which is not extremely uncommon, Dr. Matthews Duncan has given the name of remote parametritis.

Results and Symptoms.—The onset of the disease

is acute in the great majority of cases, and a decided
rigor and elevation of temperature (often reaching or
exceeding 103° or 104°) more generally occur than in
the case of pelvic peritonitis. The fever is accompanied or quickly followed by pain, which is not
always very acute, and often depends in great measure
upon implication of the peritoneum. Vesical tenesmus and pain on defecation are frequently added.
If menstruation is present at the time of onset, the
flow may be increased, except in cases in which the
inflammation is complicated by any considerable degree
of acute endometritis or metritis, the effect of which
is usually to arrest either the lochial or the menstrual
discharge. In some puerperal cases the inflammation
commences more gradually, and is only kindled into
activity after the patient leaves her bed. More rarely
the symptoms are limited to slight pelvic pain, and
trouble in micturition, with feverishness and debility,
and the exudation may only be discovered on examination at a considerable interval after delivery. More
frequently a swelling appears within a few days in the
groin or iliac region, or extending to the hypogastrium.
Flexion and adduction of the thigh, which are often
enforced in consequence of the pressure of the exudation, are characteristic symptoms. In the course of a
few weeks, in the majority of cases, the disease either
ends in suppuration, or resolution has commenced.
Thickening in the cellular tissue, however, is only
slowly absorbed, and a certain amount of induration
may be permanent. Lameness on the affected side is
often slow in disappearing. The uterus may be drawn
to one side by contraction of the inflamed tissue in a
late stage; but complete fixation of the uterus, with
sterility, and other permanent sequelæ, when they
occur, are commonly due to associated peritonitis.
Suppuration probably most frequently commences
within a few days, when it takes place at all, but
the period of bursting of the abscess commonly varies
from two weeks to three months.

Thrombosis of the veins is a common result of inflammation in the cellular tissue surrounding them, and involves a risk of pulmonary embolism. This is one reason why protracted rest should be enforced after even a slight attack of cellulitis. If thrombosis extends to the iliac or femoral veins and lymphatics, phlegmasia dolens may be a sequel, especially in puerperal, but sometimes also in non-puerperal cases. In some instances an abscess burrows extensively in the pelvic cellular tissue. Suppuration may then be protracted, especially if the abscess opens by a long fistulous track, or an opening exists in two directions simultaneously. The patient may thus be greatly reduced by hectic fever, and even a fatal result follow. In very rare cases there is extensive sloughing of areolar tissue. The mortality, however, of uncomplicated pelvic cellulitis is, in general, small, and much less than that of pelvic peritonitis.

Diagnosis.—In the typical form of pelvic cellulitis, namely, that of phlegmon of the broad ligament, in which the cellular tissue of the broad ligament is the chief focus of inflammation, the diagnosis is generally easy. On vaginal examination, a considerable and immovable swelling, shading off into the pelvic wall, is felt on one side of the cervix, and rather low down. The cervix itself is pushed toward the opposite side, and its mobility, although diminished, is often not entirely destroyed. Some thickening may also extend round the front and back of the uterus. The lateral swelling can be reached by the external hand above the groin, and on bimanual examination is felt as a considerable mass between the fingers. The thigh on the affected side is frequently retracted. Unless the extent of inflammation is very limited, it forms a swelling in the inguinal and iliac region, either prominent and readily tangible, or, at any rate, sufficient to give rise to a feeling of resistance, and partial or complete dulness on percussion. A cellulitic swelling, however, rarely extends higher than two or three inches above Poupart's ligament, or is liable to be mistaken for a tumour, except in the rare case of a large abscess between uterus and bladder, which may rise as high as half-way between pubes and umbilicus. For the differential diagnosis of a swelling in the abdomen due to peritonitis, *see* p. 346. The symptom of retraction of the thigh, with pain upon any attempt to extend it, may persist for a long time after delivery, and may be the only local sign to indicate the presence of inflammation or abscess about the psoas and iliacus muscles, when no swelling can be detected.

When cellulitis affects the tissue in front of, or behind, the uterus, the induration caused by it is very difficult to distinguish from that due to peritonitis, and it is often complicated by that affection. When, however, the induration extends low down upon the walls of vagina or rectum, or when it chiefly affects the base of the bladder, the diagnosis of cellulitis becomes positive. Diagnosis is often assisted by the fact of parturition, abortion, or some operation upon uterus or vagina having preceded.

From a fibroid or ovarian tumour, a cellulitic swelling is distinguished by its fixity, and by the fact that no sign of tumour had existed before the onset of inflammatory symptoms. For the differential diagnosis of extra-uterine fœtation, *see* p. 311, and for that of hæmatocele, *see* p. 362.

Treatment.—The local and general treatment is similar to that of pelvic peritonitis (*see* p. 347). The use of leeches is, however, less frequently desirable, since the affection often occurs from a septic cause in anæmic patients, or those debilitated by hæmorrhage. If used at all, they should only be employed quite at the onset. From the frequent presence of a septic element, quinine, or other internal antiseptic, in large doses, combined with opiates, is often given with advantage until the temperature is reduced.

It is not desirable to interfere at an early stage to

evacuate pus. When, however, an abscess is commencing to point externally, so that fluctuation is distinct, and there is no fear of opening the peritoneal cavity in making the incision, it may be evacuated with advantage. It is desirable that this should be done under carbolic spray, and the wound dressed with antiseptic precautions, a drainage tube being introduced. When the pointing takes place by vagina or rectum, artificial evacuation requires more caution, and is more frequently superfluous. If a distinctly fluctuating spot is felt from the vagina, aspiration with a fine trocar may first be employed, and a larger opening afterwards made, if necessary. The puncture or incision should be made in the manner already described for the case of a peritonitic abscess (*see* p. 349).

Some authorities, as Sir James Simpson, Dr. Savage, and Sir Spencer Wells, recommend puncture as soon as the existence of an abscess is distinctly ascertained. When grave constitutional symptoms referable to an abscess arise, it is certainly worth while to venture on puncture with an aspirator in the earlier stage. If the abscess fails to close for a long period, and a large abscess cavity is found to exist, it should be treated by the use of a large drainage tube, antiseptic irrigation, and pressure as already described (*see* p. 350) ; or in exceptional cases a second opening may be made at the most dependent point.

PELVIC HÆMATOCELE.

By pelvic or periuterine hæmatocele, in its wider sense, is understood a limited collection of blood wholly or partially in the pelvis, whether within the peritoneal cavity, or in the cellular tissue outside it. An effusion of blood while still free within the peritoneum should not receive the name of hæmatocele, though its causes may be the same, and though it may form an antecedent stage to that affection.

Causation.—The *immediate mechanism* of the production of hæmatocele may be, for the intra-peritoneal
variety—(1) Reflux of menstrual blood through the
Fallopian tubes, due either to atresia or obstruction of
the cervix or vagina, to a morbid condition of the tubes
themselves, or to excessive menstruation ; (2) excessive
hæmorrhage on rupture of a Graafian follicle; (3) rupture of a vessel in the broad ligament or elsewhere ;
(4) hæmorrhage from inflamed peritoneum, or from
vascular pseudo-membranes ; (5) rupture of a cyst in
the ovary or broad ligament ; (6) rupture of the distended Fallopian tube ; (7) rupture of the sac of an
early extra-uterine fœtation, or a fœtation in a rudimentary uterine cornu.   Of these, the first four, which
are generally the menstrual kinds of hæmatocele, form
the least severe varieties, while in the fifth, sixth, or
seventh the hæmorrhage is more frequently excessive,
and is apt to prove fatal before the blood becomes
encysted. In the extra-peritoneal variety the mechanism
is generally that of the rupture of a vessel, either into
the surrounding tissue, or into a pre-existing cyst. This,
like the first four varieties of the intra-peritoneal effusion, is generally a menstrual form of hæmatocele.

The *predisposing causes* are the presence of menstruation ; active or passive hyperæmia of the uterus and
adjoining parts, by whatever cause produced ; previous
disease within the pelvis, especially obstruction of the
cervix uteri or vagina, morbid conditions of the Fallopian tubes or ovaries, or varicose distension of veins ;
the hæmorrhagic diathesis ; and diseased conditions of
blood, such as those produced by zymotic diseases,
jaundice, purpura, or scurvy.

The *exciting cause* is most frequently external
violence ; muscular strain ; coitus, especially during
menstruation ; or the effect of cold or mental emotion
in producing a sudden increase of hyperæmia during
the same condition.

Pathological Anatomy.—Pelvic hæmatocele is not
excessively rare, but yet undoubted instances of it form

a very small proportion to cases of pelvic cellulitis or peritonitis. In the great majority of fatal cases in which an autopsy has been made, the effusion of blood has been reported as being intra-peritoneal, although it has often proved extremely difficult to determine positively the true position of the peritoneum. It is probable, however, that the extra-peritoneal variety is relatively much commoner among those cases which end in recovery. It is also probable that, in a considerable number of cases which are not distinguished from pelvic cellulitis or peritonitis, the starting-point of inflammation may have been a slight or moderate effusion of blood.

Blood effused into the peritoneal cavity tends to gravitate into the retro-uterine fossa, but does not form a tangible swelling there while it remains fluid. When clotting has taken place, there may be a mass to be felt behind the cervix; but the uterus will not be displaced more than it is when the lymph effused in pelvic peritonitis gravitates into the same position. An induration of this kind, but no more, may be formed by gravitation into the pelvis of blood effused, not within the pelvis itself, but elsewhere in the peritoneal cavity. When, however, the amount of blood effused is not sufficient to cause death, it is soon enclosed by false membranes, which separate it from the intestines which it has displaced from the retro-uterine fossa. If further hæmorrhage now takes place within the enclosed space, the uterus is pushed forward and upward, the rectum flattened against the sacrum, and a retro-uterine tumour formed, which may extend upwards as high as the umbilicus (Fig. 75, p. 360). This condition, which constitutes the most typical and recognizable form of hæmatocele, and the one which specially deserves the name of retro-uterine hæmatocele, thus necessarily implies (unless it be of the extra-peritoneal variety) either a slow and gradual hæmorrhage or one repeated at intervals.

An intra-peritoneal hæmatocele situated in front of

the uterus has occasionally been observed, but it is scarcely possible for it to be confined to that position, unless the retro-uterine fossa has previously been occluded by false membranes. The extra-peritoneal variety of hæmatocele chiefly occurs in the broad ligaments, more rarely in the cellular tissue in front of the

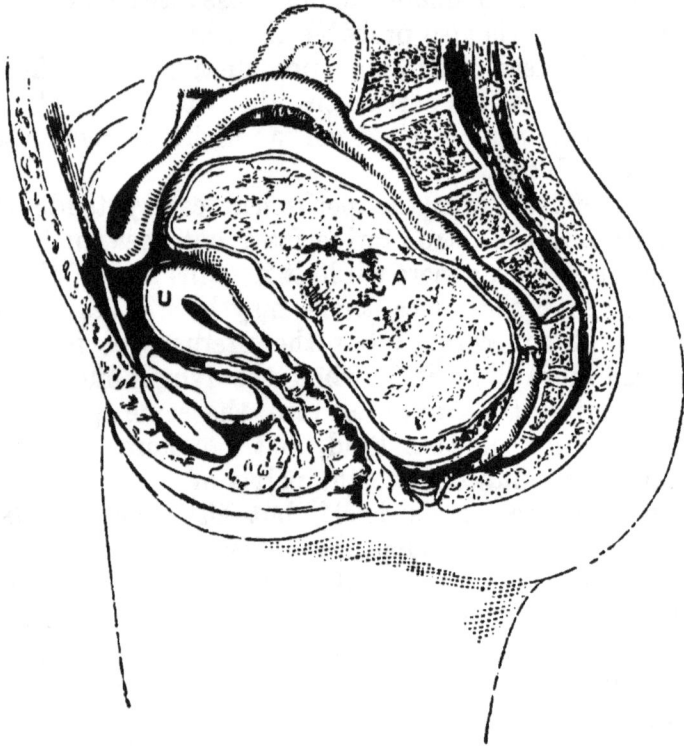

Fig. 75.—Retro-uterine Hæmatocele (after BARNES). U, The uterus pushed forward. A, The hæmatocele filling the cavity of the sacrum, bounded above by plastic effusions and the small intestines. R, The rectum compressed by the hæmatocele.

uterus. It is also possible for blood-effusion, like cellulitis, to occur in the cellular tissue behind the cervix. Retro-uterine hæmatocele, however, so far as a conclusion can be drawn from records of autopsies, appears to be almost invariably intra-peritoneal.

A frequent and important variety of hæmatocele is that which arises from reflux of blood through the Fallopian tubes. This is especially likely to occur in' cases of dysmenorrhœa with menorrhagia dependent upon obstruction of the cervix from stenosis or flexion. The Fallopian tubes are then often unduly patent, and may themselves abnormally pour out blood from their mucous membrane in menstruation. Under the influence of a sudden contraction of the uterus or of the tube itself, the outlet not being free, the blood escapes from the ostium abdominale of the tube. The same may occur even without obstruction to the outlet if the flow of blood is very excessive. In cases of complete atresia of cervix or vagina, the Fallopian tubes eventually become distended by retained blood, which may escape by reflux, or, more frequently, by ulceration or rupture. The rupture of the sac of a tubal fœtation at an early stage, often so early that no pregnancy has been suspected, is probably not a very uncommon cause of the more severe forms of hæmatocele. If, however, the fœtation is rather more advanced, so that a positive diagnosis has become possible, the amount of blood lost is usually sufficient to destroy life before it can become encysted. Virchow has maintained that the usual source of hæmorrhage is vascular peritoneal adhesions, and that therefore peritonitis is an almost invariable antecedent of hæmatocele. General experience does not lead to the conclusion that this is often the case; but instances have been recorded of hæmorrhage into a pseudo-cyst, formed by the adhesions of peritonitis (*see* p. 346).

Results and Symptoms.—In a marked case of hæmatocele, where the effusion of blood is considerable, a patient, generally during a period of profuse menstruation, and often from the effect of one of the exciting causes before mentioned, is suddenly attacked by pain, which is quickly followed by faintness, and often collapse, with nausea or vomiting. The loss of blood may be sufficient to produce pallor. The external menstrual hæmorrhage is generally considerably diminished, or may be arrested altogether, although frequently it continues to some extent. In other cases, the onset of the attack takes place while menstruation is imminent, or after its suppression, or after partial or temporary suppression, from the effect of cold or emotion. After a while, symptoms of pressure in the pelvis arise—a feeling as of the presence of a foreign body— with vesical and rectal tenesmus. A swelling may also appear in the hypogastrium, and extend upward toward the umbilicus. At first the temperature may be subnormal, but within two or three days a febrile reaction generally occurs, with symptoms of pelvic or general peritonitis, which may be of greater or less intensity. When, however, the hæmorrhage is slight or gradual, the onset of the attack may be little marked. When the cause is a rupture of the sac of an extra-uterine fœtation, the occurrence more frequently takes place apart from menstruation, and the signs of hæmorrhage are generally more severe.

A recurrence of hæmorrhage at succeeding menstrual periods is not unfrequent, and in this way the tumour may undergo repeated increase in size; otherwise it diminishes and becomes harder by absorption of the serum. Sometimes its contents again become softened in consequence of decomposition or suppuration, and constitutional symptoms of septicæmia may then supervene. In some cases spontaneous evacuation, before or after suppuration, takes place by rectum, or, more rarely, by vagina. In others the tumour becomes very slowly and gradually absorbed. In rare cases rupture into the general peritoneal cavity occurs. Death may occur from this cause, or from septicæmia or peritonitis; and the prognosis is always grave when the effusion is of enormous size or when decomposition takes place in it.

Diagnosis.—The diagnosis is easy in a typical case of retro-uterine hæmatocele when the onset has been sudden and well marked. A characteristic history, such as that already described, and a recently acquired appearance of anæmia, afford valuable evidence. On vaginal examination a large mass is felt, pressing down the recto-vaginal septum by distension of the pouch of Douglas, and encroaching upon vagina and rectum. The cervix is displaced forwards, and generally upwards, much more considerably than in pelvic peritonitis or cellulitis. The fundus uteri is pushed forward against the abdominal wall, so as to be much more readily tangible than usual (*see* Fig. 75, p. 360). On bimanual examination, a mass is felt behind, and generally above, the fundus uteri, continuous with that behind the cervix, and sometimes reaching as high as the umbilicus. Such a tumour may be recognized within two days of the first hæmorrhage. The mass is at first somewhat soft and yielding, but becomes gradually very hard and nodular, though it may afterwards again soften.

When the history is not clear, when the amount of effusion is moderate, or when the case is only seen at a late period, there may be much difficulty in distinguishing hæmatocele from other masses which may exist behind the uterus. Such masses may be formed by a retroflexed fundus uteri, pregnant or not, by pelvic peritonitis or cellulitis, fibroid tumours, ovarian tumours, parovarian cysts, hydatid cysts, dermoid tumours, extrauterine fœtation, a distended Fallopian tube, malignant disease, outgrowths from the pelvic wall, or fæcal accumulations. The distinction is most likely to be difficult between hæmatocele and cystic or dermoid tumours of the ovary, fibroid tumours, or extra-uterine fœtation. The sudden appearance of the tumour of hæmatocele is its chief distinction. For the distinctive signs of extra-uterine fœtation *see* p. 311. In estimating the value of enlargement of the uterus as a sign of this affection, it must be remembered that a uterus may be elongated when adherent to a hæmatocele. A retroflexed uterus will be distinguished by bimanual examination and the use of the sound if necessary.

When the hæmatocele is of small size, and does not

descend low in the pelvis, or when it is situated laterally
or anteriorly, especially if it is of the extra-peritoneal
variety, the diagnosis may be very difficult, and it may
be impossible to distinguish it from pelvic peritonitis
or cellulitis, except by means of exploratory puncture,
a proceeding generally not to be recommended. In such
cases a positive diagnosis is of little consequence as
regards treatment. Sometimes the case is cleared up by
spontaneous evacuation. In very rare cases an extra-
peritoneal hæmatocele has been closely attached to the
uterus, and movable to some extent with it, thus simu-
lating a fibroid tumour.

Treatment.—The question of immediate gastrotomy
in the case of rupture of the sac of an extra-uterine fœtation will not be considered here, since it belongs rather
to obstetrics. In cases where life does not appear to
be immediately threatened by loss of blood, as soon as
symptoms of the primary hæmorrhage are detected,
perfect rest in the horizontal position, or with the pelvis
somewhat raised, should be immediately secured, and
ice may be applied over the hypogastrium. The best
hæmostatic, and at the same time stimulant, is a hypodermic injection of morphia. Ergotin and gallic acid
may also be given in the form of pill, or ergot may be
administered subcutaneously (*see* p. 236), if the heart is
not too feeble. Alcohol and ether should be absolutely
avoided, unless there appears to be imminent risk of
death from syncope.

When the febrile reaction occurs, the case should be
treated like one of pelvic peritonitis (*see* p. 347), except
that there will generally be no occasion for leeching.
Special precautions should be taken at recurring
menstrual periods, particularly by the observance of
absolute rest. When there is recurrence of pain and
tenderness, it may sometimes be useful to apply leeches
to the hypogastrium or groin. All early surgical interference in the way of puncture or evacuation of blood
is undesirable, and is especially dangerous before there
has been time for the effusion to be shut off completely

by adhesion from the peritoneal cavity. There is also
risk that fresh hæmorrhage may occur if the blood is
drawn off early. At a later stage, however, if suppuration or softening has occurred, and symptoms of septicæmic fever appear, the hæmatocele should be evacuated. A free opening should be made, by the vagina
if possible, and clots may be cleared out so far as they
can be reached by the finger, care being taken not to
break down the limiting adhesions. The safest way to
avoid hæmorrhage in making the opening is to use the
galvanic or benzoline cautery. It will generally be
desirable to wash out the cavity repeatedly with disinfectants, and a drainage tube may sometimes be useful.
If diminution of the tumour is taking place, however
slowly, interference should be avoided. In general,
apart from decomposition or suppuration, the contents
should be evacuated only when the tumour is of such
enormous size as to cause great inconvenience by pressure, and shows no tendency to become absorbed. If
comparatively early evacuation is demanded in order
to relieve extreme pressure, the aspirator, or a trocar
and canula, may be used. If spontaneous perforation
takes place, the evacuation may generally be left to
nature.

CHAPTER XI.

DISEASES OF THE VAGINA AND VULVA.

VAGINITIS.

INFLAMMATION of the mucous membrane of the vagina is called vaginitis, or, with stricter etymological propriety, "colpitis."

Causation.—Acute catarrhal inflammation of the vagina most frequently arises from gonorrhœal contagion. Vaginitis may also be produced by cold, sexual excess, parturition, the presence of a pessary or other traumatic cause, too hot or too cold injections, the irritation of acrid uterine discharges, or may arise in the course of zymotic diseases, as measles or scarlatina. It is promoted by want of cleanliness. Occasionally a simple vaginitis, produced by one of these causes, is so severe as to be indistinguishable from the specific form, and it may then resemble it in its power of carrying contagion to the other sex, or exciting purulent ophthalmia if any of the secretion comes in contact with the eye. Chronic catarrh is most frequently the sequel of more acute inflammation, or the result of irritating uterine leucorrhœa. It may also arise from any of the causes already mentioned, acting in a less acute degree, and is especially liable to exist in debilitated women, or those of strumous, gouty, or rheumatic diathesis.

Pathological Anatomy.—At the onset of acute

catarrhal inflammation, the mucous membrane is swollen and congested, and its secretion diminished. After a day or two the secretion is increased and becomes purulent or sero-purulent. There is then great injection of the mucous membrane, especially upon the summits of its folds; and small ecchymoses may be formed in its substance, or superficial abrasions upon its surface. The gonorrhœal form of vaginitis is more frequently limited to the lower portion of the canal, and is more apt to extend to the urethra and vulvovaginal glands. In chronic catarrh, the secretion contains a large quantity of epithelial cells, with a variable proportion of mucous and pus corpuscles. When there is any admixture of pus, the "trichomonas vaginalis," an infusorium possessed of one long cilium, is generally present. From long-continued catarrh, the vaginal walls become relaxed, and the mucous membrane thickened. The term "*granular vaginitis*" is given to a chronic form of inflammation, in which the mucous membrane feels rough to the finger from the existence of numerous minute elevations. These are more frequently due to hypertrophy of papillæ than to enlargement of mucous follicles, since mucous glands are so scarce in the vagina that many observers have failed to find any. As to their existence, however, positive testimony ought to outweigh negative. In both acute and chronic vaginitis the vulva commonly takes part in the inflammation, and very frequently the redness of the mucous membrane is greater at the lower part of the vagina and at the vulva than in the upper part of the canal. This may often depend upon the secretion becoming more irritating through exposure to air.

In some cases of very severe inflammation of the mucous membrane, as those produced by pessaries or other foreign bodies, by highly acrid discharges, by exposure and violent friction, in consequence of prolapse, or more especially in septicæmic conditions following any lesion, the epithelium may be thrown

off, and adherent diphtheroid exudations may be formed
upon the surface. Adhesion of the vaginal walls or
cicatricial contraction is then apt to follow. The
vagina may also be affected by true diphtheria.

Results and Symptoms.—In acute catarrh there
may be some febrile disturbance. Burning, aching,
and throbbing are felt in the vagina. After the first
day or two the discharge is profuse, and yellow or
greenish. Often it is offensive, and by its acrid quality
excoriates the vulva and surrounding parts. Generally
there is vesical tenesmus, and smarting on micturition.
The vulva and vagina are very tender, so that even the
careful introduction of a single finger produces much
pain. In the chronic form there may be the same
symptoms in milder degree, or the presence of discharge
may be the only one noticed.

Diagnosis.—The degree of inflammation of the
mucous membrane is best judged of by inspection
with the aid of the speculum, or without it, if there
is so much tenderness as to render its introduction
painful. The speculum will also show how much of
the discharge is coming from the cervix. If necessary,
microscopic examination will distinguish the epithelial
cells, or epithelial cells mixed with pus, of the vaginal
discharge from the mucoid or muco-purulent secretion
of the uterus. The chief difficulty in diagnosis is to
distinguish gonorrhœal from simple inflammation. The
chief characters of gonorrhœa are its sudden onset;
the markedly yellow or greenish colour, offensive smell,
and irritating quality of the discharge; the smarting on
micturition produced by extension of inflammation to
the urethra; the occurrence of inflammation or abscess
in the vulvo-vaginal glands, the ducts of which often
become distinguishable as injected points just in front
of the hymen or its remnant; and the communication
of contagion to the male. The occurrence of marked
œdema of the vulva, buboes, or consequent peritonitis,
furnish still stronger evidence of gonorrhœa. A conclusion based upon all these signs, or the majority of

them, would be right in ninety-nine cases out of a hundred ; but, since a simple vaginitis may possibly have the same characters, it is never safe to pronounce an absolute affirmative opinion as to origin from gonorrhœal contagion. On the other hand, a chronic or recurrent gonorrhœa often presents no sign, except its contagious quality, by which it can be distinguished from an ordinary form of simple inflammation.

Treatment.—In the very acute stage, warm hip-baths, and injections with emollient and sedative fluids, as decoction of poppies, or a weak decoction of linseed or starch, with the addition of a drachm of laudanum to the pint, should be used, the patient being placed in the dorsal position for the injections. If the patient can bear it, these medicated injections may be preceded by the injection of a large quantity of hot water between 100° and 110°, as hot as the patient can comfortably bear it, either by the Higginson's syringe or irrigator, in the manner described at page 188. The hot-water injections may be repeated at intervals of a few hours. Complete rest in bed also affords relief.

A little later a warm solution of borax, chloride of ammonium, bicarbonate of soda, or acetate of lead (ʒj. ad Oj.), or the liquor plumbi subacetatis dilutus, may be used. Tampons of cotton-wool or oakum soaked in glycerine containing a small proportion of carbolic acid (1—200) may be placed in the vagina in the intervals, and the vulva may be protected from the irritating effect of the discharge by vaseline or cold cream. After subsidence of the more acute symptoms, the injections may be made more astringent by alum, tannic acid, or sulphate of zinc (ʒj.—ij. ad Oj.). A lotion containing forty grains of carbolic acid and the same quantity of sulphate of zinc, or forty to sixty grains of sulpho-carbolate of zinc to the pint, is also very useful, especially in gonorrhœal forms of inflammation. In gonorrhœa, other antiseptic lotions are also useful, especially chloride of zinc (gr. xx.—xl. ad Oj.), per-

B B

chloride of mercury (gr. iij.—vj. ad Oj.), and liquor carbonis detergens (Ʒss.—ij. ad Oj.). All these should be used very weak at first, and afterwards increased in strength. Of these injections the tannic acid and lead lotions have the disadvantage of often staining linen indelibly. For the mode of administering injections effectually, *see* pp. 187—191.

In the acute stage laxatives and salines, especially the citrate or acetate of potash, should be given. With these may be combined drachm doses of tincture of hyoscyamus in camphor water, or infusion of uva ursi, if there is any urethral or bladder inflammation. Alcohol and spices must be absolutely avoided. If the inflammation is becoming chronic, or injections fail to relieve it, a tampon of cotton-wool, large enough to keep the vaginal wall separate, and soaked in glycerine containing acetate of lead, sulphate of zinc, or tannic acid (gr. xx—lx. ad Ʒj.), may be introduced from time to time, and left from twenty-four to forty-eight hours. Suppositories containing the same drugs are also useful (*see* p. 192). In the intervals, warm water injections should be freely used to wash away secretions, and warm water should also be employed before using the medicated lotions. It is also serviceable to apply occasionally to the whole vaginal walls a solution of nitrate of silver containing ten or twenty grains to the ounce, or, in more obstinate cases, to apply, at longer intervals, one containing from thirty to sixty grains to the ounce. The most convenient mode of doing this is to pour the solution into Ferguson's speculum, while the patient is in the dorsal position. By altering the direction of the speculum, and finally withdrawing it very slowly, the liquid is brought into contact with every part of the vagina. When the outlet is reached, the fluid is poured out by tilting up the speculum, or mopped up by a tampon of absorbent cotton. Carbolic acid (Ʒij.—iv. ad glycerini Ʒj.), or even in obstinate cases the strong carbolic acid, may also be applied with a mop of cotton to the whole of the vaginal walls through Ferguson's speculum. In using the stronger applications, the sensitive structures of the vulva should be avoided. But when the weaker ones (such as nitrate of silver gr. x.—xxx. ad ʒj.) are used, and there is vulvitis as well as vaginitis, the solution may afterwards be applied thoroughly with a swab of cotton to the vulva. Meantime, any cause of passive hyperæmia which tends to promote excessive secretion should be removed as far as possible. In debilitated or anæmic patients, tonic remedies, and especially iron, should be given. Chronic forms of vaginitis can frequently only be cured by treating the cervical or corporeal endometritis which keeps them up.

MALFORMATIONS, DISPLACEMENTS, AND ATRESIA OF THE VAGINA have been considered in connection with the corresponding conditions of the uterus.

CICATRICES OF THE VAGINA, producing contractions or partial atresia, are generally the result of injury in labour, sloughing after parturition, or the incautious use of caustics. If they cause great inconvenience superficial incisions should be made in them, followed by dilatation. The vagina should be plugged for a few hours immediately after the incisions, or, if possible, a Sims' dilator of glass (Fig. 78, p. 411) should be at once introduced, and worn either continuously, or for some hours daily. Care should be taken to use frequent antiseptic injections, and to keep the patient in bed for a few days. To prevent the tendency to subsequent contraction, a Hodge's pessary may often be used with effect, if the cicatrices affect the posterior cul-de-sac, or upper part of the canal. If the upper part is free, and the lower part only contracted, a Sims' dilator of vulcanite may be substituted for that of glass, after the incisions have healed, and worn daily for at least some hours.

FIBROUS OR SARCOMATOUS GROWTHS occur in rare cases in the vaginal walls, or assume the form of polypi. In the latter case they may easily be removed by the écraseur.

VAGINAL CYSTS are also occasionally found. They generally contain a clear, glairy fluid, and, in most cases, are formed by dilatation of the mucous glands which exist very sparingly in the vaginal walls. If they cause inconvenience they should be freely incised, and tincture of iodine applied to the cavity.

PRIMARY CANCER OF THE VAGINA is very much more rare than that of the cervix uteri or vulva. It may have the form either of carcinoma or epithelioma. The former sometimes appears in an infiltrating form in old women, commencing most commonly at the anterior vaginal wall, and producing contraction of the canal, with induration of its walls. Epithelioma may occur in comparatively young women, and more frequently commences in the posterior vaginal wall. The symptoms are similar to those of cancer of the cervix, but hæmorrhage is not usually so considerable. In infiltrating carcinoma difficulty in micturition and lancinating pain may be the chief symptoms. In epithelioma, or ulcerating forms of carcinoma, the first symptom is frequently pain and hæmorrhage on coitus. In an early stage the disease may possibly be confounded with syphilitic ulceration. Cancer is distinguished by its friable surface, with hard base and edges, by its greater proneness to bleed on touching, and also by its resisting syphilitic remedies.

Treatment.—Epithelioma in the early stage should be removed, if possible, by the knife or the galvanic or benzoline cautery, or it may be excised with the knife or scissors, and the cautery applied afterwards. In general, however, the disease rapidly spreads in the loose cellular tissue beneath the vaginal wall, and extirpation becomes impossible. In the more advanced stage, if there is much hæmorrhage or fœtid discharge, some relief may be afforded by the use of the sharp spoons, cautery, or caustics, in the mode described under the head of cancer of the cervix uteri (*see* p. 264).

CYSTIC DILATATION OF THE URETHRA occasionally

forms a swelling, projecting from the anterior vaginal wall, which causes inconvenience mechanically, and may even project at the vulva. The urine retained in the pouch also becomes irritating from decomposition, and is liable to be discharged involuntarily from time to time. The best *treatment* is to dissect off a portion of vaginal mucous membrane over the pouch, and bring the edges together by sutures of silkworm gut, or silver wire. Sometimes the pouch attains a considerable size, and may involve a portion of the base of the bladder. The only radical cure in this case is to excise a portion of the wall of the pouch completely, and immediately close by sutures the urethro-vaginal, or vesico-urethro-vaginal fistula so produced.

VULVITIS.—Catarrhal inflammation of the vulva, gonorrhœal or simple, is commonly associated with a similar inflammation of the vagina, and has been already considered in connection with vaginitis. When the vulvitis forms the prominent part of the affection, it is useful, except at the very acutest stage of the inflammation, to keep a sedative and astringent lotion in constant contact with the inflamed parts by means of a dossil of lint placed between the labia.* The vulva may also be painted with a weak solution of nitrate of silver (gr. x. ad ℨj.) every other day; or, in obstinate cases, with a stronger solution (gr. xl.—lx. ad ℨj.) at longer intervals.

Either after subsidence of vaginal gonorrhœa or other forms of acute vaginitis, as a sequel of marriage, or, occasionally, even in virgins, a very chronic and obstinate inflammation of the vulva may exist, generally most acute at its posterior part, affecting especially the hymen or its remnant, and extending to the four-chette. It may be associated with superficial excoriations or fissures, and is the condition which most commonly gives rise to the symptoms of vaginismus (*see* p. 409). In its treatment, care should first be taken

---

* The following is a useful formula:—Ext. Opii, gr. iv. ; Glycerini, ℨj. ; Liq. Plumb. Subacet. dilut. ad ℨj.

to cure any irritating uterine or vaginal discharge. When this has been done, the solution of nitrate of silver applied at intervals to the vulva, as already described, is often effectual, but if milder means fail, the mucous membrane may be brushed over with equal parts of strong carbolic acid and glycerine. Dr. Matthews Duncan describes, as one cause of vaginismus, a form of obstinate and recurrent superficial excoriation, which he regards as analogous to lupus, finding it occasionally to be associated with small tubercles. This he finds to be curable only by application of the actual cautery, or strong caustics, such as nitric acid. In chronic vulvitis, constitutional treatment, especially by saline purgatives, abstinence from alcohol, and a somewhat sparing diet, are of much importance. This is especially so in the case of gouty subjects, who are liable to an obstinate form of the complaint. A somewhat severe form of vulvitis may be the result of diabetes, and it is important to look out for the presence of this disease, especially in women rather beyond middle life.

The form of purulent catarrh common in weakly or strumous children, which sometimes gives rise to a suspicion of contagion, is usually confined to the vulva. It is often promoted by uncleanliness or the irritation of thread-worms. It should be treated by frequent ablutions, and mild astringent lotions, or an ointment containing acetate of lead. At the same time, good diet and tonics, especially cod-liver oil and iron, should be given.

FOLLICULAR VULVITIS is a chronic form of inflammation, in which either the mucous or sebaceous glands of the vulva may be inflamed and enlarged. In the former case, the internal surface of the nymphæ and vestibule are chiefly affected ; in the latter, the enlarged follicles are seen most on the external surface of the nymphæ and internal surface of the labia majora, and the parts may be covered with an offensive cheesy secretion. This affection may be the cause of severe pruritus or vaginismus. The *treatment*, local and constitutional, is similar to that of chronic catarrhal vulvitis. An ointment, made with vaseline, and containing acetate of lead (gr. x.—xxx. ad ℥j.), to which hydrocyanic acid or morphia may be added, is often useful.

GANGRENE OF THE VULVA occurs in cachectic children in the form of noma, and also occasionally appears in some forms of puerperal septicæmia, or in severe zymotic diseases. Sporadic cases in adults, of doubtful causation, have also been recorded.

CYSTIC DILATATION OF THE VULVO-VAGINAL GLANDS arises from occlusion of the duct of the gland (the opening of which is situated just in front of the hymen), and is generally the consequence of vulvitis. A fluctuating swelling is thus formed, which may enlarge to the size of a small hen's egg, and distend the labium majus. It contains a clear, glairy fluid. The chief symptom is usually that of pain or inconvenience on coitus. The *treatment* is to incise the cyst freely, and plug it with lint soaked in tincture of iodine, or the strong solution of perchloride of iron. If the swelling recurs, the cyst may be dissected out. Cysts at the vulva may occasionally also be formed by obstruction of an ordinary mucous gland.

INFLAMMATION AND ABSCESS OF THE VULVO-VAGINAL GLAND is commonly a sequel of gonorrhœa, but may arise from simple vulvitis, especially when combined with want of cleanliness. The *treatment* consists in rest, the application of poultices, and free incision from the mucous surface as soon as fluctuation is discovered. Inflammation and suppuration may also extend from the gland to the areolar tissue of the labium majus, or arise there from the direct effect of violence.

VARICOSE DILATATION OF THE VEINS OF THE VULVA is generally the result of pregnancy, but may occur apart from that condition, or persist afterwards. The *treatment* should generally be limited to bathing with cold water, administration of laxatives, and rest. Fatal hæmorrhage may occur from puncture of these veins by a sharp instrument, or rupture by a blow or kick. Rupture has even occurred in coitus, or in straining at stool. If the hæmorrhage is detected, it may always be arrested by pressure.

HÆMATOMA, OR THROMBUS OF THE LABIUM, is chiefly of importance in relation to pregnancy and parturition, but may result from violence or puncture by a pointed instrument, even in the non-pregnant condition. In non-puerperal cases it is rarely necessary to evacute the swelling. This should not be done unless decomposition or suppuration occurs in it, or its size is so enormous that no progress is made in its absorption.

ERUPTIONS.—Of the eruptions which may occur about the vulva, as elsewhere, the most frequent are lichen, acne, furuncles, and especially *eczema*. Eczema of the vulva is often the source of extreme distress, from the soreness or pruritus which it occasions. It usually commences on the outer surface of the labia majora, and extends to the adjoining skin of the thighs and abdomen, as well as to the mucous membrane of the vulva. When chronic, it causes loss of hair, and considerable thickening of the skin and mucous membrane. The point chiefly to be noted about eczema in this situation is its frequent association with the presence of sugar in the urine, often without any loss of flesh or general symptoms of diabetes sufficient to attract attention. The eruption is not solely due to local irritation from the urine, since, as Dr. Braxton Hicks has pointed out, eczema not unfrequently occurs in other parts of the body in the same cases. Eczema also occurs from the irritation of a leucorrhœal discharge, from incontinence of urine in gouty subjects, or from the excoriation consequent upon excess of fat.

When the urine is saccharine, constitutional *treatment* suitable to diabetes should be employed, and the genitals should be washed with water after micturition. The local and constitutional treatment is otherwise similar to that of eczema in other parts. In obstinate cases it may be necessary to modify the condition of the skin by brushing over it caustic fluids, as a solution of nitrate of silver (3j. ad 3j.), strong carbolic acid, or a solution of caustic potash (3ss. ad 3j.), or by rubbing over it the solid nitrate of silver.

VASCULAR CARUNCLE OF THE URETHRA is a growth of connective tissue, springing from just within the orifice of the urethra, generally at its lower or lateral border. Its size may be from that of a pin's head up to that of a hazel nut, or more rarely that of a cherry, and it is frequently pedunculated. In most cases the growth is very abundantly supplied with vessels and nerves, covered by an extremely thin epithelium, so that it is excessively sensitive, and readily bleeds. It is sometimes single, but not infrequently there are a number of small growths extending some distance within the urethral orifice. The more sensitive variety of caruncle has a bright cherry-red colour, and the tendency to bleed is generally in proportion to the sensitiveness. It is usually so friable that it can scarcely be grasped by forceps. The less sensitive variety of caruncle may be in colour like the surrounding mucous membrane, and is not so friable.

The *causation* is obscure, but the growth may sometimes originate in inflammation of the vulva and urethra ; any cause of passive hyperæmia also tends to promote it. It is more common in married women, but is found not very infrequently even in young virgins, and is not rare in the old. The *symptoms* are generally pain on micturition, which is often extreme, and excessive tenderness to any sort of contact, so that coitus is generally impossible or very painful, and even walking may give distress. Hence it is always desirable to examine visually the orifice of the urethra when great hyperæsthesia at the vulval outlet is found on digital examination. Sometimes bleeding occurs in micturition or at other times. Frequently severe hysterical symptoms are the result of the affection, and the mind sometimes becomes affected by serious

depression. The *treatment* is to administer an anæsthetic, and, if the growth is pedunculated, to snip it off
with scissors, and apply to its base nitric acid, the
solid nitrate of silver, or, what is preferable, the actual
or benzoline cautery. Lead lotion, with the addition
of opium or morphia, may afterwards be applied. If
the growths are sessile, they should be destroyed by
cautery. When they extend up the urethra, Mr.
Bryant's urethral speculum dilator of ivory (Fig. 76)
may be used with great advantage, both to expose
them and to allow convenient access. The use of the
cautery appears to be the most effectual means of
guarding against recurrence of the caruncle, to which
there is a strong tendency.

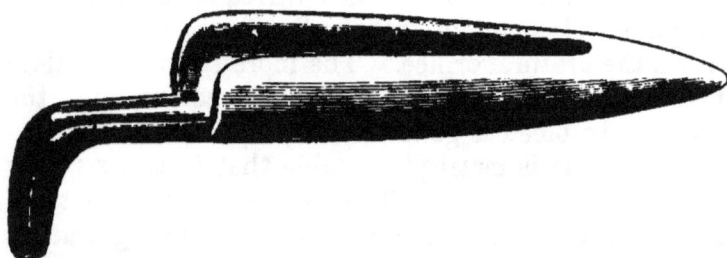

Fig. 76.—BRYANT'S Urethral Speculum Dilator (Actual Size).

Granular inflammation of the urethral outlet, or
extending some distance up the urethral canal, sometimes persists after removal of a caruncle, or may exist
independently of any caruncle, especially in old women.
The surface is then intensely red, may have the same
extreme sensitiveness as a caruncle, and often readily
bleeds. This condition may be treated by the application, with the aid of Playfair's probe (Fig. 60, p. 193),
of equal parts of carbolic acid and glycerine, or a strong
solution of nitrate of silver (gr. xl.—lx. ad ℥j.), or by
the repeated application of the undiluted liquor plumbi
subacetatis at intervals of two or three days.

HYPERPLASIA OF THE CLITORIS is generally in whole
or in part congenital. The hypertrophy is usually
unconnected with masturbation. If much inconvenience

is caused, amputation of the organ may be called for, and is most conveniently performed by means of the galvanic écraseur.

HYPERPLASIA OF THE NYMPHÆ.—The nymphæ may be elongated into long flaps, either congenitally or from the effect of masturbation. They may then form an impediment in coitus, or may become irritated from the contact of the clothes in walking. If necessary, they may be partially or wholly removed.

ELEPHANTIASIS OF THE VULVA is very rare except in Eastern countries. It generally commences in one labium majus, and may form an enormous pedunculated tumour. If the growth is pedunculated or localized, it may either be excised, and its vessels tied or twisted, or it may be amputated by the galvanic écraseur, or by the knife of the benzoline cautery. Syphilitic hypertrophy of the vulva may take a form approximating to elephantiasis, or a syphilitic taint may predispose to the latter affection.

FIBROID OR SARCOMATOUS TUMOURS in rare cases have their origin in the labia.

CANCER OF THE VULVA is not unfrequent, especially at the clitoris and margin of the labia. It generally commences in the form of epithelioma. At an early stage it may be *treated* by free excision, in the same way as cancer of the vagina, and with more hopefulness. The knife of the benzoline cautery is generally the best instrument to use. An ulceration due to tertiary syphilis may, in some instances, somewhat resemble cancer. In a doubtful case, it will be distinguished by its yielding to syphilitic remedies. The so-called *rodent ulcer* in this situation is probably a superficial form of epithelioma.

LUPUS may occur on the mons veneris, on the cutaneous surface of the labia, or on their mucous surface; and, in the last case, it may extend to the vagina. It more frequently appears before the age of thirty, and is characterized by a superficial ulceration, sometimes accompanied by tubercular elevations, which spreads in one direction, while it heals in another. Its course may be one of many years. When it affects the mucous surface it is apt to be accompanied by hypertrophy and induration, with contraction of surrounding tissue. The disease is rare in this situation. It may be *treated* by local applications of the cautery, or of nitric acid, and by the same constitutional remedies as lupus elsewhere.

RUPTURE OF THE PERINEUM in almost all cases occurs in parturition, although, in a few instances, it may be produced in the extraction per vaginam of a large tumour, such as a fibroid. Cases of rupture of the perineum may be divided into two great classes : first, incomplete ruptures, in which the sphincter ani is not divided; secondly, complete ruptures, in which the sphincter ani is divided, and therefore more or less of the recto-vaginal septum destroyed. In both cases the primary operation ought always to be performed at the time of the rupture, and is much easier than the secondary operation, since no freshening of surfaces is required. This primary operation will not, however, be considered here, since it is described in text-books of midwifery.

The effect of incomplete rupture of the perineum is to deprive the anterior vaginal wall in its lower part of the support which it normally receives from the perineal body (*see* Fig. 25, p. 69), and so to lead to the production of prolapse of the vagina, and consequently of the uterus. The use of the vagina in coitus may also be impaired, from the laxity at the outlet which is so produced. When the rupture is at all extensive, reaching up to, or nearly up to, the sphincter ani, it is desirable to operate for its cure without waiting for prolapse to be produced. The time for such operation, supposing the primary operation to have failed, or not to have been performed, should not be less than two months after delivery, so that the effects of the puerperal state may have completely passed away. It is also convenient, if the infant can be weaned before the operation, that the patient may not have the disturbance of suckling while the union is taking place; but this is not absolutely essential. The mode of performing the operation for incomplete rupture has already been described in the section on prolapse of the uterus and vagina (p. 130).

The effect of complete rupture is, in addition, to destroy or impair the power of retaining the contents of the bowel. This may vary from complete incontinence of fæces to a diminished power of retaining flatus or liquid motions when the bowels are loose. The main object of operation in this case is to restore the functions of the sphincter, and the operation is a failure if this is not attained, however strong a perineum may be produced. When the sphincter ani is torn through, its two ends separate, and, instead of being a circle, it becomes nearly a straight line in the position E F (Fig. 77, p. 382). Thus, in such cases, the radiating folds of skin indicating the sphincter are seen at the lower margin only of the bowel orifice, and the sphincter itself can be felt by the finger under the skin as a straight or nearly straight ridge, the ends of which have retracted away somewhat from the edges of the cicatrix at E F. The most important point in the operation is so to regulate the freshening and placing of sutures that the ends of the sphincter are brought together again.

Operation for Complete Rupture of Perineum.— The following is the mode in which I generally perform the operation. Beforehand the rectum must be washed out by enema, and, at the time of operating a sponge, tied by a tape, is passed just within the bowel, to prevent fæcal matter coming down. The thighs are then secured by Clover's crutch (*see* p. 130), and the fingers of assistants put the mucous membrane on a stretch by drawing the skin of the thigh outward near C and D (Fig. 77, p. 382). A point B in the median line of the vagina, a sufficient distance above the apex of the rent in the septum, is taken, and an incision through the mucous membrane is made from B to G, and from G to E and F along the edges of the septum, between the rectal mucous membrane and the cicatrix. Incisions are also made through the

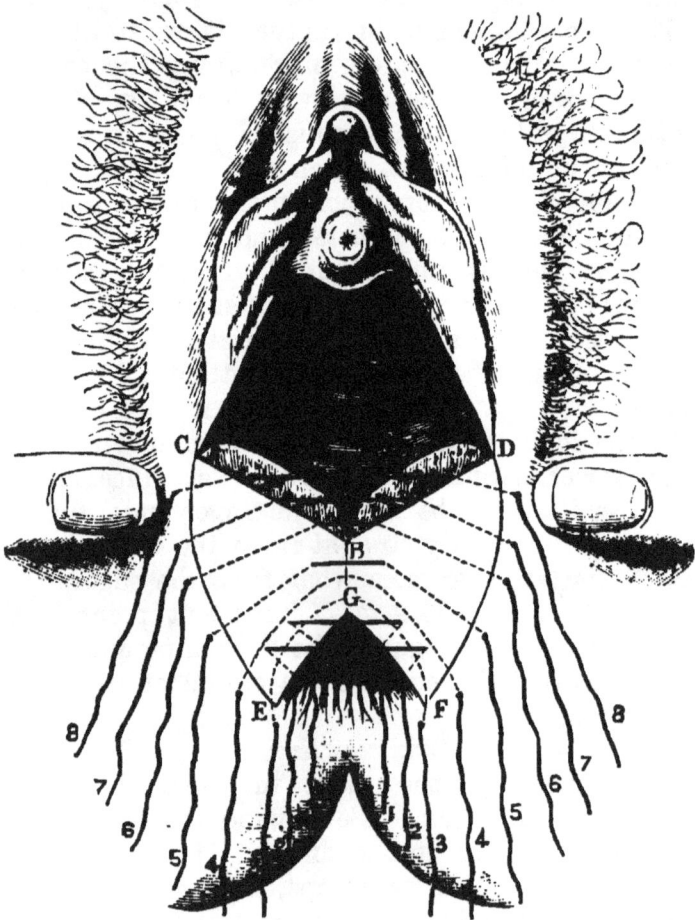

Fig. 77.—Operation for Complete Rupture of Perineum.

skin from E to C and F to D, so that the freshened surface may extend somewhat beyond the limits of the cicatrix left by the rent, C and D not to be higher than the lower extremities of the nymphæ. The quadrilateral flap E G B C is then seized at E by dissecting forceps, and dissected up with the knife from the angle E, and afterwards from the angle G, towards the base B C. While this is done, the parts are kept on the stretch by an assistant drawing down the skin below E with a tenaculum. The flap is then cut away with scissors, except an upturned border, which is left along B C. The flap F G B D is treated in a similar manner. If, as is usual, the ends of the sphincter at E and F have retracted from the margins of the cicatrix, it is well to cut away with the scissors a narrow strip of rectal mucous membrane, generally somewhat everted, a short distance from E and F toward G, so as to bring the freshened surface up to the ends of the sphincter.

Sutures of silkworm gut (*see* p. 132) are then applied in the following manner :—First, rectal sutures, either two or three, according to the extent of rent in the septum, are applied. These are destined to be tied in the rectum, and the ends left projecting through the anus.* They are best applied with a half-curved needle, held in a needle-holder. The needle is passed in a little distance from the margin of the rent, and brought out almost at the very edge of the rectal mucous membrane, on the line G F. The needle is then threaded at the other end of the suture, and that is drawn through in the same way from without inward, on the margin E G. Next two sutures at least are passed completely round through the remnant of the septum, by means of a curved needle, not too large, mounted in a handle. This is passed unthreaded, and draws the suture back with it on withdrawal. The first of these (3, Fig. 77) is passed in somewhat behind and below the angle F, so as to take up, if possible, or at least go quite close to, the end of the divided sphincter, and is brought out in a similar position near E. Thus, when tightened, it brings together the

* The use of rectal sutures has been adopted by Dieffenbach, Simon, and Bantock.

ends of the sphincter, drawing it into a circle; but it often brings into apposition, not so much the freshened surfaces above as the unfreshened rectal mucous membrane. This serves as a barrier to keep out fæcal matter, while the next suture (4, Fig. 77) aids the rectal sutures in uniting the freshened surfaces. The remaining sutures are passed as shown in the figure (5—8, Fig. 77) by a slightly curved needle mounted in a handle, in the same way as in the operation for incomplete rupture (see p. 132). The needle, unthreaded, is passed in pretty close to the edge c e or f d, is brought out (except in the case of suture 5, Fig. 77) on the line where the margin c b or d b is turned up, and draws one end of the suture back with it, the other end being afterwards drawn through in the same way. The effect is, that, when the sutures are tightened, the margins b c, b d, are turned up into a slight ridge toward the vagina, and afterwards fall over and cover any portion of the vaginal border which does not unite quite up to the edge. Suture 5 (Fig. 77) may either be buried throughout, or brought out for a very short space near the median line b g.

When all the sutures are in place, the sponge is removed from the rectum, and the rectal sutures are tied first. Care must be taken to draw up the whole of the slack in the centre, and bring the edges e g, f g perfectly together. This will approximate the ends of the sphincter to a great extent, and the approximation is completed by tightening suture 3. The remaining sutures are then tied in the order of the numbers, care being taken to allow no clots or blood to remain between, and to tighten them just enough to bring the surfaces into contact. The ends of the rectal sutures may be left moderately long, to distinguish them, the rest cut pretty short.

The perineal sutures are removed in seven days. The rectal sutures may be left from ten to fourteen days longer, till the perineum is consolidated. They are then removed through a small rectal speculum, care being taken not to break down any of the union in passing it. By this operation the anus is generally much more completely restored than by the use of quilled sutures, or the plan of making deep lateral incisions to relieve tension. If there is much resistance to bringing the surfaces together, the only thing required is to use more numerous sutures, so as to diminish the tension on each.

In some cases, by the primary operation after labour, only superficial union is secured, and a recto-vaginal fistula is left close to the part united. The best plan is then to cut through the bridge of union with scissors at the time of the operation, and then proceed as in the case of complete rupture. This is the only way to secure a firm and thick perineum, and is less likely to fail than an operation on the fistula alone.

CHAPTER XII.

FUNCTIONAL AND SYMPTOMATIC DISORDERS.

AMENORRHŒA.

AMENORRHŒA, or the absence of the menstrual flow within the limits of age during which it should naturally continue, is to be distinguished from occlusion of the genital canal, and consequent retention of menstrual fluid, which gives rise to an apparent only, and not a real, amenorrhœa. Amenorrhœa, besides being a natural physiological condition in pregnancy and lactation, is a result common to a large number of constitutional and local pathological conditions. It has already been mentioned as a symptom of absence or imperfect development of uterus and ovaries, and of cystic or other form of degeneration affecting both ovaries ; also as a sequel of severe inflammation of the pelvic organs, especially of acute ovaritis or pelvic peritonitis. The chief varieties are *primary amenorrhœa*, in which menstruation has never appeared at all ; and *secondary amenorrhœa*, or suppression of menstruation.

The age at which menstruation commences may vary in different persons by a considerable number of years without calling for any special medical interference ; but the longer its onset is deferred beyond the normal age, the more likely is constitutional disturbance to attend the change. The difference depends partly on the general vigour and development of the whole body, partly on the relative development and activity of ovaries and uterus. Thus in girls of deficient intellect puberty is commonly much retarded. The occurrence of any serious illness within a few years before the natural date for the commencement of menstruation often has the effect of considerably deferring its appearance. Primary amenorrhœa may be due to absence or imperfect development of uterus and ovaries, and imperfect development of either or both organs strongly predisposes to the production of secondary amenorrhœa, or a premature menopause, by comparatively slight causes. Sudden suppression of menstruation during the period of flow may be produced by cold or by mental emotions, even when the suppression is not a symptom of actual inflammation ; and this may be the starting-point of secondary amenorrhœa of considerable duration. Long-protracted and even permanent amenorrhœa may be the sequel of acute diseases or strong depressing emotions, or the menopause may come on prematurely without obvious cause. Sometimes superinvolution of the uterus after labour is a starting-point. Towards the natural period of cessation, it is common for considerable periods of amenorrhœa to alternate with an occasional and sometimes excessive flow. Any chronic and wasting disease, and more especially phthisis, may induce primary or secondary amenorrhœa, according to the age at which it makes its appearance. Again, amenorrhœa may be produced by repeated loss of blood, as from hæmorrhoidal tumours. The same effect may result from a simple anæmia and failure of nutrition, due to insufficient diet, indigestion, or a too sedentary life. A sudden change in the mode of life, such as often occurs in the case of girls on going to school, is especially likely to interrupt menstruation, when combined with any other of the above-mentioned causes. Amenorrhœa also sometimes comes on shortly after marriage, even without any pregnancy ; and is still more likely to occur after illicit intercourse, when there is a strong reason to dread the possibility of pregnancy.

Among all the causes of amenorrhœa there is none

more frequent or more important than *chlorosis*, the
relation of which to menstruation is a somewhat com-
plex one. The important significance of this relation
is shown by the fact that the disease is almost limited
to the female sex, and to an age not far removed from
that of puberty. Chlorosis is a disease largely dependent upon congenital predisposition, and frequently
associated with imperfect development of the heart and
narrowness of large arteries. It has also a close relation
to the nervous system, for it is often characterized by
the symptoms of nervous depression or irritability, and
frequently owes its origin to a powerful depressing
emotion, such as disappointment in love or bereave-
ment. As regards the condition of the blood, chlorosis
differs from other forms of anæmia chiefly in the fact
that the deficiency in hæmoglobin is far more than proportionate to the deficiency in number of the red corpuscles. This circumstance accounts for the extreme
degree of the pallor of the skin, and its greenish tint.

Chlorosis may come on before the age of puberty,
and give rise to primary amenorrhœa. In other cases,
the commencement of menstruation is the starting-
point of chlorosis, the extra demand which thus arises
having proved too much for the feeble powers of the
system. In more rare instances, the same effect is
produced by a menstruation which in the first
instance was excessive, although it becomes scanty,
or is entirely interrupted, after the chlorosis is established. In general, therefore, the amenorrhœa of
chlorosis is secondary to the condition of the system
generally, and that of the blood. It is probable,
however, that in many, if not in most, cases, the
deficiency of the stimulus to nutrition furnished by
ovarian development and activity contributes to the
disease. Thus the tendency to the production of fat at
the expense of muscular tissue, so often characteristic of
ovarian torpidity, is frequently observed in chlorosis.
Again, cases are not very unfrequent in which the
amenorrhœa appears to be primary, and to be associated

at first with plethora, while anæmia and the signs of chlorosis only come on after an interval. The same inference may be drawn from the cases of chlorosis in which benefit is derived from marriage, or from direct emmenagogue treatment.

Contrasted with cases of chlorosis are those in which primary or secondary amenorrhœa is associated with an appearance of plethora and symptoms of general disturbance, similar to those which frequently attend the menopause, such as headache, flushing of the face, constipation, hepatic derangement, and a tendency to morbid nervous and mental conditions. Ovarian inactivity may then generally be inferred, and this may be due either to a congenital condition, or to a sedentary life, with too good living. To these symptoms may be added hæmorrhages from various parts, as the lungs, stomach, nose, or rectum, or even sometimes from a wound or ulcer. These are sometimes spoken of as ectopic or vicarious menstruation. It is very rarely, however, that the vicarious hæmorrhages have any monthly periodicity, but they indicate an excess of vascular pressure which does not find its natural relief.

Diagnosis.—In primary amenorrhœa it is desirable to make a local examination, if periodical pain, or any other symptom, suggests the suspicion that atresia may exist; if the appearance of menstruation is delayed many years beyond the normal time; if signs of general or local plethora coexist with amenorrhœa; or if marriage is projected. In secondary amenorrhœa special care must be taken to decide the question as to the possibility of pregnancy. If, in a healthy-looking young woman, menstruation, having been previously normal, has ceased suddenly without the occurrence of any illness, pregnancy is naturally the first cause which suggests itself. In a suspicious case, an inspection of the breasts will often indicate the necessity for a more complete examination. Special care should also be taken to seek for signs of any bygone inflammation of the uterus or surrounding parts, especially in the form of pelvic peritonitis. Chlorosis is generally manifest in a patient's face. Even in the slighter degrees of anæmia, there are usually characteristic symptoms in the shortness of breath, debility, neuralgic pains, or indigestion, while anæmic murmurs are often to be heard over the heart and large arteries. If there is no manifest chlorosis, or other sufficient cause, signs of phthisis or other constitutional disease should be carefully searched for. The diagnosis of the conditions of uterus and ovaries associated with amenorrhœa has already been considered (*see* pp. 43, 277).

Treatment.—If amenorrhœa is a symptom of any constitutional disease, such as phthisis, the treatment should be directed solely to the primary disease; and, if it is the sequel of pelvic inflammation, the inflammation must be treated in the first place. In all forms of anæmia, but especially in chlorosis, iron is the great remedy, and in chlorosis it should be given in large doses. It is necessary, however, in the first place, to see that the digestive organs are in a condition to bear and to assimilate the iron, and it is often desirable to give first vegetable bitters with salines, or combined with acids or alkalies, according to circumstances. Dr. Barnes recommends iodide of potassium as preparatory to, or in combination with, the iron. The syrup of the iodide of iron may often be used with advantage. If digestion is weak, the iron should be given in the most easily assimilable form, as the liquor ferri dialysatus, the ferrum redactum, or one of the vegetable salts. It is often of use to combine it with aloes, especially if any tendency to constipation exists. The aloes and iron may be given in pill, or the decoctum aloes co. may be combined in a mixture with the ferri et ammoniæ citras. Permanganate of potash, given in pill, in doses of two or three grains, has recently been praised as a remedy for functional amenorrhœa. Other tonic medicines, as quinine, strychnia, and especially arsenic, also sometimes prove useful. Cod-liver oil is beneficial, except in cases where there is a tendency to corpulence. Hygienic treatment is still more important than medicinal. It should comprise nourishing diet, especially an ample allowance of fresh meat, abundance of fresh air, judiciously regulated exercise (the most effectual form of which is riding on horseback), cold fresh, or still better, salt water baths, and change of air and scene. A stay at the seaside or watering-place with chalybeate springs is especially useful. If the appearance of menstruation be deferred several years beyond the usual time, it is of special importance to guard against a too sedentary mode of life, overmuch study, or unsuitable diet; for if the commencement of ovarian activity be too long deferred, the natural development of the pelvis at puberty may fail (*see* p. 277), and menstruation itself is more subject to disturbance when it commences much too late. In all cases of amenorrhœa associated with anæmia, especially in the young, careful watch should be kept for the appearance of any sign indicating the onset of phthisis. A warm seaside residence in winter, when circumstances allow it, has often a beneficial effect on the menstrual functions, even apart from any question of delicacy of chest. In amenorrhœa or scanty menstruation associated with apparent plethora rather than anæmia, the diet, while nourishing, should be rather sparing, and should consist more of the nitrogenous than of the fat-forming elements of food. A greater amount of exercise is desirable than in anæmic cases, and occasional purgatives are often called for. In all cases in which the development of uterus or ovaries, and not the general health, is at fault, marriage generally has a beneficial effect, especially when it is ovarian activity which is defective ; and, if pregnancy occurs, menstruation is usually afterwards more natural.

If menstruation is arrested by cold or any other cause in the midst of a period, without the occurrence of actual inflammation, and the arrest is followed by headache or other symptoms of general congestion, an

attempt should be made to restore it by the use of hot
hip-baths or foot-baths, with the addition of mustard,
by hot applications to the hypogastrium, and the
administration of acetate of ammonia with ether, or
(with caution and moderation) of the domestic remedy
of gin in hot water.   Similar treatment should be
repeated at ensuing periods, if menstruation does not
come on normally.   In all cases of amenorrhœa not
dependent upon anæmia, but associated with general
or local congestive symptoms, a similar mode of stimulation may be employed for three or four consecutive
days in several succeeding months, either at the period
of menstrual nisus, if that is revealed by any sign, or
at intervals of about four weeks.   The hip-bath may
be taken at night, followed by a hot linseed or bran
poultice to the hypogastrium, and the hot foot-bath
with mustard may be used in the morning, while a pill
of aloes and myrrh is taken every night.   Stimulating
liniments may be employed to the inner surfaces of the
thighs; and hot vaginal injections or enemata may
also be tried.   It is often of use also to apply about
the same time three or four leeches to the inner
surfaces of the knees or thighs, or, when there is pain
indicating local congestion, to the labia, or, in married
women, to the cervix uteri.   This measure tends to
induce a periodical fluxion towards the pelvic region,
and is especially indicated in primary amenorrhœa of
long standing, associated with signs of plethora, or when
vicarious hæmorrhage has occurred.

There are some cases in which, after full trial of
measures of this kind, it may be desirable to use direct
means of stimulus to the uterus or ovaries.   Such
treatment should generally be limited to cases in which
there is no constitutional condition to account for the
amenorrhœa, but an imperfect development of the
uterus, not too extreme in degree, is discovered, or
deficient development of the ovaries is inferred, and
in which, also, either there is reason to believe that
the absence of menstruation is affecting the health

injuriously, or vicarious hæmorrhages occur. It is to be remembered that women themselves are very apt to attach an exaggerated importance to amenorrhœa, and that, in the absence of any evidence of injurious effect from plethora, they may be advised not to concern themselves too much about this condition. Supposing that a sufficiently urgent reason exists for adopting local treatment, a Faradic current may be passed through the uterine and ovarian regions every day or every other day. The electrodes may be placed, one over the sacrum, the other over the ovarian regions alternately, or one rheophore may even be introduced into the uterus. The uterine sound, or metallic dilating bougies, may be used from time to time as a mechanical stimulus, or, with due precautions (see p. 31), the cervix may be dilated from time to time by a tent. The most powerful means of all, however, is the introduction of an intra-uterine stem, and especially of the galvanic stem of Simpson, the upper half of which is made of zinc, the lower of copper. The effect of this is rather that of a chemical than an electrical stimulus, owing to the constant slow production of chloride of zinc, although doubtless a weak galvanic current over the surface of the uterine mucous membrane is produced. A modified kind of galvanic pessary, in which the zinc and copper are arranged side by side, in the form of a spiral coil of wire, is more readily tolerated, since it allows the uterus more mobility. The stem pessary is more likely to be of avail when the uterus is in fault, but it may also have some reflex stimulating influence upon the ovaries. It is scarcely necessary to say that it is only in very exceptional cases that a mode of treatment which is not entirely free from risk should be thought of. It should be an indispensable condition also that the general health be such as to make it quite certain that the cause of amenorrhœa is solely local, that there has been no previous inflammation, and that the patient can be kept completely under control. A large proportion of patients with imperfectly developed ovaries may with advantage be left alone. Among cases in which treatment by means of a stem might be tried after failure of all milder means, may be mentioned those in which vicarious hæmorrhage occurs in dangerous situations, or in which great distress is caused by unrelieved menstrual nisus, the uterus being somewhat deficient in development. As in the case of uterine flexion, this treatment should not be ventured on, except by those specially experienced in the diseases of women. The precautions already detailed as to the use of intra-uterine stems (*see* p. 108) must be observed; and, on account of the corrosion and consequent roughening of the zinc, the stem should not usually be left more than about three weeks at a time without removal. When the galvanic stem is not tolerated, either an expanding (Fig. 38, p. 110) or simple straight stem may be borne. In some cases of decidely imperfect development of the uterus, not too extreme in degree, the prolonged use of a stem has been recorded as having led to gradual enlargement of the organ. When amenorrhœa or scanty menstruation is the result of pelvic peritonitis, cellulitis, or acute ovaritis, the use of the sound or any other local treatment to the uterus must be avoided. When atrophy of the ovaries is inferred to have taken place, a cautious trial of the milder kind of local treatment may sometimes be desirable, if the condition is recent, but should not be prolonged if not soon successful.

Besides aloes, which influences the uterus from the sympathy of that organ with the rectum, some other drugs have the repute of being direct emmenagogues. Of these the most effective appear to be oil of savine, in doses of from five to ten minims, and the tinctura hellebori, in doses of twenty or thirty minims. Ergot is also reputed to act as an emmenagogue in certain cases, as well as a hæmostatic in excessive menstruation. All these drugs, however, are apt to prove disappointing, and can hardly be expected to produce any effect when the development of Graafian follicles is altogether wanting. In amenorrhœa or scanty menstruation resulting from chronic metritis, or from periuterine inflammation, tincture of iodine, in doses of from five to ten minims, sometimes acts as an emmenagogue.

In chlorosis and other forms of anæmia, direct emmenagogues should not be used until full trial has been given to treatment by iron, with other tonics, and hygienic measures. In obstinate cases, stimulation by heat and external applications every four weeks may be tried, or the Faradic current may be passed through the ovarian regions.

SCANTY MENSTRUATION generally depends upon causes similar to those which produce amenorrhœa, but acting in lesser degree. It is to be treated in a similar manner.

MENORRHAGIA AND METRORRHAGIA.—By the term menorrhagia is meant an excessive loss of blood from the uterus at menstrual periods; by the term metrorrhagia, a loss during the intervals, or of such an irregular kind that no monthly periodicity can be detected. The following are the main causes of menorrhagia and metrorrhagia :—(1) A morbid condition of blood, such as is found in Bright's disease, in some forms of simple malnutrition, and in febrile affections, especially those of a zymotic kind. (2) A general undue relaxation of the vessels or diseased condition of their walls, the result either of hæmophilia, of general debility, of the effects of a hot climate, or any other cause. (3) General active hyperæmia, the result of constitutional plethora or excessive arterial pressure. (4) Passive hyperæmia, whether general, as from obstructive heart, lung, or liver disease, or local, as from the pressure of a tumour, or from displacement of the uterus. (5) Want of tone in the muscular walls of the uterus, by the contraction of which the circulation through the organ is normally regulated and controlled. This may result from defective general nutrition, or from a morbid local condition. (6) Local active hyperæmia. This may depend upon the retention of a portion of placenta or membranes within the uterus; upon inflammation of, or the presence of new growths in, the uterus itself, whether body or cervix, the ovaries, or adjoining parts; upon ovarian irritability or congestion; or upon mental or mechanical causes, such as sexual excitement or sexual excess. (7) Increased surface of the mucous membrane, resulting from enlargement of the body of the uterus. (8) A diseased condition of the uterine mucous membrane, whether due to inflammation, villous or glandular degeneration, or to new growths, especially to those in a state of ulceration.

Another practically useful classification of menorrhagia and metrorrhagia is to divide them into those forms due to a general systemic cause, and those depending upon some morbid condition of the sexual organs. The first of these classes comprises the first, second, third, and a great part of the fourth and fifth of the above-mentioned divisions, while the second includes the remainder.

The amount of blood lost in menstruation varies considerably in different individuals, the difference depending in great measure upon the development and activity of the ovaries. When the ovaries are more active than usual, menstruation commences early in life and continues late, while the flow is considerable in amount, sexual feelings are strong, and there is a liability to menorrhagia or metrorrhagia, especially soon after the first establishment of the menstrual function, as well as to active hyperæmia of the sexual organs.

Diagnosis.—As a general rule, the symptom of menorrhagia or metrorrhagia is one which calls for local examination. In the case, however, of menorrhagia of only moderate degree in an unmarried girl soon after the age of puberty, such as is a common result of a somewhat excessive ovarian activity, it may be desirable in the first place to try the effect of general treatment. In investigating the cause of the disorder all available means of examination should be used, not only vaginal touch, but bimanual examination, the sound, and, except in the case of virgins, the speculum. If the source of the affection is not otherwise discoverable, and if it does not yield readily to treatment, the cervix should be dilated, to allow exploration of the cavity of the uterus.

Treatment.—The curative treatment of the various disorders of the sexual organs, of which menorrhagia and metrorrhagia are symptoms, has already been considered. It remains only to speak of the immediate and palliative treatment, and of the management of those cases in which no local cause is discoverable. Menorrhagia has occasionally actually produced a fatal result in cases in which no morbid condition could be detected even at an autopsy, and hence the primary indication is often simply to arrest the hæmorrhage.

In the first place, all systemic causes should be treated, as far as possible, and any general passive hyperæmia relieved, especially by saline purgatives if any constipation is present (*see* p. 167). If hæmorrhage is at all severe, perfect rest in the horizontal position, or with the pelvis raised, should be secured, and all hot drinks or alcohol must be avoided. The most efficient hæmostatic is ergot. Half-drachm or drachm doses of the liquid extract, or of Richardson's liquor secalis ammoniatus, may be given in cases of moderate severity. In more serious ones, drachm doses of the powdered ergot, in the form of fresh infusion, are to be preferred, or subcutaneous injections may be given, either of Savory and Moore's discs, or of some other preparation of ergotin (*see* p. 236), especially if a rapid effect is required. Next to ergot in value come digitalis, given in rather full doses (such as half a drachm of the tincture), and strychnia, either of which may be combined with the ergot. Quinine acts as a hæmostatic if given in very large doses. Cannabis indica, in doses of fifteen or twenty minims of the tincture, is also useful, espe-

cially when the hæmorrhage is associated with pain.
Full doses of bromide of potassium are of value, particularly when there is excessive ovarian activity. In
very severe hæmorrhage, full doses of opium should be
given, and cold applied, except within the first three
days of menstruation, at which period the latter means
should be avoided as a rule. For the application of
cold, sponging with cold water may be employed, or,
what is more effectual, ice may be applied to the hypogastrium or within the vagina. Injections of hot water
at a temperature of 110° or 115° F., into the vagina,
or, by means of an irrigator, into the uterus after
dilatation of the cervix, may also be tried, and these
are preferable to the use of cold when there is very
great depression. If the loss is alarming, the vagina
should be plugged, or, by preference, the os uteri
should be plugged by a sponge tent. In plugging the
vagina, it is best to use long strips of lint, moistened
with glycerine or carbolized oil, a piece of tape being
attached to those first introduced to facilitate their
removal. The strips are to be introduced, one by one,
through a Sims' or cylindrical speculum, the speculum
being gradually withdrawn meanwhile, until the vagina
is fully distended. The plug should not be left more
than twenty-four hours, but may be renewed if necessary. If bleeding still recurs after dilatation of the
cervix with a tent, the cavity of the uterus should be
swabbed with a styptic fluid, such as the tincture of
iodine, the liquor ferri perchloridi, or liquor ferri subsulphatis.* If even this fails, styptic intra-uterine injec-

---

* The liquor ferri subsulphatis of the United States Pharmacopœia, or Monsell's solution, is prepared in a similar way to
the liquor ferri perchloridi of the British Pharmacopœia, but
the ingredients are so proportioned that the result is a basic
ferric oxysulphate. The proportions are (by weight)—Sulphate
of iron, 5,760 grains; sulphuric acid, 510 grains; nitric acid,
780 grains. Water is added to make up 12 fluid ounces.
A less irritating fluid than the liquor ferri perchloridi
may also be made by dissolving the solid perchloride in
water.

tions may be used as a last resort, and with due precautions (*see* p. 219).

After relief of the hæmorrhage, special precautions, particularly with regard to rest, are to be used for several ensuing periods, while any local cause of hæmorrhage should receive suitable treatment. Cold bathing during the inter-menstrual intervals is generally beneficial. In menorrhagia dependent on debility or an impaired quality of blood, or that associated with anæmia, provided that there is no active pelvic engorgement or tenderness, prolonged administration of iron in an astringent form, such as the tincture of the perchloride of iron, in combination with ergot, is often of value. In mild cases when no organic lesion is discoverable, the administration of mineral acids, with cinchona or quinine, may complete the cure.

DYSMENORRHŒA.—The old division of dysmenorrhœa, or painful menstruation, into the neuralgic, congestive, and obstructive forms is still the most useful and comprehensive. Most cases of dysmenorrhœa, however, do not belong exclusively to one or other of these classes, but partake in some degree of the character of two, or even all three of them. The names, therefore, should only be understood as indicating the preponderating element in each case. The neuralgic form probably never exists without some basis in the shape either of undue congestion, or obstructed outflow of menstrual fluid; a small degree, however, of either of these conditions, which, if the nervous system were healthy, would pass entirely unnoticed, or cause the slightest possible inconvenience, may in some persons not only produce severe pelvic pain, but be the source of irritation which gives rise to distant and reflex pain, as headache, pain in the breasts, in the intercostal nerves, or extending down the sciatic or anterior crural nerves. In most cases of very severe dysmenorrhœa, which completely incapacitates a woman during the period, this hyperæsthetic or neuralgic element plays an important part. In such cases the state of nervous system may be the most important condition, and that chiefly calling for treatment. Hence the name of neuralgic dysmenorrhœa may be retained on this understanding, although the congestive or obstructive element may be in all cases the primary one. The morbid state of nervous system which leads to this result may be a peculiarity natural to the individual, especially in those of hysterical predisposition, or it may result from deficient nutrition and deteriorated blood, or from the impression on the nervous system produced by constant or repeated pain, especially the pain of a dysmenorrhœa having some adequate basis in congestion or obstruction.

Of obstructive dysmenorrhœa it is to be remembered that it does not depend only on the absolute size or straightness of the canal, but largely on the formation of clots or casting off of shreds of membrane, and also that the available canal of the cervix may be diminished by swelling of the mucous membrane or by spasm of the internal os. Obstructive dysmenorrhœa is also commonly complicated by the addition of congestion or inflammation set up by the irritation of the uterine mucous membrane from retained clots or secretions, with the addition sometimes, in the case of flexions, of interference with the uterine circulation. Thus it is not uncommon in primary dysmenorrhœa for pain to be limited at first to the period of flow, while there is afterwards added pain which commences some days beforehand, or continues in some measure through the intervals.

By some authors separate classes are made of inflammatory, ovarian, membranous, and spasmodic dysmenorrhœa. Inflammatory dysmenorrhœa should, however, rather be regarded as a variety or subdivision of congestive dysmenorrhœa, since in inflammation the increase of pain at the period is mainly due to the concomitant congestion, and it is frequently impossible to draw an absolute line between congestion and inflammation. Ovarian dysmenorrhœa, again, is a variety of congestive or inflammatory dysmenorrhœa, and membranous dysmenorrhœa is one form of obstructive dysmenorrhœa, commonly associated with signs of congestion (*see* p. 220). Three other forms of inflammatory dysmenorrhœa specially deserve attention—(1) that in which the pain in menstruation is a symptom of fresh and active inflammation, indicated by febrile disturbance, for then it is the inflammation which calls for treatment; (2) that in which the dysmenorrhœa is the sequel of periuterine inflammation, for here any active mechanical treatment is generally contra-indicated; (3) that in which it is the result of corporeal endometritis, and often forms the most prominent symptom which calls attention to the existence of that malady.

By spasmodic dysmenorrhœa is meant that form in which the pain comes on in recurrent paroxysms, sometimes very severe, which are attributed, with much probability, to painful contractions of the uterus. In this there is often a large addition of the hyperæsthetic or neuralgic element to a basis of obstruction, or, possibly, in some cases, only of congestion. The periodic contractions of the uterus, which are manifest in pregnancy, doubtless occur in some degree in the unimpregnated organ, and to a greater extent during menstruation, when they serve to expel the menstrual fluid. They are liable to be increased if any clots are formed or shreds of membrane detached, or if the outflow through the cervix is less free than normal, but the degree to which they become painful is greatly dependent upon the state of the nervous system. Dysmenorrhœa characterized by intermittent pain limited to the time of the flow is generally alleviated for the time, if not cured, by adequate dilatation of the cervix, whether by graduated bougies, tents, two-bladed or three-bladed dilators, or incisions; and this fact is a strong argument in favour of the view that some impediment to outflow is an element in the causation, though the absolute size of the cervical canal may not be less than the average. There appears to be no evidence to show that a continuous pain during the period can be solely due to a tonic spasm of the uterus.

Diagnosis.—In congestive dysmenorrhœa the pain commences some time before the onset of the flow, generally at an interval of from one or two days to a week, and frequently some degree of pain exists also throughout the inter-menstrual intervals. When there is no complicating obstruction, the pain is generally relieved, more or less, soon after the commencement of the flow, or, at any rate, towards its termination. In some cases it recurs after the flow has ceased. If the quantity of blood lost at different periods varies, pain is greatest when the amount of flow is least. In purely obstructive dysmenorrhœa, the pain does not commence more than a few hours before the appearance of the flow, unless there is very extreme stenosis. It is often intermittent in character, being dependent in part upon painful uterine contractions, but continuous pain may be produced by the irritation of a retained clot or shred of membrane. Frequently clots are formed from retention of blood within the uterus, even though its quantity is not excessive, from lack of the usual admixture with the vaginal mucus. Hence a scanty flow, associated with the expulsion of clots (not shreds of membrane), affords the best proof which any symptom can give that freedom of outflow through the cervix is deficient. Sometimes paroxysmal pain is noticed to be coincident with the passage of a clot. In the intervals symptoms are absent, provided that there is no complication with congestion or inflammation. When pain is markedly increased after the commencement of the flow, although existing for some time previously, it may be inferred that an obstructive element is present in addition to congestion, since congestion is generally at its height just before the onset of menstruation. A congestive dysmenorrhœa may be suspected to be ovarian if pain and tenderness are localized in the iliac region, especially when they are associated with reflex pain in the thigh, inter-

costal nerves, or mamma, and with hysterical symptoms, but physical examination alone justifies a
positive diagnosis. If, however, the pain commences
regularly at a certain interval before menstruation,
and ceases before the flow begins, it may be almost
certainly attributed to difficult ovulation. Some idea
of the importance of the neurotic element in any case
of dysmenorrhœa may often be obtained from the
amount of hyperæsthesia noticed on local examination.

The physical diagnosis of the cause of dysmenorrhœa is merged in the diagnosis of inflammation, congestion, displacement, or other morbid condition of the
uterus, ovaries, and adjoining parts, or stenosis of the
cervical canal or vagina—combined with a judgment
as to the neurotic or hyperæsthetic disposition of the
patient.

Treatment.—Palliative treatment alone remains to
be spoken of here, since curative treatment consists
in the treatment of the various causes. An essential
point is to enjoin the avoidance of all exertion, and, if
pain is severe, the horizontal position should be maintained during the period. In congestive dysmenorrhœa
saline purgatives should be given just before the period,
at which time there is often a tendency to constipation,
and full doses of bromide of potassium are useful. In
all cases the hot hip-bath, or the whole bath, in which
the patient should remain for as much as half-an-hour,
affords much relief. Hot applications to the hypogastrium have a similar effect, and hot water with
mustard to the feet, followed by rest and warmth in
bed, is also useful. Cold should always be avoided, and
the wearing of woollen drawers is generally desirable,
if the patient is not kept to her bed. Considerable
alleviation may also be procured by diffusible stimulants and the milder sedatives. Among the former
may be mentioned ether and ammonia, one or both of
which may be given with the liquor ammoniæ acetatis.[*]

[*] R. Sp. Ætheris Sulphurici, ♏xxx. ; Sp. Chloroformi,
♏xv. ; Liq. Ammoniæ Acetatis, ℥ss. ; Aq. ad ℥j.

One of the most efficacious is the favourite domestic remedy of gin in hot water, which tends to increase the flow, as well as to diminish pain, but, for obvious reasons, much caution is necessary in recommending it. Essence of ginger in hot water may be used as a substitute. Among the most useful sedatives are hyoscyamus or belladonna, hydrocyanic acid, chloral, camphor in five or ten-grain doses, or bromide of camphor in capsules. Cannabis indica is very useful, especially when the flow is profuse, but it has the disadvantage of being uncertain in its quality and effects. Small doses should, therefore, be given at first, and increased up to about thirty minims of the tincture, or two grains of the extract. Assafœtida may be given by enema, if there is much hysteria. Opium and its alkaloids are, of course, the most effectual remedies for the pain, but they should be avoided, as far as possible, in all chronic conditions. They are required, however, in severe cases, especially when dysmenorrhœa is the sequel of peritonitis. They may often conveniently be given in the form of suppository or enema.

The tendency to neuralgia and nervous hyperæsthesia should be treated in the intervals of menstruation by tonic and hygienic remedies, with cold bathing, air and exercise, and sufficient occupation. The same treatment tends to promote the formation of a more healthy menstrual decidua if menstruation is scanty as well as painful. Marriage is generally beneficial in primary congestive dysmenorrhœa, the result of ovarian irritation without any serious organic lesion. Mild forms of obstructive dysmenorrhœa are also often cured by marriage, followed by parturition. If, however, the marriage prove sterile, as is likely to be the case when the obstruction is considerable, the condition is frequently rendered worse.

The treatment of spasmodic dysmenorrhœa (*see* p. 401) by dilatation of the cervix with metallic bougies (*see* p. 60) has been supposed by some to be of use, not by rendering the canal more patent, but merely by producing an effect on the nervous system, and thereby diminishing the tendency to spasmodic contraction. It appears much more probable that the benefit really arises in great part from the canal being made more free, and this view is largely confirmed by the fact that, by such treatment in cases of spasmodic dysmenorrhœa, not only is the dysmenorrhœa relieved, but sterility appears to be not unfrequently cured. This can hardly be explained by any mere impression upon the nervous system.

CLIMACTERIC DISTURBANCES.—The cessation of menstruation at the menopause is frequently accompanied by constitutional disturbances of a well-known character, which often last over a period of several years. These are to be attributed, not only to the cessation of the periodical active hyperæmia and discharge of blood to which the system has been accustomed for some thirty-five years, but to that of the expenditure of nervous energy in a particular direction. The chief phenomena, therefore, are signs of plethora, with transient vascular disturbances, inducing flushings of the face, or feelings of heat, chilliness, or sinking in the epigastric region and other parts. Vicarious hæmorrhages from the nose and rectum are frequent; the liability to cerebral hæmorrhage is also increased. Any previously existing congestion or inflammation of pelvic organs is liable to undergo a temporary aggravation, after which, as a rule, it tends to subside. In many cases, especially when any previous uterine disturbance has existed, the diminution of menstruation is not gradual and progressive, but long periods of amenorrhœa are interrupted by profuse and often prolonged hæmorrhage, which may arouse a suspicion of the existence of cancer.

Irregular discharges of nervous energy are usual, and may take the form of headaches, of epileptiform or apoplectiform attacks, or of hysterical manifestations, in those predisposed to that disorder. In other cases the nervous disturbance takes the shape of irritability or depression, which, when there is a constitutional

proclivity, sometimes develops into insanity. Sometimes, again, women seek refuge in alcohol from low
spirits, or from the pain produced by pelvic disorders
or by indigestion, and the foundation of intemperance
is not unfrequently laid about this time of life. With
the diminution of sexual activity is associated a tendency to corpulence, and to deposit of fat about
internal organs, which is apt to lead to neglect of outdoor exercise. To this cause are partly to be ascribed
the digestive disturbances which often form the most
prominent feature of the general condition. They
consist mainly of constipation, inactivity of liver, and
distension of the abdomen by flatus, with frequent
spasmodic and painful contractions of the intestines.

Treatment.—No emmenagogue treatment should be
adopted, unless the menopause appear to be coming on
at a period very long anterior to the normal age ; nor,
on the other hand, should the intercurrent hæmorrhages,
which often afford a natural relief, be checked too
suddenly, unless signs of anæmia appear. Local examination should, however, always be made in case of
undue hæmorrhage, lest there should be commencing
cancer, or an erosion of the cervix, which at such an
age is more likely to form the starting-point of cancer,
and therefore calls the more urgently for treatment.
Diet should be rather sparing, and patients should be
urged to take a due amount of outdoor exercise. The
allowance of alcohol should be diminished ; beer, porter,
and spirits are to be avoided, and claret or other light
wine alone taken. Occasional venesection has proved
useful, but is hardly an available remedy in the present
state of popular opinion. If, however, epileptiform or
apoplectiform attacks occur, or very severe headache is
associated with an appearance of plethora, leeches to
the temples, or cupping, may be employed. Occasional
mercurial purgatives are often useful, and saline laxatives, especially the Hunyadi Janos, or other mineral
water, are to be taken daily if required. For the
nervous disorders, the bromides form the most useful

remedies. For the digestive disturbances, alkalies, with a vegetable bitter, taken before meals, or ammonia with aromatics, and a small dose of rhubarb,* are most generally useful. When, however, there is much general debility, mineral acids, combined with tonics, are to be preferred; and nux vomica is often of value for stimulating the muscular walls of the intestines (*see* formula, p. 186).

PSEUDO-CYESIS.—By the term pseudo-cyesis is denoted spurious or imaginary pregnancy. This is not uncommonly one of the neuroses of the climacteric period, and its starting-point is then the enlargement of the abdomen by fat and flatus, combined with the arrest of menstruation. The movements of the distended intestines are often mistaken for the movements of a child, even by women who have the experience of former pregnancies to guide them. The mental condition may be of any degree, from that of a not unnatural mistake, which is at once dispelled by a medical opinion, to that of a delusion amounting to monomania, which is proof against all assurances, and may even persist for a far longer period than the normal duration of pregnancy. The delusion may also occur at other times, especially soon after marriage, or after illicit intercourse. It may be entertained even though menstruation continues normal, or is merely diminished in quantity, the mistake in such cases being generally based upon corpulent or flatulent enlargement of the abdomen, and imaginary fœtal movements. The breasts, in some cases, are actually developed, and secrete a mucoid fluid, as in pregnancy, though in others the supposed enlargement is simply due to fat. The apparent enlargement of the abdomen is often increased by arching of the back and rigidity of the abdominal muscles.

The diagnosis is generally easily made by the recog-

---

* Ammon. Carb. gr. ij.; Tinct. Rhei, ℔ xxx.; Sodæ Bicarb. gr. x.; Syrup. Zingiberis, ℥j.; Aq. Menth. Pip. ad ℥j.—ter quotidie.

nition of the small size of the uterus on bimanual examination, and by resonance of the abdomen, although this may be, in some measure, diminished by great thickness of fat. The administration of an anæsthetic will clear up any doubt, and the formality of this proceeding, combined with that of a consultation, is often of use in dispelling the patient's illusion.

DYSPAREUNIA AND VAGINISMUS.—By the word dyspareunia is signified pain or difficulty in sexual intercourse. This symptom is frequently that which leads patients to seek for medical relief, although often they do not mention it until questioned on the subject. It is, therefore, generally desirable in the case of married women to make inquiry on the point when any condition is discovered likely to lead to such a result. A vaginal examination will generally reveal whether any obstruction exists due to a rigid or imperfectly ruptured hymen, to narrowness, cicatricial contraction, or spasm of the vagina; also whether the vagina is unduly short, or the uterus displaced, and whether the tenderness which causes the symptoms is situated at the vulval outlet, the urethra, the vagina, the cervix or body of the uterus, or the ovaries, or is due to periuterine inflammation or tumour. If the tenderness is found to be at the vulval outlet, a careful visual examination with a good light should be made as to its cause, which may be found in some vulvitis, erosion, fissure, or urethral caruncle, which would escape detection by the finger. The treatment will depend upon the cause which may be discovered, and should, in almost all cases, include abstinence from any attempt at intercourse for a considerable period. It should be especially remembered that any partial retroversion of the uterus, by which its canal is brought nearly into a line with that of the vagina, exposes the cervix to a direct impact to which it is not normally liable, and that this effect is increased if there is any concomitant prolapse.

The word vaginismus denotes a spasmodic contraction of the sphincter vaginæ, or pubo-coccygeus muscle, which takes place upon any attempt at intercourse, and renders intromission difficult or impossible. The same effect is frequently produced by the introduction of the index finger, or even by touching the vulval outlet with a camel-hair brush. In severe cases the spasm does not only affect the sphincter, but the muscles of the whole body are thrown into intense energy of resistance by the mere idea of any contact with the vulva, so that intercourse could only be accomplished by absolute violence. In some women who have an excessive nervous dread of the consummation of marriage, even the first attempt at intercourse may be thus prevented. In the great majority of cases, however, two causes are present—a local cause of irritation at the vulva, and a hyperæsthesia of the nerves of that region, almost invariably associated with an extreme general reflex susceptibility, dependent upon, or closely allied to, the hysterical temperament. Women who suffer in this way are often by no means destitute of sexual desire, but rather the opposite. The spasm of vaginismus is indeed an exaggeration of the contraction of the sphincter vaginæ normally produced by sexual excitement or pressure on the clitoris, but associated with painful sensitiveness of the mucous membrane on which the spasm causes pressure. The most characteristic cases of vaginismus are those which show themselves from the commencement of married life. The mental distress which is apt to follow often leads to great depression of spirits and impairment of general health. Usually there is no suffering apart from coitus, but sometimes the vulva becomes so hyperæsthetic that sitting and walking are painful, and the patient is reduced to a complete invalid life. The condition most commonly found as a cause of vaginismus is a vulvitis, most intense towards the posterior part of the vulval outlet, affecting especially the anterior surface of the hymen or its remnant, and sometimes associated with fissures or erosions. This may arise from some original disproportion of parts or awkwardness on the part of the husband; not unfrequently there is in addition a communication of contagion from a latent gonorrhœa; while in other cases there was a vaginitis or vulvitis existing before marriage, often dependent on the irritation of a uterine leucorrhœa.

Vaginismus may also be set up by follicular vulvitis, by lacerations of the vaginal or vulval outlet resulting from parturition, by urethral caruncles or other growths at the vulva, or by granular inflammation at the meatus urethræ. The form of superficial ulceration described by Dr. Matthews Duncan as a cause of the affection has been already mentioned (see p. 374). Mr. Lawson Tait describes, as a frequent cause of vaginismus in women over forty, a local atrophy of the mucous membrane, producing red spots which are excessively tender in consequence of the exposure of nerve-fibres through atrophy of the other tissues. This is said to be incurable, to lead to gradual contraction of the vulva, and to be only capable of palliation by occasional applications of strong carbolic acid.

Treatment.—In all cases of vaginismus, and in many of those of dyspareunia even without vaginismus, the introduction either of cold cream or other oily substance or of glycerine into the vagina before intercourse is a valuable palliative, since whenever there is pain on intercourse, or even a mere absence of sexual feeling, the natural lubricating secretion, poured out abundantly under the influence of emotion, is apt to fail. A little glycerine of starch is perhaps better than any oily preparation, since it increases the natural secretion of the glands, and mixes with it. In mild cases this plan, together with temporary sexual abstinence, and the treatment of vulvitis or vaginitis by the methods already mentioned (see pp. 369, 373) will prove sufficient. In a more severe case, the patient should be placed under anæsthesia, and unless the vaginal outlet is then found to be capacious, it should be fully stretched by the fingers, and a full-sized Sims' vaginal dilator of glass (Fig. 78) should be introduced. This should at first be worn all day, if possible, the patient being kept in bed ; afterwards it may be worn for some hours each day, while the treatment of any vulvitis, fissures, or erosions is continued. If this plan fails to effect a cure, after all erosions or lacerations have healed, a careful examination by probe or camel-hair brush should be made, to discover which are the sensitive points. If these prove to be chiefly the remnants of the hymen, Sims' operation should be performed. This consists in dissecting away completely with scissors the whole circuit of the hymen, whether

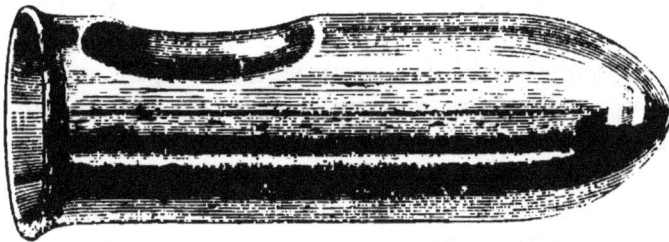

Fig. 78.—SIMS' Vaginal Dilator.

inflamed or not. After the operation the glass dilator should be immediately introduced, and worn for some time. In severe cases of vaginismus, when the hymen is inflamed and very sensitive, it is well to have recourse to this operation without delay. After the removal of the hymen, the vaginal secretion escapes more freely, and any vaginitis is more readily cured. The most hopeless cases are those in which little or no inflammation can be discovered, but the condition is one of *hyperæsthesia of the vulva*, affecting not only the hymen, but the nymphæ, vestibule, and clitoris, and apparently depending upon a perverted character of the nerves of sexual feeling. This state may produce incurable dyspareunia, even when there is no vaginismus, and may resist, not only Sims' operation, but repeated parturition, the application of strong caustics, and even the dissecting off of the whole of the sensitive mucous membrane. Generally those cases are curable in which the vaginismus arises from inflammation, or in which the hyperæsthesia is limited to the hymen, although many months of treatment and of sexual abstinence may be required. Sometimes parturition may effect a cure by effectually dilating the vagina, but severe forms of the affection generally persist notwithstanding.

ABSENCE OF SEXUAL FEELING, apart from any dyspareunia, is so dependent upon emotional conditions as to be little amenable to medical treatment. When primary it is often the result of individual peculiarity, but treatment is more frequently sought when it is secondary and acquired. If not associated with a premature menopause, it is more often dependent upon constitutional debility, anæmia, anxiety, or overwork, than upon local causes, and the only remedy is to be found in the removal of these conditions as far as practicable. Among local causes may be mentioned an undue relaxation of vagina, due to subinvolution of that canal after delivery, or to rupture of perineum or prolapse of uterus or vagina; and a want of muscular tone in its walls, associated with chronic leucorrhoea. These causes may be capable of removal.

STERILITY.—For the occurrence of conception with the greatest possible facility, it is necessary not only that there should be no deficiency on the part of the male, no dyspareunia or vaginismus, that the ovum should be properly formed and conveyed by the Fallopian tube, and that the uterine mucous membrane should be in a fit state to receive it, but also that the cervical canal should have its normal patency, straightness, and relative direction, should not be obstructed by a plug of mucus too tenacious to be displaced during coitus, and that neither the vaginal nor uterine secretion should have undergone any change rendering it adverse to the life of the spermatozoa. As a rule, the spermatozoa live only for a few hours in the acid vaginal secretion, while in that of the cervix or body of the uterus they may remain alive for a considerable number of days. The direction of the cervical canal should be nearly at right angles to that of the penis, and it is probable that normally the semen makes its way for a considerable distance into the cervix almost at the moment of ejaculation, not through any active suction by the cervix, nor by exact apposition of the os uteri to the male urethra, nor even by the force of ejaculation, but through the intermittent pressure on the cervix while it is turgid and tense with blood, so that its canal is probably rendered more circular than usual. Hence the occasional failure of vaginal injections after coitus as a prophylactic against pregnancy. If the direction of the os is changed, as in the case of retroversion or cervical anteflexion, so that it does not dip into the pool of semen in the posterior cul-de-sac, or if the external os is very narrow, or the canal plugged by mucus too tenacious to be displaced by pressure, this mechanism is interfered with. It has been asserted that there is a wide gaping of the os and cervix in the sexual orgasm, but this is not the fact.

These causes of sterility, as well as stenosis or flexion near the internal os, do not render conception impossible, but only diminish its probability. In some cases spermatozoa have effected impregnation by making their way even from outside the vulva through an intact and narrow hymen, or by passing an almost complete atresia of the vagina. If, however, they have not free access to the cervix, there is much greater probability of their perishing in the vagina before they can enter it, especially if the vaginal secretion is more adverse than usual to their life. The occurrence of the sexual orgasm on the part of the woman is not necessary to conception, but probably favours the entrance of the semen into the cervical canal. Thus

I have known an instance of a woman who had been married for many years to two husbands in succession, and who, when over forty years old, experienced the sexual orgasm for the first and only time, and became then for the first time pregnant.

It has been said that a narrow cervix cannot cause sterility because the spermatozoa have to pass through the uterine orifice of the Fallopian tube, which is normally much narrower. It is not known whether the Fallopian tube expands in the sexual orgasm, though the probability appears to be against its doing so, the uterine wall being at that time tense. But even if it does not, the spermatozoa, maintaining their life and activity within the uterus, have ample time to pass the Fallopian tube singly through their own movements. There is quite space enough to allow this, but the probability of its happening must be much greater if semen penetrates *en masse* into the safe refuge of the uterus, than if each spermatozoon has to travel from the vagina or from some still more external part. As a rule, acid solutions are injurious to the vitality of spermatozoa, and saline and very weakly alkaline solutions may tend to promote it, while of injurious fluids few have a more fatal influence than plain water in sufficient quantity. Hence, if the vaginal secretion has an acrid quality, a vaginal injection before intercourse of a solution containing 1 per cent. of common salt and $\frac{1}{10}$ per cent. of caustic soda or potash may tend to promote conception.

Among the commonest conditions associated with primary sterility is an imperfectly developed uterus, with a small external os and cervical anteflexion. In this case the sterility often persists even after the canal has been made patent, a result which is probably due to some other congenital imperfection less easily remedied than the shape of the cervix. Other frequent causes are vaginismus, or any other form of severe dyspareunia, displacements of the uterus, fibroid tumours, stenosis of external or internal os, vaginitis, and cervical and corporeal endometritis. One of the most important causes of incurable sterility is distortion, obstruction, or atresia of the Fallopian tubes, due to adhesions resulting from pelvic peritonitis. Gonorrhœa has thus a very important influence as an indirect cause of sterility through the medium of peritonitis as well as through that of vaginitis and endometritis. The views of Noeggerath as to the incurable character and important sequelæ of gonorrhœa have been already mentioned (*see* p. 205). Besides the causes of sterility depending upon the wife, it may happen that spermatozoa are absent in the semen, even though there is no apparent impotence in the male, and probably the vitality of spermatozoa may vary in different cases. There is also evidence to show that there may be a relative sterility between husband and wife, each being capable of procreation with another person.

The frequent failure of attempts to cure sterility appears to indicate that, in a large proportion of cases, it depends, not upon any mechanical or other obvious cause, but on some inscrutable imperfection in ova or spermatozoa, or in their relation to each other, depending on want of vigour in either parent, or on an unknown cause. This is confirmed by the analogy of animals and plants. Animals in confinement, or in an uncongenial climate, may be sterile, even without any other sign of want of vigour. Plants may be so also, under domestication, or in unfavourable localities. In the human race it has been noticed that heiresses, who are often only children or members of small families, are more apt to be sterile than the average of women. The imperfect fertility which results in animals from breeding in-and-in, or from self-fertilization in plants, is another instance of the obscure causes which influence propagation. Another instance of sterility not due to any discoverable local conditions, but rather to the general state of the system, is the usual cessation of child-bearing as women advance in years, some considerable time before the cessation of menstruation, and, probably, before that of ovulation.

Treatment.—The treatment will generally, in this country, be limited to the removal, when possible, of any of the above-mentioned curable causes, the existence of which may be detected, especially vaginitis, cervical and corporeal endometritis, displacements of the uterus, or stenosis of the cervical canal. Inquiry should also be made as to any sign of impotence on the part of the husband. If children are desired, coitus should not take place too often, and any vaginal injection, except that of a saline solution like that already mentioned, should be avoided for some days afterwards. Conception may follow insemination at any period of the menstrual cycle, but is believed to be most probable either just before or shortly after a period ; which of these occasions is most favourable is not yet determined.

For a complete investigation of causes, it would be necessary to adopt the method of Marion Sims, namely, to examine microscopically the cervical mucus for spermatozoa on the day following coitus, and, if none are then found alive, to repeat the examination at shorter intervals, and, if necessary, immediately after that act. In this way may be established the absence of spermatozoa, their immediate expulsion by the vagina, their rapid death either in the vagina or in the cervix, or their failure to penetrate into the cervix. The method of intra-uterine injection of semen has not had sufficient success to outweigh its difficulties and other obvious drawbacks. Out of fifty-five trials made by Marion Sims on six women, conception followed in one case only.

The treatment of sterility is not very hopeful unless the patient comes under observation while still young, and within a few years from the time of marriage, for, at a later stage, the cause of sterility, whatever it may be, is apt to have led to other alterations, not easily removed. It must be admitted that the treatment of sterility is the least successful part of gynæcological therapeutics, since a complete study of its origin is generally forbidden by a sense of delicacy. Those causes of sterility which consist in some obscure condition of the general system, impairing its reproductive vigour, lie almost entirely outside the domain of therapeutics, although a large proportion of the cases may be influenced by them. Prolonged change of air and scene, however, sometimes appears to have a favourable influence on conception, and pregnancy has sometimes occurred unexpectedly after a long separation between husband and wife.

PRURITUS VULVÆ.—Either associated with vulvitis, simple or follicular, or with any morbid condition of the uterus or vagina, an irritation, itching, or burning at the vulva is a frequent and often a very distressing symptom. Sometimes it is combined with a general hyperæsthesia, especially of cutaneous nerves. It is not uncommon in pregnancy or as a symptom of commencing cancer, and is promoted by all causes of active or passive hyperæmia of the sexual organs. It is also common in diabetes, or may be dependent upon a gouty diathesis. Any eruption around the vulva, pediculi about the pubes, or thread-worms wandering from the rectum, are also among its causes. The itching may extend to the vagina, the anus, and the adjoining skin. It is generally much aggravated by warmth, and hence may render sleep at night almost impossible, while the effect upon the nervous system of the constant or frequently recurring torment is often very severe. In other cases the sexual irritability which results is a source of great annoyance, or may give rise to the habit of masturbation. The scratching excited by the itching aggravates the malady, and is apt to produce vulvitis, if none existed previously. Pruritus vulvæ thus partakes, in great measure, of the character of a neurosis, but in a considerable proportion of cases it is excited by some discharge, either uterine or vaginal, which either sets up actual vulvitis, or at least irritates the terminations of the nerves.

E E

The **Treatment** consists in the discovery and removal of the cause. Endometritis is to be especially sought for, if no other is readily discoverable. In all cases, diet should be sparing, alcohol and spices should be avoided, and extreme cleanliness observed. As a temporary palliative, warm hip-baths at intervals of a few hours are of use. If uterine or vaginal leucorrhœa is the exciting cause, very frequent vaginal injections, used in an effectual manner (see p. 187) are to be recommended, and a tampon of cotton-wool soaked in glycerine containing acetate of lead or borax may be kept constantly in the vagina. The vulva, if not itself much inflamed, may be protected by unctuous applications, of which the best is vaseline, to which may be added acetate of lead (ʒj. ad ʒj.), with acetate of morphia (gr. x. ad ʒj.), chloroform (ʒss. ad ʒj.), or dilute hydrocyanic acid (ʒj. ad ʒj.). If there is actual inflammation of the vulva, it is preferable to keep between the labia a pledget of lint soaked in a lotion containing glycerine (ʒj. ad ʒj.) and carbolic acid (gr. iv. ad ʒj.), to which may be added acetate of morphia (gr. ij. ad ʒj.) or dilute hydrocyanic acid (ʒss. ad ʒj.), or a combination of the two. Carbolic acid and glycerine may also be combined with the liquor plumbis subacetatis dilutus. A solution of perchloride of mercury (gr. iv. ad ʒj.) is also a remedy of repute. Any constitutional disorder must be treated, saline laxatives being generally of use. Bromide of potassium in full doses is often valuable, and opiates, or chloral, must be given to secure sleep in severe cases. When the neurotic element is predominant, quinine or arsenic may be of service.

COCCYGODYNIA.—By coccygodynia is meant pain in the situation of the coccyx. It is generally accompanied by tenderness, and is greatly increased by any movement of the sacro-coccygeal joint, or the muscles attached to the coccyx. It is thus usually most acute on defecation, and on sitting down, or rising from the sitting posture. Sometimes pain is also felt on walking or while sitting. Coccygodynia is either a symptom of disease of the coccyx or of its articulation, or it may be, like pruritus vulvæ, a neurosis depending on any source of irritation in the sexual organs, anus, or rectum. In the former case it is generally either the result of injury during parturition, or one received from without, or of horse exercise. In the form of a neurosis, the affection is not uncommon in single women. For diagnosis, the coccyx should be explored between one finger in the rectum and another used externally. The detection of actual inflammation of the coccyx itself or of its articulation respectively will be assisted by the degree of tenderness of the bone itself on pressure, or of the pain produced by moving it. When there is no history of any cause likely to have produced inflammation, careful search should be made for a source of reflex irritation.

Treatment.—In the neurotic form of the affection, the chief object is to cure the primary cause. Subcutaneous injections of morphia over the coccyx afford relief. When any local inflammation is diagnosed, leeches may be applied over the seat of pain, followed by repeated counter-irritation. In very obstinate cases a tenotomy knife may be introduced at the tip of the coccyx, and the bone severed from its attachments posteriorly and along its lateral border by subcutaneous incision. If this fails, the whole bone may be excised. The latter plan is preferable if the pain is a sequel of actual dislocation, fracture, or ankylosis of the bone. It is only in exceptional cases, however, that surgical interference is desirable.

THE END.

INDEX.

J. & A. CHURCHILL'S
MEDICAL CLASS BOOKS.

ANATOMY.

BRAUNE.—An Atlas of Topographical Anatomy, after Plane Sections of Frozen Bodies. By WILHELM BRAUNE, Professor of Anatomy in the University of Leipzig. Translated by EDWARD BELLAMY, F.R.C.S., and Member of the Board of Examiners ; Surgeon to Charing Cross Hospital, and Lecturer on Anatomy in its School. With 34 Photo-lithographic Plates and 46 Woodcuts. Large Imp. 8vo, 40s.

FLOWER.—Diagrams of the Nerves of the Human Body, exhibiting their Origin, Divisions, and Connexions, with their Distribution to the various Regions of the Cutaneous Surface, and to all the Muscles. By WILLIAM H. FLOWER, F.R.C.S., F.R.S. Third Edition, containing 6 Plates. Royal 4to, 12s.

GODLEE.—An Atlas of Human Anatomy : Illustrating most of the ordinary Dissections and many not usually practised by the Student. By RICKMAN J. GODLEE, M.S., F.R.C S., Assistant-Surgeon to University College Hospital, and Senior Demonstrator of Anatomy in University College. With 48 Imp. 4to Coloured Plates, containing 112 Figures, and a Volume of Explanatory Text, with many Engravings. 8vo, £4 14s. 6d.

HEATH.—Practical Anatomy : a Manual of Dissections. By CHRISTOPHER HEATH, F.R.C.S., Holme Professor of Clinical Surgery in University College and Surgeon to the Hospital. Sixth Edition, revised by RICKMAN J. GODLEE, M.S. Lond., F.R.C.S., Demonstrator of Anatomy in University College, and Assistant Surgeon to the Hospital. With 24 Coloured Plates and 274 Engravings. Crown 8vo, 15s.

ANATOMY—*continued.*

HOLDEN.—A Manual of the Dissection of the
Human Body. By LUTHER HOLDEN, F.R.C.S., Consulting-Surgeon to
St. Bartholomew's Hospital. Fifth Edition, by JOHN LANGTON,
F.R.C.S., Surgeon to, and Lecturer on Anatomy at, St. Bartholomew's
Hospital. With 208 Engravings. 8vo, 20s.

By the same Author.

Human Osteology : comprising a Descrip-
tion of the Bones, with Delineations of the Attachments of the
Muscles, the General and Microscopical Structure of Bone
and its Development. Sixth Edition, revised by the Author and
JAMES SHUTER, F.R.C.S., late Assistant-Surgeon to St. Bartholomew's Hospital. With 61 Lithographic Plates and 89 Engravings.
Royal 8vo, 16s.

ALSO,

Landmarks, Medical and Surgical. Fourth
Edition. *[In the press.*

MORRIS.—The Anatomy of the Joints of Man.
By HENRY MORRIS, M.A., F.R.C.S., Surgeon to, and Lecturer on Ana-
tomy and Practical Surgery at, the Middlesex Hospital. With 44
Plates (19 Coloured) and Engravings. 8vo, 16s.

The Anatomical Remembrancer; or, Com-
lete Pocket Anatomist. Eighth Edition. 32mo, 3s. 6d.

WAGSTAFFE.—The Student's Guide to Human
Osteology. By WM. WARWICK WAGSTAFFE, F.R.C.S., late Assistant-
Surgeon to, and Lecturer on Anatomy at, St. Thomas's Hospital.
With 23 Plates and 66 Engravings. Fcap. 8vo, 10s. 6d.

WILSON — BUCHANAN — CLARK. — Wilson's
Anatomist's Vade-Mecum : a System of Human Anatomy. Tenth
Edition, by GEORGE BUCHANAN, Professor of Clinical Surgery in the
University of Glasgow, and HENRY E. CLARK, M.R.C.S., Lecturer on
Anatomy in the Glasgow Royal Infirmary School of Medicine. With
450 Engravings, including 26 Coloured Plates. Crown 8vo, 18s.

BOTANY.

BENTLEY.—A Manual of Botany. By Robert
BENTLEY, F.L.S., M.R.C.S., Professor of Botany in King's College
and to the Pharmaceutical Society. With 1185 Engravings. Fourth
Edition. Crown 8vo, 15s.

By the same Author.

The Student's Guide to Structural,
Morphological, and Physiological Botany. With 660 Engravings.
Fcap. 8vo, 7s. 6d.

ALSO,

The Student's Guide to Systematic
Botany, including the Classification of Plants and Descriptive
Botany. With 357 Engravings. Fcap. 8vo, 3s. 6d.

BENTLEY AND TRIMEN.—Medicinal Plants:
being descriptions, with original Figures, of the Principal Plants
employed in Medicine, and an account of their Properties and Uses.
By ROBERT BENTLEY, F.L.S., and HENRY TRIMEN, M.B., F.L.S.
In 4 Vols., large 8vo, with 306 Coloured Plates, bound in half
morocco. gilt edges, £11 11s.

CHEMISTRY.

BERNAYS.—Notes for Students in Chemistry;
being a Syllabus of Chemistry compiled mainly from the Manuals of
Fownes-Watts, Miller, Wurz, and Schorlemmer. By ALBERT J. BERNAYS,
Ph.D., Professor of Chemistry at St. Thomas's Hospital. Sixth
Edition. Fcap. 8vo, 3s. 6d.

BLOXAM.—Chemistry, Inorganic and Organic;
with Experiments. By CHARLES L. BLOXAM, Professor of Chemistry in
King's College. Fifth Edition. With 292 Engravings. 8vo, 16s.

By the same Author.

Laboratory Teaching; or, Progressive
Exercises in Practical Chemistry. Fourth Edition. With 83
Engravings. Crown 8vo, 5s. 6d.

CHEMISTRY—continued.

BOWMAN AND BLOXAM.—Practical Chemistry,
including Analysis. By JOHN E. BOWMAN, and CHARLES L. BLOXAM, Professor of Chemistry in King's College. Eighth Edition. With 90 Engravings. Fcap. 8vo, 5s. 6d.

BROWN.—Practical Chemistry: Analytical
Tables and Exercises for Students. By J. CAMPBELL BROWN, D.Sc. Lond., Professor of Chemistry in University College, Liverpool. Second Edition. 8vo, 2s. 6d.

CLOWES.—Practical Chemistry and Qualita-
tive Inorganic Analysis. Adapted for use in the Laboratories of Schools and Colleges. By FRANK CLOWES, D.Sc. Lond., Professor of Chemistry in University College, Nottingham. Fourth Edition. With Engravings. Post 8vo, 7s. 6d.

FOWNES.—Manual of Chemistry.—See WATTS.

FRANKLAND AND JAPP.—Inorganic Chemistry.
By EDWARD FRANKLAND, Ph.D., D.C.L., F.R.S., and F. R. JAPP, M.A. Ph.D., F.I.C. With 2 Lithographic Plates and numerous Wood Engravings. 8vo, 24s.

TIDY.—A Handbook of Modern Chemistry,
Inorganic and Organic. By C. MEYMOTT TIDY, M.B., Professor of Chemistry and Medical Jurisprudence at the London Hospital, 8vo, 16s.

VACHER.—A Primer of Chemistry, including
Analysis. By ARTHUR VACHER. 18mo, 1s.

VALENTIN.—Chemical Tables for the Lecture-
room and Laboratory. By WILLIAM G. VALENTIN, F.C.S. In Five large Sheets, 5s. 6d.

VALENTIN AND HODGKINSON.—A Course of
Qualitative Chemical Analysis. By W. G. VALENTIN, F.C.S. Sixth Edition by W. R. HODGKINSON, Ph.D. (Wurzburg), Senior Demonstrator of Practical Chemistry in the Science Schools, South Kensington, and H. M. CHAPMAN, Assistant Demonstrator. With Engravings and Map of Spectra. 8vo, 8s. 6d.

The Tables for the Qualitative Analysis of
Simple and Compound Substances, with Map of Spectra, printed separately on indestructible paper. 8vo, 2s. 6d.

CHEMISTRY—continued.

WATTS.—Physical and Inorganic Chemistry.
By HENRY WATTS, B.A., F.R.S. (being Vol. I. of the Thirteenth Edition of Fownes' Manual of Chemistry). With 150 Wood Engravings, and Coloured Plate of Spectra. Crown 8vo, 9s.

By the same Author.

Chemistry of Carbon - Compounds, or
Organic Chemistry (being Vol. II. of the Twelfth Edition of Fownes' Manual of Chemistry). With Engravings. Crown 8vo, 10s.

CHILDREN, DISEASES OF.

DAY.—A Manual of the Diseases of Children.
By WILLIAM H. DAY, M.D., Physician to the Samaritan Hospital for Women and Children. Second Edition. Crown 8vo, 12s. 6d.

ELLIS.—A Practical Manual of the Diseases
of Children. By EDWARD ELLIS, M.D., late Senior Physician to the Victoria Hospital for Sick Children. With a Formulary. Fourth Edition. Crown 8vo, 10s.

GOODHART.—The Student's Guide to Diseases
of Children By JAMES F. GOODHART, M.D., F.R.C.P., Assistant Physician to Guy's Hospital; Physician to the Evelina Hospital for Sick Children. Fcap. 8vo, 10s. 6d.

SMITH.—On the Wasting Diseases of Infants
and Children. By EUSTACE SMITH, M.D., F.R.C.P., Physician to H.M. the King of the Belgians, and to the East London Hospital for Children. Fourth Edition. Post 8vo, 8s. 6d.

By the same Author.

A Practical Treatise on Disease in Children. 8vo, 22s.

STEINER.—Compendium of Children's Diseases; a Handbook for Practitioners and Students. By JOHANN STEINER, M.D. Translated by LAWSON TAIT, F.R.C.S., Surgeon to the Birmingham Hospital for Women, &c. 8vo, 12s. 6d.

DENTISTRY.

GORGAS. — Dental Medicine : a Manual of
Dental Materia Medica and Therapeutics, for Practitioners and
Students. By FERDINAND J. S. GORGAS, A.M., M.D., D.D.S., Professor
of Dentistry in the University of Maryland; Editor of "Harris's
Principles and Practice of Dentistry," &c. Royal 8vo, 14s.

HARRIS. — The Principles and Practice of
Dentistry; including Anatomy, Physiology, Pathology, Therapeutics,
Dental Surgery, and Mechanism. By CHAPIN A. HARRIS, M.D., D.D.S.
Eleventh Edition, revised and edited by FERDINAND J. S. GORGAS,
A.M., M.D., D.D.S. With 750 Illustrations. 8vo, 31s. 6d.

SEWILL. — The Student's Guide to Dental
Anatomy and Surgery. By HENRY E. SEWILL, M.R.C.S., L.D.S., late
Dental Surgeon to the West London Hospital. Second Edition.
With 78 Engravings. Fcap. 8vo, 5s. 6d.

STOCKEN. — Elements of Dental Materia Medica
and Therapeutics, with Pharmacopœia. By JAMES STOCKEN, L.D.S.R.C.S.,
late Lecturer on Dental Materia Medica and Therapeutics and Dental
Surgeon to the National Dental Hospital; assisted by THOMAS GADDES,
L.D.S. Eng. and Edin. Third Edition. Fcap. 8vo, 7s. 6d.

TOMES (C. S.). — Manual of Dental Anatomy,
Human and Comparative. By CHARLES S. TOMES, M.A., F.R.S.
Second Edition. With 191 Engravings. Crown 8vo, 12s. 6d.

TOMES (J. and C. S.). — A Manual of Dental
Surgery. By JOHN TOMES, M.R.C.S., F.R.S., and CHARLES S. TOMES,
M.A., M.R.C.S., F.R.S. ; Lecturer on Anatomy and Physiology at the
Dental Hospital of London. Third Edition. With many Engravings,
Crown 8vo. [*In the press.*

EAR, DISEASES OF.

BURNETT. — The Ear: its Anatomy, Physio-
logy, and Diseases. A Practical Treatise for the Use of Medical
Students and Practitioners. By CHARLES H. BURNETT, M.D., Aural
Surgeon to the Presbyterian Hospital, Philadelphia. Second Edition.
With 107 Engravings. 8vo, 18s.

DALBY. — On Diseases and Injuries of the Ear.
By WILLIAM B. DALBY, F.R.C.S., Aural Surgeon to, and Lecturer on
Aural Surgery at, St. George's Hospital. Third Edition. With
Engravings. Crown 8vo. 7s. 6d.

EAR, DISEASES OF—*continued.*

JONES.—A Practical Treatise on Aural Sur-
gery. By H. MACNAUGHTON JONES, M.D., Professor of the Queen's
University in Ireland, late Surgeon to the Cork Ophthalmic and Aural
Hospital. Second Edition. With 63 Engravings. Crown 8vo, 8s. 6d.

By the same Author.

Atlas of the Diseases of the Membrana
Tympani. In Coloured Plates, containing 59 Figures. With Explanatory Text. Crown 4to, 21s.

FORENSIC MEDICINE.

ABERCROMBIE. — The Student's Guide to
Medical Jurisprudence. By JOHN ABERCROMBIE, M.D., Senior
Assistant to, and Lecturer on Forensic Medicine at, Charing Cross
Hospital. Fcap 8vo, 7s. 6d.

OGSTON.—Lectures on Medical Jurisprudence.
By FRANCIS OGSTON, M.D., late Professor of Medical Jurisprudence
and Medical Logic in the University of Aberdeen. Edited by FRANCIS
OGSTON, Jun., M.D., late Lecturer on Practical Toxicology in the
University of Aberdeen. With 12 Plates. 8vo, 18s.

TAYLOR.—The Principles and Practice of
Medical Jurisprudence. By ALFRED S. TAYLOR, M.D., F.R.S.
Third Edition, revised by THOMAS STEVENSON, M.D., F.R.C.P., Lec-
turer on Chemistry and Medical Jurisprudence at Guy's Hospital;
Examiner in Chemistry at the Royal College of Physicians; Official
Analyst to the Home Office. With 188 Engravings. 2 Vols. 8vo, 31s. 6d.

By the same Author.

A Manual of Medical Jurisprudence.
Tenth Edition. With 55 Engravings. Crown 8vo, 14s.

ALSO,

On Poisons, in relation to Medical Jurisprudence and Medicine. Third Edition. With 104 Engravings.
Crown 8vo, 16s.

TIDY AND WOODMAN.—A Handy-Book of
Forensic Medicine and Toxicology. By C. MEYMOTT TIDY, M.B.; and
W. BATHURST WOODMAN, M.D., F.R.C.P. With 8 Lithographic Plates
and 116 Wood Engravings. 8vo, 31s. 6d.

HYGIENE.

PARKES.—A Manual of Practical Hygiene.
By EDMUND A. PARKES, M.D., F.R.S. Sixth Edition by F. DE CHAUMONT, M.D., F.R.S., Professor of Military Hygiene in the Army Medical School. With 9 Plates and 103 Engravings. 8vo, 18s.

WILSON.—A Handbook of Hygiene and Sanitary Science. By GEORGE WILSON, M.A., M.D., F.R.S.E., Medical Officer of Health for Mid Warwickshire. Fifth Edition. With Engravings. Crown 8vo, 10s. 6d.

MATERIA MEDICA AND THERAPEUTICS.

BINZ AND SPARKS.—The Elements of Therapeutics; a Clinical Guide to the Action of Medicines. By C. BINZ, M.D., Professor of Pharmacology in the University of Bonn. Translated and Edited with Additions, in conformity with the British and American Pharmacopœias, by EDWARD I. SPARKS, M.A., M.B., F.R.C.P. Lond. Crown 8vo, 8s. 6d.

LESCHER.—Recent Materia Medica. Notes on their Origin and Therapeutics. By F. HARWOOD LESCHER, F.C.S., Pereira Medallist. Second Edition. 8vo, 2s. 6d.

OWEN.—A Manual of Materia Medica; incorporating the Author's "Tables of Materia Medica." By ISAMBARD OWEN, M.D., F.R.C.P., Lecturer on Materia Medica and Therapeutics to St. George's Hospital. Crown 8vo, 6s.

ROYLE AND HARLEY.—A Manual of Materia Medica and Therapeutics. By J. FORBES ROYLE, M.D., F.R.S., and JOHN HARLEY, M.D., F.R.C.P., Physician to, and Joint Lecturer on Clinical Medicine at, St. Thomas's Hospital. Sixth Edition. With 139 Engravings. Crown 8vo, 15s.

THOROWGOOD. — The Student's Guide to Materia Medica and Therapeutics. By JOHN C. THOROWGOOD, M.D., F.R.C.P., Lecturer on Materia Medica at the Middlesex Hospital. Second Edition. With Engravings. Fcap. 8vo, 7s.

WARING.—A Manual of Practical Therapeutics. By EDWARD J. WARING, C.I.E., M.D., F.R.C.P. Fourth Edition. Crown 8vo.   [*In the press.*

MEDICINE.

BARCLAY.—A Manual of Medical Diagnosis.
By A. WHYTE BARCLAY, M.D., F.R.C.P., late Physician to, and
Lecturer on Medicine at, St. George's Hospital. Third Edition. Fcap.
8vo, 10s. 6d.

CHARTERIS.—The Student's Guide to the
Practice of Medicine. By MATTHEW CHARTERIS, M.D., Professor of
Materia Medica, University of Glasgow ; Physician to the Royal In-
firmary. With Engravings on Copper and Wood. Third Edition.
Fcap. 8vo, 7s.

FAGGE.—The Principles and Practice of Medi-
cine. By the late C. HILTON FAGGE, M.D., F.R.C.P., Edited by P. H.
PYE-SMITH, M.D., F.R.C.P., Physician to, and Lecturer on Medicine
at, Guy's Hospital. 2 Vols. 8vo. [*In the press.*

FENWICK.—The Student's Guide to Medical
Diagnosis. By SAMUEL FENWICK, M.D., F.R.C.P., Physician to the
London Hospital. Fifth Edition. With 111 Engravings. Fcap. 8vo, 7s.

By the same Author.

The Student's Outlines of Medical Treatment. Second Edition. Fcap. 8vo, 7s.

FLINT.—Clinical Medicine : a Systematic Trea-
tise on the Diagnosis and Treatment of Disease. By AUSTIN FLINT,
M.D., Professor of the Principles and Practice of Medicine, &c., in
Bellevue Hospital Medical College. 8vo, 20s.

SANSOM.—Manual of the Physical Diagnosis
of Diseases of the Heart, including the use of the Sphygmograph
and Cardiograph. By A. E. SANSOM, M.D., F.R.C.P., Assistant-
Physician to the London Hospital. Third Edition. With 47 Woodcuts.
Fcap. 8vo, 7s. 6d.

WARNER.—The Student's Guide to Clinical
Medicine and Case-Taking. By FRANCIS WARNER, M.D., F.R.C.P.,
Assistant-Physician to the London Hospital. Second Edition. Fcap.
8vo, 5s.

WEST.—How to Examine the Chest : being a
Practical Guide for the Use of Students. By SAMUEL WEST, M.D.,
F.R.C.P., Physician to the City of London Hospital for Diseases of
the Chest, &c. With 42 Engravings. Fcap. 8vo, 5s.

11, *NEW BURLINGTON STREET.*

9

MEDICINE—*continued.*

WHITTAKER.—Student's Primer on the Urine.
By J. TRAVIS WHITTAKER, M.D., Clinical Demonstrator at the Royal Infirmary, Glasgow. With Illustrations, and 16 Plates etched on Copper. Post 8vo, 4s. 6d.

MIDWIFERY.

BARNES.—Lectures on Obstetric Operations,
including the Treatment of Hæmorrhage, and forming a Guide to the Management of Difficult Labour. By ROBERT BARNES, M.D., F.R.C.P., Obstetric Physician to, and Lecturer on Diseases of Women, &c., at, St. George's Hospital. Third Edition. With 124 Engravings. 8vo, 18s.

BURTON.—Handbook of Midwifery for Mid-
wives. By JOHN E. BURTON, M.R.C.S., L.R.C.P., Surgeon to the Liverpool Hospital for Women. Second Edition. With Engravings. Fcap 8vo, 6s.

RAMSBOTHAM.—The Principles and Practice
of Obstetric Medicine and Surgery. By FRANCIS H. RAMSBOTHAM, M.D., formerly Obstetric Physician to the London Hospital. Fifth Edition. With 120 Plates, forming one thick handsome volume. 8vo, 22s.

REYNOLDS.—Notes on Midwifery: specially
designed to assist the Student in preparing for Examination. By J. J. REYNOLDS, L.R.C.P., M.R.C.S. Fcap. 8vo, 4s.

ROBERTS.—The Student's Guide to the Practice
of Midwifery. By D. LLOYD ROBERTS, M.D., F.R.C.P., Lecturer on Clinical Midwifery and Diseases of Women at Owen's College, Physician to St. Mary's Hospital, Manchester. Third Edition. With 2 Coloured Plates and 127 Engravings. Fcap. 8vo, 7s. 6d.

SCHROEDER.—A Manual of Midwifery; includ-
ing the Pathology of Pregnancy and the Puerperal State. By KARL SCHROEDER, M.D., Professor of Midwifery in the University of Erlangen. Translated by CHARLES H. CARTER, M.D. With Engravings. 8vo, 12s. 6d.

SWAYNE.—Obstetric Aphorisms for the Use of
Students commencing Midwifery Practice. By JOSEPH G. SWAYNE, M.D., Lecturer on Midwifery at the Bristol School of Medicine. Eighth Edition. With Engravings. Fcap. 8vo, 3s. 6d.

MICROSCOPY.

CARPENTER.—The Microscope and its Revela-
tions. By WILLIAM B. CARPENTER, C.B., M.D., F.R.S. Sixth Edition.
With 26 Plates, a Coloured Frontispiece, and more than 500 Engravings.
Crown 8vo, 16s.

LEE. — The Microtomist's Vade-Mecum ; a
Handbook of the Methods of Microscopic Anatomy. By ARTHUR
BOLLES LEE. Crown 8vo, 8s. 6d.

MARSH. — Microscopical Section-Cutting : a
Practical Guide to the Preparation and Mounting of Sections for the
Microscope, special prominence being given to the subject of Animal
Sections. By Dr. SYLVESTER MARSH. Second Edition. With 17
Engravings. Fcap. 8vo, 3s. 6d.

MARTIN.—A Manual of Microscopic Mounting.
By JOHN H. MARTIN, Member of the Society of Public Analysis, &c.
Second Edition. With several Plates and 144 Engravings. 8vo, 7s. 6d.

OPHTHALMOLOGY.

HARTRIDGE.—The Refraction of the Eye. By
GUSTAVUS HARTRIDGE, F.R.C.S., Assistant Surgeon to the Royal
Westminster Ophthalmic Hospital. With 87 Illustrations, Test Types,
&c. Crown 8vo, 5s.

HIGGENS.—Hints on Ophthalmic Out-Patient
Practice. By CHARLES HIGGENS, F.R.C.S., Ophthalmic Surgeon to,
and Lecturer on Ophthalmology at, Guy's Hospital. Second Edition.
Fcap. 8vo, 3s.

JONES.—A Manual of the Principles and
Practice of Ophthalmic Medicine and Surgery. By T. WHARTON JONES,
F.R.C.S., F.R.S., late Ophthalmic Surgeon and Professor of Ophthalmo-
logy to University College Hospital. Third Edition. With 9 Coloured
Plates and 173 Engravings. Fcap. 8vo, 12s. 6d.

MACNAMARA.—A Manual of the Diseases of
the Eye. By CHARLES MACNAMARA, F.R.C.S., Surgeon to, and Lecturer
on Surgery at, the Westminster Hospital. Fourth Edition. With
4 Coloured Plates and 66 Engravings. Crown 8vo, 10s. 6d.

OPHTHALMOLOGY—continued.

NETTLESHIP.—The Student's Guide to Diseases
of the Eye. By EDWARD NETTLESHIP, F.R.C.S., Ophthalmic Surgeon to, and Lecturer on Ophthalmic Surgery at, St. Thomas's Hospital. Third Edition. With 157 Engravings, and a Set of Coloured Papers illustrating Colour-blindness. Fcap. 8vo, 7s. 6d.

TOSSWILL.—Diseases and Injuries of the Eye
and Eyelids. By LOUIS H. TOSSWILL, B.A., M.B. Cantab., M.R.C.S., Surgeon to the West of England Eye Infirmary, Exeter. Fcap. 8vo, 2s. 6d.

WOLFE.—On Diseases and Injuries of the Eye:
a Course of Systematic and Clinical Lectures to Students and Medical Practitioners. By J. R. WOLFE, M.D., F.R.C.S.E., Senior Surgeon to the Glasgow Ophthalmic Institution, Lecturer on Ophthalmic Medicine and Surgery in Anderson's College. With 10 Coloured Plates, and 120 Wood Engravings, 8vo, 21s.

PATHOLOGY.

JONES AND SIEVEKING.—A Manual of Patho-
logical Anatom By C. HANDFIELD JONES, M.B., F.R.S., and EDWARD H. SIEVEKING M.D., F.R.C.P. Second Edition. Edited, with considerable enlargement, by J. F. PAYNE, M.B., Assistant-Physician and Lecturer on General Pathology at St. Thomas's Hospital. With 195 Engravings. Crown 8vo, 16s.

LANCEREAUX.—Atlas of Pathological Ana-
tomy. By Dr. LANCEREAUX. Translated by W. S. GREENFIELD, M.D., Professor of Pathology in the University of Edinburgh. With 70 Coloured Plates. Imperial 8vo, £5 5s.

VIRCHOW. — Post-Mortem Examinations: a
Description and Explanation of the Method of Performing them, with especial reference to Medico-Legal Practice. By Professor RUDOLPH VIRCHOW, Berlin Charité Hospital. Translated by Dr. T. B. SMITH. Second Edition, with 4 Plates. Fcap. 8vo, 3s. 6d.

PSYCHOLOGY.

BUCKNILL AND TUKE.—A Manual of Psycho-
logical Medicine: containing the Lunacy Laws, Nosology, Ætiology, Statistics, Description, Diagnosis, Pathology, and Treatment of Insanity, with an Appendix of Cases. By JOHN C. BUCKNILL, M.D., F.R.S., and D. HACK TUKE, M.D., F.R.C.P. Fourth Edition with 12 Plates (30 Figures). 8vo, 25s.

PSYCHOLOGY—*continued.*

CLOUSTON. — Clinical Lectures on Mental
Diseases. By THOMAS S. CLOUSTON, M.D., and F.R.C.P. Edin.; Lecturer on Mental Diseases in the University of Edinburgh. With
8 Plates (6 Coloured). Crown 8vo, 12s. 6d.

MANN.—A Manual of Psychological Medicine
and Allied Nervous Disorders. By EDWARD C. MANN, M.D., Member
of the New York Medico-Legal Society. With Plates. 8vo, 24s.

PHYSIOLOGY.

CARPENTER.—Principles of Human Physio-
logy. By WILLIAM B. CARPENTER, C.B., M.D., F.R.S. Ninth Edition.
Edited by Henry Power, M.B., F.R.C.S. With 3 Steel Plates and
377 Wood Engravings. 8vo, 31s. 6d.

DALTON.—A Treatise on Human Physiology :
designed for the use of Students and Practitioners of Medicine. By
JOHN C. DALTON, M.D., Professor of Physiology and Hygiene in the
College of Physicians and Surgeons, New York. Seventh Edition.
With 252 Engravings. Royal 8vo, 20s.

FREY.—The Histology and Histo-Chemistry of
Man. A Treatise on the Elements of Composition and Structure of the
Human Body. By HEINRICH FREY, Professor of Medicine in Zurich.
Translated by ARTHUR E. BARKER, Assistant-Surgeon to the University
College Hospital. With 608 Engravings. 8vo, 21s.

PYE-SMITH.—Syllabus of a Course of Lectures
on Physiology. By PHILIP H. PYE-SMITH, B.A., M.D., F.R.C.P.,
Physician to Guy's Hospital. With Diagrams, and an Appendix of
Notes and Tables. Crown 8vo, 5s.

SANDERSON.—Handbook for the Physiological
Laboratory : containing an Exposition of the fundamental facts of the
Science, with explicit Directions for their demonstration. By J.
BURDON SANDERSON, M.D., F.R.S.; E. KLEIN, M.D., F.R.S.; MICHAEL
FOSTER, M.D., F.R.S., and T. LAUDER BRUNTON, M.D., F.R.S. 2 Vols.,
with 123 Plates. 8vo, 24s.

YEO.—A Manual of Physiology for the Use of
Junior Students of Medicine. By GERALD F. YEO, M.D., F.R.C.S.,
Professor of Physiology in King's College, London. With 301 Engravings. Crown 8vo, 14s.

SURGERY.

BELLAMY.—The Student's Guide to Surgical
Anatomy; an Introduction to Operative Surgery. By EDWARD
BELLAMY, F.R.C.S., and Member of the Board of Examiners ; Surgeon
to, and Lecturer on Anatomy at, Charing Cross Hospital. Third
Edition. With 80 Engravings. Fcap. 8vo, 7s. 6d.

BRYANT.—A Manual for the Practice of
Surgery. By THOMAS BRYANT, F.R.C.S., Surgeon to, and Lecturer on
Surgery at, Guy's Hospital. Fourth Edition. With 750 Illustra-
tions (many being coloured), and including 6 Chromo-Lithographic
Plates. 2 Vols. Crown 8vo, 32s.

CLARK AND WAGSTAFFE. — Outlines of
Surgery and Surgical Pathology. By F. LE GROS CLARK, F.R.C.S.,
F.R.S., Consulting Surgeon to St. Thomas's Hospital. Second Edition.
Revised and expanded by the Author, assisted by W. W. WAGSTAFFE,
F.R.C.S., Assistant Surgeon to St. Thomas's Hospital. 8vo, 10s. 6d.

DRUITT.—The Surgeon's Vade-Mecum ; a
Manual of Modern Surgery. By ROBERT DRUITT, F.R.C.S. Eleventh
Edition. With 369 Engravings. Fcap. 8vo, 14s.

FERGUSSON.—A System of Practical Surgery.
By Sir WILLIAM FERGUSSON, Bart., F.R.C.S., F.R.S., late Surgeon and
Professor of Clinical Surgery to King's College Hospital. With 463
Engravings. Fifth Edition. 8vo, 21s.

HEATH.—A Manual of Minor Surgery and
Bandaging, for the use of House-Surgeons, Dressers, and Junior Practi-
tioners. By CHRISTOPHER HEATH, F.R.C.S., Holme Professor of
Clinical Surgery in University College and Surgeon to the Hospital.
Seventh Edition. With 129 Engravings. Fcap. 8vo, 6s.

By the same Author.

A Course of Operative Surgery: with
Twenty Plates (containing many figures) drawn from Nature by
M. LÉVEILLÉ, and Coloured. Second Edition. Large 8vo, 30s.

ALSO,

The Student's Guide to Surgical Diagnosis. Second Edition. Fcap. 8vo, 6s. 6d.

SURGERY—*continued.*

SOUTHAM.—Regional Surgery : including Surgical Diagnosis. A Manual for the use of Students. BY FREDERICK A. SOUTHAM, M.A., M.B. Oxon, F.R.C.S., Assistant-Surgeon to the Royal Infirmary, and Assistant-Lecturer on Surgery in the Owen's College School of Medicine, Manchester.

Part I. The Head and Neck. Crown 8vo, 6s. 6d.
  ,,   II. The Upper Extremity and Thorax. Crown 8vo, 7s. 6d.

TERMINOLOGY.

DUNGLISON.—Medical Lexicon : a Dictionary of Medical Science, containing a concise Explanation of its various Subjects and Terms, with Accentuation,. Etymology, Synonyms, &c. By ROBERT DUNGLISON, M.D. New Edition, thoroughly revised by RICHARD J. DUNGLISON, M.D. Royal 8vo, 28s.

MAYNE.—A Medical Vocabulary : being an Explanation of all Terms and Phrases used in the various Departments of Medical Science and Practice, giving their Derivation, Meaning, Application, and Pronunciation. By ROBERT G. MAYNE, M.D., LL.D., and JOHN MAYNE, M.D., L.R.C.S.E. Fifth Edition. Crown 8vo, 10s. 6d.

WOMEN, DISEASES OF.

BARNES.—A Clinical History of the Medical and Surgical Diseases of Women. By ROBERT BARNES, M.D., F.R.C.P., Obstetric Physician to, and Lecturer on Diseases of Women, &c., at, St. George's Hospital. Second Edition. With 181 Engravings. 8vo, 28s.

COURTY.—Practical Treatise on Diseases of the Uterus, Ovaries, and Fallopian Tubes. By Professor COURTY, Montpellier. Translated from the Third Edition by his Pupil, AGNES M'LAREN, M.D., M.K.Q.C.P. With Preface by Dr. MATTHEWS DUNCAN. With 124 Engravings. 8vo, 24s.

DUNCAN.—Clinical Lectures on the Diseases of Women. By J. MATTHEWS DUNCAN, M.D., F.R.C.P., F.R.S.E., Obstetric Physician to St. Bartholomew's Hospital. Second Edition, with Appendices. 8vo, 14s.

EMMET. — The Principles and Practice of Gynæcology. By THOMAS ADDIS EMMET, M.D., Surgeon to the Woman's Hospital of the State of New York. Third Edition. With 150 Engravings. Royal 8vo, 24s.

WOMEN, DISEASES OF—*continued.*

GALABIN.—The Student's Guide to the Diseases of Women. By ALFRED L. GALABIN, M.D., F.R.C.P., Obstetric Physician to, and Lecturer on Obstetric Medicine at, Guy's Hospital. Third Edition. With 78 Engravings. Fcap. 8vo, 7s. 6d.

REYNOLDS.—Notes on Diseases of Women. Specially designed to assist the Student in preparing for Examination. By J. J. REYNOLDS, L.R.C.P., M.R.C.S. Second Edition. Fcap. 8vo 2s. 6d.

SAVAGE.—The Surgery of the Female Pelvic Organs. By HENRY SAVAGE, M.D., Lond., F.R.C.S., one of the Consulting Medical Officers of the Samaritan Hospital for Women. Fifth Edition, with 17 Lithographic Plates (15 Coloured), and 52 Woodcuts. Royal 4to, 35s.

WEST AND DUNCAN.—Lectures on the Diseases of Women. By CHARLES WEST, M.D., F.R.C.P. Fourth Edition. Revised and in part re-written by the Author, with numerous additions by J. MATTHEWS DUNCAN, M.D., F.R.C.P., F.R.S.E., Obstetric Physician to St. Bartholomew's Hospital. 8vo, 16s.

ZOOLOGY.

CHAUVEAU AND FLEMING.—The Comparative Anatomy of the Domesticated Animals. By A. CHAUVEAU, Professor at the Lyons Veterinary School; and GEORGE FLEMING, Veterinary Surgeon, Royal Engineers. With 450 Engravings. 8vo, 31s. 6d.

HUXLEY.—Manual of the Anatomy of Invertebrated Animals. By THOMAS H. HUXLEY, LL.D., F.R.S. With 156 Engravings. Post 8vo, 16s.

By the same Author.

Manual of the Anatomy of Vertebrated Animals. With 110 Engravings. Post 8vo, 12s.

WILSON.—The Student's Guide to Zoology: a Manual of the Principles of Zoological Science. By ANDREW WILSON, Lecturer on Natural History, Edinburgh. With Engravings. Fcap. 8vo, 6s. 6d.